SpringerWienNewYork

CISM COURSES AND LECTURES

Series Editors:

The Rectors
Friedrich Pfeiffer - Munich
Franz G. Rammerstorfer - Wien
Jean Salençon - Palaiseau

The Secretary General
Bernhard Schrefler - Padua

Executive Editor
Paolo Serafini - Udine

The series presents lecture notes, monographs, edited works and proceedings in the field of Mechanics, Engineering, Computer Science and Applied Mathematics.
Purpose of the series is to make known in the international scientific and technical community results obtained in some of the activities organized by CISM, the International Centre for Mechanical Sciences.

INTERNATIONAL CENTRE FOR MECHANICAL SCIENCES

COURSES AND LECTURES - No. 532

NUMERICAL MODELING OF CONCRETE CRACKING

EDITED BY

GÜNTER HOFSTETTER
UNIVERSITY OF INNSBRUCK, AUSTRIA

GÜNTHER MESCHKE
RUHR UNIVERSITY BOCHUM, GERMANY

SpringerWienNewYork

This volume contains 185 illustrations

This work is subject to copyright.
All rights are reserved,
whether the whole or part of the material is concerned
specifically those of translation, reprinting, re-use of illustrations,
broadcasting, reproduction by photocopying machine
or similar means, and storage in data banks.
© 2011 by CISM, Udine
Printed in Italy
SPIN 80073533

All contributions have been typeset by the authors.

ISBN 978-3-7091-0896-3 SpringerWienNewYork

PREFACE

Reliable model-based prognoses of the initiation and propagation of cracks in concrete plays an important role for the durability and integrity assessment of concrete and reinforced concrete structures. To this end, a large number of material models for concrete cracking based on different theories (e.g., damage mechanics, fracture mechanics, plasticity theory and combinations of the mentioned theories) as well as advanced finite element methods suitable for the representation of cracks (e.g., the Extended Finite Element Method and Embedded Crack Models) have been developed in recent years.

The focus of the Advanced School on "Numerical Modeling of Concrete Cracking" at the International Centre for Mechanical Sciences (CISM) at Udine in May 2009 was laid on numerical models for describing crack propagation in concrete and their applications to numerical simulations of concrete and reinforced concrete structures. The lectures of this course formed the basis for this book. Its aim is to impart fundamental knowledge of the underlying theories of the different approaches for modelling cracking of concrete and to provide a critical survey of the state-of-the-art in computational concrete mechanics.

This book covers a relatively broad spectrum of topics related to modelling of cracks, including continuum-based and discrete crack models, meso-scale models, advanced discretization strategies to capture evolving cracks based on the concept of finite elements with embedded discontinuities and on the extended finite element method, respectively, and, last but not least, extensions to coupled problems such as hygro-mechanical problems as required in computational durability analyses of concrete structures.

*Innsbruck and Bochum,
March 2011,*

*Günter Hofstetter
University of Innsbruck
Austria*

*Günther Meschke
Ruhr-University Bochum
Germany*

CONTENTS

Damage and smeared crack models
by M. Jirásek.. 1

Cracking and fracture of concrete at meso-level using zero-thickness interface elements
by I. Carol, A. Idiart, C. López and A. Caballero........ 51

Crack models with embedded discontinuities
by A. E. Huespe and J. Oliver............................ 99

Plasticity based crack models and applications
by G. Hofstetter, C. Feist, H. Lehar, Y. Theiner,
B. Valentini and B. Winkler.............................. 161

Crack models based on the extended finite element method
by N. Moës... 221

Smeared crack and X-FEM models in the context of poromechanics
by G. Meschke, S. Grasberger, C. Becker and S. Jox...... 265

Damage and Smeared Crack Models

Milan Jirásek

Czech Technical University in Prague, Czech Republic

1 Isotropic Damage Models

Continuum damage mechanics is a constitutive theory that describes the progressive loss of material integrity due to the propagation and coalescence of microcracks, microvoids, and similar defects. These changes in the microstructure lead to a degradation of material stiffness observed on the macroscale. The term "continuum damage mechanics" was first used by Hult in 1972 but the concept of damage had been introduced by Kachanov already in 1958 in the context of creep rupture (Kachanov, 1958) and further developed by Rabotnov (1968); Hayhurst (1972); Leckie and Hayhurst (1974). The simplest version of the isotropic damage model considers the damaged stiffness tensor as a scalar multiple of the initial elastic stiffness tensor, i.e., damage is characterized by a single scalar variable. A general isotropic damage model should deal with two scalar variables corresponding to two independent elastic constants of standard isotropic elasticity. More refined theories take into account the anisotropic character of damage; they represent damage by a family of vectors (Krajcinovic and Fonseka, 1981), by a second-order tensor (Vakulenko and Kachanov, 1971) or, in the most general case, by a fourth-order tensor (Chaboche, 1979). Anisotropic formulations can be based on the principle of strain equivalence (Lemaitre, 1971), or on the principle of energy equivalence (Cordebois and Sidoroff, 1979) (the principle of stress equivalence is also conceptually possible but is rarely used).

In the present chapter, we will focus on isotropic damage models and on smeared crack models, which incorporate anisotropy in a simplified way. Anisotropic damage models based on tensorial description of damage will are treated e.g. in Lemaitre and Desmorat (2005).

1.1 One-Dimensional Damage Model

Damage models work with certain internal variables that characterize the density and orientation of microdefects. To introduce the basic concepts, we start from the case of uniaxial stress. For the present purpose, the material is idealized as a

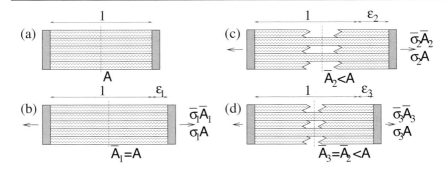

Figure 1. Representation of a uniaxial damage model as a bundle of parallel elastic fibers breaking at different strain levels

bundle of fibers parallel to the direction of loading (Fig. 1a). Initially, all the fibers respond elastically, and the stress is carried by the total cross section of all fibers, A (Fig. 1b). As the applied strain is increased, some fibers start breaking (Fig. 1c). Each fiber is assumed to be perfectly brittle, which means that the stress in the fiber drops down to zero immediately after a critical strain level is reached. However, since the critical strain is different for each fiber, the effective area $\bar{A}$ (i.e., the area of unbroken fibers that can still carry stress) decreases gradually from $\bar{A} = A$ to $\bar{A} = 0$. We have to make a distinction between the *nominal stress* σ, defined as the force per unit initial area of the cross section, and the *effective stress* $\bar{\sigma}$, defined as the force per unit effective area. The nominal stress enters the Cauchy equations of equilibrium on the macroscopic level, while the effective stress is the "true" stress acting in the material microstructure.[1] From the condition of equivalence, $\sigma A = \bar{\sigma} \bar{A}$, we obtain

$$\sigma = \frac{\bar{A}}{A} \bar{\sigma} \qquad (1)$$

The ratio of the effective area to the total area, $\bar{A}/A$, is a scalar characterizing the *integrity* of the material. In damage mechanics it is customary to work with the *damage variable* defined as

$$D = 1 - \frac{\bar{A}}{A} = \frac{A - \bar{A}}{A} = \frac{A_d}{A} \qquad (2)$$

where $A_d = A - \bar{A}$ is the damaged part of the area. An intact (undamaged) material is characterized by $\bar{A} = A$, i.e., by $D = 0$. Due to propagation of microdefects and

---

[1] Of course, detailed micromechanical analysis would reveal local oscillations of the stress fields dependent on the specific defect geometry, and the representation of the actual stress distribution by one averaged value—the effective stress—is just a simplification for modeling purposes.

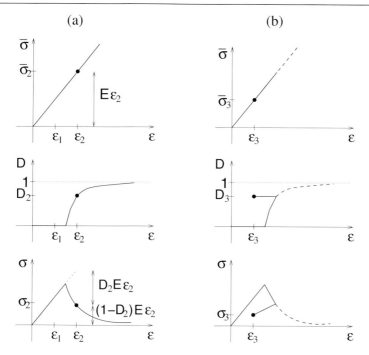

Figure 2. Evolution of effective stress $\bar{\sigma}$, damage variable D and nominal stress σ under a) monotonic loading, b) non-monotonic loading

their coalescence, the damage variable grows and at late stages of the degradation process it attains or asymptotically approaches the limit value $D = 1$, corresponding to a completely damaged material with effective area reduced to zero. In the simplest version of the model, each fiber is supposed to remain linear elastic up to the strain level at which it breaks.[2] Consequently, the effective stress $\bar{\sigma}$ is governed by Hooke's law,

$$\bar{\sigma} = E\varepsilon \qquad (3)$$

Combining (1)–(3) we obtain the constitutive law for the nominal stress,

$$\sigma = (1 - D)E\varepsilon \qquad (4)$$

Damage evolution can be characterized by the dependence of the damage variable on the applied strain,

$$D = g(\varepsilon) \qquad (5)$$

---

[2] In general, the fictitious "fibers" can obey any (nonlinear) constitutive law, which provides one possible framework for coupling of damage with other dissipative phenomena, such as plasticity.

Function g affects the shape of the stress-strain diagram and can be directly identified from a uniaxial test. The evolution of the effective stress, damage variable, and nominal stress in a material that remains elastic up to the peak stress is shown in Fig. 2a. This description is valid only for monotonic loading by an increasing applied strain ε. When the material is first stretched up to a certain strain level ε_2 that induces damage $D_2 = g(\varepsilon_2)$ and then the strain decreases (Fig. 1d), the damaged area remains constant and the material responds as an elastic material with a reduced Young's modulus $E_2 = (1 - D_2)E$. This means that, during unloading and partial reloading, the damage variable in (4) must be evaluated from the largest previously reached strain and not from the current strain ε. It is convenient to introduce an internal variable κ characterizing the maximum strain level reached in the previous history of the material up to a given time t, i.e., to set

$$\kappa(t) = \max_{\tau \leq t} \varepsilon(\tau) \tag{6}$$

where t is not necessarily the physical time—it can be any monotonically increasing parameter controlling the loading process. The damage evolution law (5) is then replaced by equation

$$D = g(\kappa) \tag{7}$$

that remains valid not only during monotonic loading but also during unloading and reloading. The evolution of the effective stress, damage variable, and nominal stress in a non-monotonic test is shown in Fig. 2b. Note that, upon a complete removal of the applied stress, the strain returns to zero (due to elasticity of the yet unbroken fibers), i.e., the pure damage model does not take into account any permanent strains. Nevertheless, the material state is different from the initial virgin state, because the damage variable is not zero and the stiffness and strength mobilized in a new tensile loading process are smaller than their initial values. The loading history is reflected by the value of the damage variable D.

To gain further insight, we rewrite the constitutive law (4) in the form $\sigma = E_s \varepsilon$ where $E_s = (1 - D)E$ is the apparent (damaged) modulus of elasticity. Instead of defining the variable κ through (6), we introduce a loading function $f(\varepsilon, \kappa) = \varepsilon - \kappa$ and postulate the loading-unloading conditions in the Kuhn-Tucker form,

$$f \leq 0, \quad \dot{\kappa} \geq 0, \quad \dot{\kappa} f = 0 \tag{8}$$

The first condition means that κ can never be smaller than ε, and the second condition means that κ cannot decrease. Finally, according to the third condition, κ can grow only if the current values of ε and κ are equal.

The basic ingredients of the uniaxial damage theory are summarized as follows:
- the stress-strain law in the secant format,

$$\sigma = E_s \varepsilon \tag{9}$$

- the equation relating the apparent stiffness to the damage variable,

$$E_s = (1 - D)E \tag{10}$$

- the law governing the evolution of the damage variable,

$$D = g(\kappa) \tag{11}$$

- the loading function

$$f(\varepsilon, \kappa) = \varepsilon - \kappa \tag{12}$$

specifying the elastic domain $\mathcal{E}_\kappa = \{\varepsilon \mid f(\varepsilon, \kappa) < 0\}$, i.e., the set of states for which damage does not grow, and
- the loading-unloading conditions (8).

1.2 Damage Models with Strain-Based Loading Functions

Simple Models with One Damage Variable. The simplest extension of the uniaxial damage theory to general multiaxial stress states is achieved by the isotropic damage model with a single scalar variable. Isotropic damage models are based on the simplifying assumption that the stiffness degradation is isotropic, i.e., stiffness moduli corresponding to different directions decrease proportionally, independently of the direction of loading. Since an isotropic elastic material is characterized by two independent elastic constants, a general isotropic damage model should deal with two damage variables. The model with a single variable makes use of an additional assumption that the relative reduction of all the stiffness coefficients is the same, in other words, that the Poisson ratio is not affected by damage. Consequently, the damaged stiffness tensor is expressed as

$$\mathbb{E}_S = (1 - D)\,\mathbb{E} \tag{13}$$

where $\mathbb{E}$ is the elastic stiffness tensor of the intact material, and D is the damage variable. In the present context, $\mathbb{E}_S$ is the secant stiffness that relates the total strain to total stress, according to the formula

$$\sigma = \mathbb{E}_S : \varepsilon = (1 - D)\,\mathbb{E} : \varepsilon \tag{14}$$

Clearly, (13) is a generalization of (10), and (14) is a generalization of (9) and (4). In terms of the *effective stress tensor*, defined as

$$\bar{\sigma} = \mathbb{E} : \varepsilon \tag{15}$$

equation (14) can alternatively be written as

$$\sigma = (1 - D)\,\bar{\sigma} \tag{16}$$

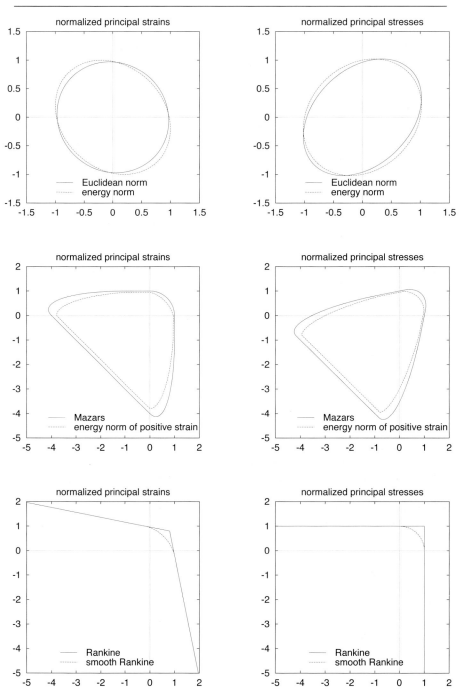

Figure 3. Loading surfaces for various definitions of equivalent strain

which is the multidimensional generalization of (1).

Similar to the uniaxial case, we introduce a loading function f specifying the elastic domain and the states at which damage grows. The loading function now depends on the strain tensor, ε, and on a variable κ that controls the evolution of the elastic domain. Physically, κ is a scalar measure of the largest strain level ever reached in the history of the material. States for which $f(\varepsilon, \kappa) < 0$ are supposed to be below the current damage threshold. Damage can grow only if the current state reaches the boundary of the elastic domain. This is described by the loading-unloading conditions (8). It is convenient to postulate the loading function in the form

$$f(\varepsilon, \kappa) = \tilde{\varepsilon}(\varepsilon) - \kappa \tag{17}$$

where $\tilde{\varepsilon}$ is the *equivalent strain*, i.e., a scalar measure of the strain level.

In some sense, the expression defining the equivalent strain plays a role similar to the yield function in plasticity, because it directly affects the shape of the elastic domain. The simplest choice is to define the equivalent strain as the Euclidean norm of the strain tensor,

$$\tilde{\varepsilon} = \|\varepsilon\| = \sqrt{\varepsilon : \varepsilon} = \sqrt{\varepsilon_{ij}\varepsilon_{ij}} \tag{18}$$

or as the energy norm,

$$\tilde{\varepsilon} = \sqrt{\frac{\varepsilon : \mathbf{E} : \varepsilon}{E}} = \sqrt{\frac{1}{E} E_{ijkl}\varepsilon_{ij}\varepsilon_{kl}} \tag{19}$$

where E_{ijkl} are the components of the elastic stiffness tensor $\mathbf{E}$ and normalization by Young's modulus E is introduced in order to obtain a strain-like quantity. Each particular definition of equivalent strain corresponds to a certain shape of the elastic domain in the strain space and can be transformed into the stress space. For illustration, Fig. 3(top) shows the elastic domains in projection onto the principal strain plane and in the principal stress plane for the case of plane stress and Poisson's ratio $\nu = 0.2$. The domains are elliptical and symmetric with respect to the origin. Consequently, there would be no difference in the response to tensile and compressive loadings.

For concrete and other materials with very different behaviors in tension and in compression, it is necessary to adjust the definition of equivalent strain. Microcracks in concrete grow mainly if the material is stretched, and so it is natural to take into account only normal strains that are positive and neglect those that are negative. This leads to the so-called Mazars definition of equivalent strain (Mazars, 1984)

$$\tilde{\varepsilon} = \|\langle\varepsilon\rangle\| = \sqrt{\langle\varepsilon\rangle : \langle\varepsilon\rangle} \tag{20}$$

or to its energetic counterpart,

$$\tilde{\varepsilon} = \sqrt{\frac{\langle\varepsilon\rangle : \mathbb{E} : \langle\varepsilon\rangle}{E}} \tag{21}$$

where McAuley brackets $\langle.\rangle$ denote the "positive part" operator. For scalars, $\langle x\rangle = \max(0, x)$, i.e., $\langle x\rangle = x$ for x positive and $\langle x\rangle = 0$ for x negative. For symmetric tensors, such as the strain tensor ε, the positive part is a tensor having the same principal directions n_I as the original one, with principal values ε_I replaced by their positive parts $\langle\varepsilon_I\rangle$. The subscript I ranges from 1 to 3 (the number of spatial dimensions) but it is not subject to Einstein's summation convention because the principal strains ε_I are not components of a first-order tensor. In terms of the spectral decomposition

$$\varepsilon = \sum_{I=1}^{3} \varepsilon_I\, n_I \otimes n_I \tag{22}$$

the positive part of ε is expressed as

$$\langle\varepsilon\rangle = \sum_{I=1}^{3} \langle\varepsilon_I\rangle\, n_I \otimes n_I \tag{23}$$

Since $(n_I \otimes n_I) : (n_J \otimes n_J) = \delta_{IJ} =$ Kronecker's delta, definition (20) can be rewritten as

$$\tilde{\varepsilon} = \sqrt{\sum_{I=1}^{3} \langle\varepsilon_I\rangle^2} \tag{24}$$

The elastic domains corresponding to (20) and (21) are shown in Fig. 3(center).

If a model corresponding to the Rankine criterion of maximum principal stress is desired, one may use the definitions

$$\tilde{\varepsilon} = \frac{1}{E} \max_{I=1,2,3} \langle\mathbb{E} : \varepsilon\rangle_I = \frac{1}{E} \max_{I=1,2,3} \langle\bar{\sigma}_I\rangle \tag{25}$$

or

$$\tilde{\varepsilon} = \frac{1}{E}\|\langle\mathbb{E} : \varepsilon\rangle\| = \frac{1}{E}\sqrt{\sum_{I=1}^{3}\langle\mathbb{E} : \varepsilon\rangle_I^2} = \frac{1}{E}\sqrt{\sum_{I=1}^{3}\langle\bar{\sigma}_I\rangle^2} \tag{26}$$

where $\langle\bar{\sigma}_I\rangle = \langle\mathbb{E} : \varepsilon\rangle_I$, $I = 1, 2, 3$, are the positive parts of principal values of the effective stress tensor (15). The former definition exactly corresponds to the Rankine criterion while the latter rounds off the corners in the octants with more than one positive principal stress; see Fig. 3(bottom).

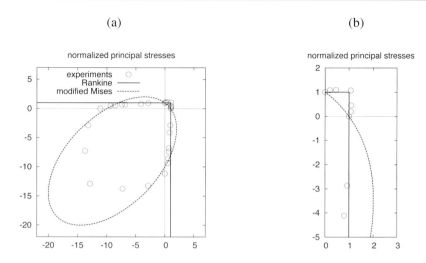

Figure 4. Biaxial strength envelope for concrete and its approximation by isotropic damage models with Rankine and modified Mises definition of equivalent strain

An alternative formula, called the modified von Mises definition (de Vree et al., 1995), reads

$$\tilde{\varepsilon} = \frac{(k-1)I_{1\varepsilon}}{2k(1-2\nu)} + \frac{1}{2k}\sqrt{\frac{(k-1)^2}{(1-2\nu)^2}I_{1\varepsilon}^2 + \frac{12kJ_{2\varepsilon}}{(1+\nu)^2}} \quad (27)$$

where

$$I_{1\varepsilon} = \mathbf{1} : \varepsilon = 3\,\varepsilon_V \quad (28)$$

is the first strain invariant (trace of the strain tensor),

$$J_{2\varepsilon} = \tfrac{1}{2}\mathbf{e} : \mathbf{e} = \tfrac{1}{2}\varepsilon : \varepsilon - \tfrac{1}{6}I_{1\varepsilon}^2 \quad (29)$$

is the second deviatoric strain invariant, and k is a model parameter that sets the ratio between the uniaxial compressive strength f_c and uniaxial tensile strength f_t. The elastic domains corresponding to the modified von Mises definition have ellipsoidal shapes but their centers are shifted from the origin along the hydrostatic axis (except for the special case with parameter $k=1$, which corresponds to the standard von Mises definition, with equivalent strain proportional to $\sqrt{J_{2\varepsilon}}$).

The uniaxial tensile strength and uniaxial compressive strength can be fitted, but the shape of the elastic domain in the tension-compression quadrant of the principal stress plane does not correspond to experimental data for concrete (Kupfer et al., 1969) and the shear strength is overestimated, see Fig. 4.

An important advantage of isotropic damage models is that the stress evaluation algorithm is usually explicit, without the need for an iterative solution of one or more nonlinear equations. The choice of a loading function in the form (17) endows the variable κ with the meaning of the largest value of equivalent strain that has ever occured in the previous deformation history of the material up to its current state; cf. (8). In other words, (6) can be generalized to

$$\kappa(t) = \max_{\tau \leq t} \tilde{\varepsilon}(\tau) \tag{30}$$

For a prescribed strain increment, the corresponding stress is evaluated simply by computing the current value of equivalent strain, updating the maximum previously reached equivalent strain and the damage variable, and reducing the effective stress according to (14). Depending on the definition of equivalent strain one may have to extract the principal strains or principal stresses. This can be done very easily, since closed-form formulas for the eigenvalues of symmetric matrices of size 2×2 or 3×3 are available.

The damaged stiffness tensor $\mathbb{E}_S = (1-D)\mathbb{E}$ introduced in (13) links the total stress to total strain and plays the role of the tangent stiffness only for unloading with constant damage ($f < 0$ or $\dot{f} < 0$). To construct the tangent stiffness tensor for loading with growing damage ($f = 0$ and $\dot{f} = 0$), we need to find the link between stress and strain increments or rates. The damage rate can be expressed in terms of the strain rate using the consistency condition $\dot{f} = 0$ with the rate of the damage loading function evaluated from (17) and combining it with the rate form of equation (11):

$$\dot{D} = \frac{\mathrm{d}g}{\mathrm{d}\kappa}\dot{\kappa} = \frac{\mathrm{d}g}{\mathrm{d}\kappa}\dot{\tilde{\varepsilon}} = \frac{\mathrm{d}g}{\mathrm{d}\kappa}\frac{\partial \tilde{\varepsilon}}{\partial \varepsilon} : \dot{\varepsilon} \tag{31}$$

For convenience, we introduce symbols g' for the derivative $\mathrm{d}g/\mathrm{d}\kappa$ of the damage function, and η for the second order tensor $\partial\tilde{\varepsilon}/\partial\varepsilon$ obtained by differentiation of the expression for the equivalent strain with respect to the strain tensor. Substituting $\dot{D} = g'\eta : \dot{\varepsilon}$ into the rate form of the stress-strain law (14) we get

$$\dot{\sigma} = (1-D)\mathbb{E} : \dot{\varepsilon} - \mathbb{E} : \varepsilon\, \dot{D} = (1-D)\mathbb{E} : \dot{\varepsilon} - \bar{\sigma}\,(g'\eta : \dot{\varepsilon}) = \mathbb{E}_{ed} : \dot{\varepsilon} \tag{32}$$

where $\bar{\sigma} = \mathbb{E} : \varepsilon$ is the effective stress and

$$\mathbb{E}_{ed} = (1-D)\mathbb{E} - g'\bar{\sigma} \otimes \eta \tag{33}$$

is the elasto-damage stiffness tensor. It is interesting to note that for a model with the equivalent strain based on the energy norm, eq. (19), the tensor η is given by

$$\eta = \frac{\partial \tilde{\varepsilon}}{\partial \varepsilon} = \frac{1}{2\sqrt{\frac{\varepsilon : \mathbb{E} : \varepsilon}{E}}} \frac{1}{E} 2\mathbb{E} : \varepsilon = \frac{\bar{\sigma}}{E\tilde{\varepsilon}} \tag{34}$$

and the resulting elasto-damage stiffness tensor

$$\mathbb{E}_{ed} = (1-D)\mathbb{E} - \frac{g'}{E\tilde{\varepsilon}}\bar{\sigma}\otimes\bar{\sigma} \qquad (35)$$

exhibits major symmetry ($E^{\text{ed}}_{ijkl} = E^{\text{ed}}_{klij}$). For other definitions of equivalent strain, this kind of symmetry is lost.

Mazars Damage Model. A popular damage model specifically designed for concrete was proposed by Mazars (Mazars, 1984, 1986). He introduced two damage variables, D_t and D_c, that are computed from the same equivalent strain (24) using two different damage functions, g_t and g_c. Function g_t is identified from the uniaxial tensile test while g_c corresponds to the compressive test. The damage variable entering the constitutive equations (14) is $D = D_t$ under tension and $D = D_c$ under compression. For general stress states the value of D is obtained as a linear combination

$$D = \alpha_t D_t + \alpha_c D_c \qquad (36)$$

where the coefficients α_t and α_c take into account the character of the stress state. In the recent implementation of Mazars model, these coefficients are evaluated as

$$\alpha_t = \left(\sum_{I=1}^{3}\frac{\varepsilon_{tI}\langle\varepsilon_I\rangle}{\tilde{\varepsilon}^2}\right)^{\beta}, \qquad \alpha_c = \left(1 - \sum_{I=1}^{3}\frac{\varepsilon_{tI}\langle\varepsilon_I\rangle}{\tilde{\varepsilon}^2}\right)^{\beta} \qquad (37)$$

where ε_{tI}, $I = 1, 2, 3$, are the principal strains due to positive stresses, i.e., the principal values of $\varepsilon_t = \mathbb{C} : \langle\mathbb{E} : \varepsilon\rangle$, in which $\mathbb{C} = \mathbb{E}^{-1}$ is the elastic compliance tensor. The exponent $\beta = 1.06$ slows down the evolution of damage under shear loading (i.e., when principal stresses do not have the same sign). In the original version of the model (Mazars, 1984), β was equal to 1.

Note that if all principal stresses are nonnegative we have $\alpha_t = 1, \alpha_c = 0$, and $D = D_t$, and if all principal stresses are nonpositive we have $\alpha_t = 0, \alpha_c = 1$, and $D = D_c$. These are the "purely tensile" and "purely compressive" stress states. For intermediate stress states the value of D is between D_t and D_c, depending on the relative magnitudes of tensile and compressive stresses. Functions characterizing the evolution of damage were originally proposed in the form (Mazars, 1984)

$$g_t(\kappa) = \begin{cases} 0 & \text{if } \kappa \leq \varepsilon_0 \\ 1 - (1-A_t)\frac{\varepsilon_0}{\kappa} - A_t\exp\left[-B_t(\kappa-\varepsilon_0)\right] & \text{if } \kappa \geq \varepsilon_0 \end{cases} \qquad (38)$$

$$g_c(\kappa) = \begin{cases} 0 & \text{if } \kappa \leq \varepsilon_0 \\ 1 - (1-A_c)\frac{\varepsilon_0}{\kappa} - A_c\exp\left[-B_c(\kappa-\varepsilon_0)\right] & \text{if } \kappa \geq \varepsilon_0 \end{cases} \qquad (39)$$

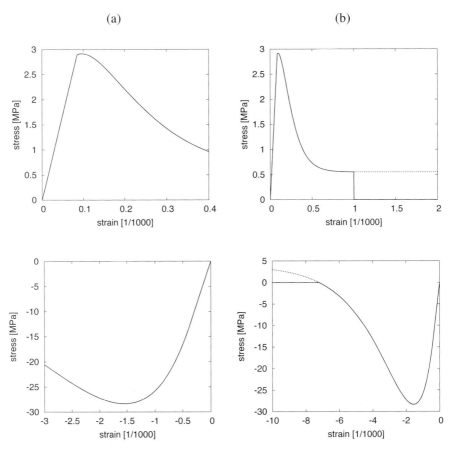

Figure 5. Stress-strain curves for Mazars damage model constructed for uniaxial tension (top) and uniaxial compression (bottom)

where ε_0 is the equivalent strain at the onset of nonlinearity, and A_t, B_t, A_c, and B_c are material parameters related to the shape of the uniaxial stress-strain diagrams. To ensure a continuous variation of slope of the compressive stress-strain curve, it is necessary to satisfy the condition $A_c B_c \varepsilon_0 = A_c - 1$, which reduces the number of independent parameters to four. A sample set of parameters used by Saouridis (1988) is $\varepsilon_0 = 10^{-4}$, $A_t = 0.81$, $B_t = 10450$, $A_c = 1.34$, and $B_c = 2537$. It is important to realize that functions (38)–(39) give a good approximation of the stress-strain curves only in the prepeak and early post-peak regime; see Fig. 5a. For large applied strains the stress level asymptotically approaches its

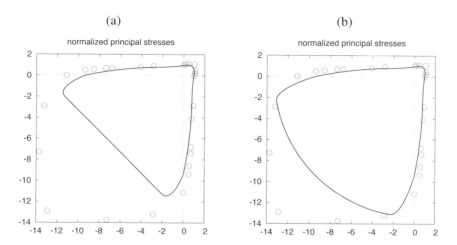

Figure 6. Biaxial failure envelope for Mazars damage model with parameters from Saouridis (1988): a) original version, b) with adjusted equivalent strain according to (41)

limit value $(1 - A_t)E\varepsilon_0$ in tension and $-(1 - A_c)E\varepsilon_0$ in compression. Typically, $A_t < 1$ and $A_c > 1$, and so the stress under uniaxial tension does not completely disappear and under uniaxial compression it changes sign from negative to positive; see the dotted curve in Fig. 5b. This can be remedied by setting $D_c = 1$ when $g_c(\kappa)$ evaluated from (39) exceeds 1, and by setting $D_t = 1$ when κ exceeds a certain limit; see the solid curve in Fig. 5b. Instead of accepting the sudden jump of tensile stress, it is more elegant to modify the definition of g_t such that the tensile stress asymptotically tends to zero. A suitable formula for g_t is e.g.

$$g_t(\kappa) = \begin{cases} 0 & \text{if} \quad \kappa \leq \varepsilon_0 \\ 1 - \dfrac{\varepsilon_0}{\kappa}\exp\left(-\dfrac{\kappa - \varepsilon_0}{\varepsilon_f - \varepsilon_0}\right) & \text{if} \quad \kappa \geq \varepsilon_0 \end{cases} \quad (40)$$

where ε_0 and ε_f are parameters.

The Mazars model allows an independent control of the tensile and compressive stress-strain curves and provides a good approximation of the biaxial failure envelope of concrete (locus of peak stress states under plane stress) under biaxial tension and under tension combined with compression. However the shape of the failure envelope is not realistic in the region of biaxial compression; see Fig. 6a.

A partial improvement of the shape of the failure envelope is obtained if the equivalent strain is adjusted by the multiplicative factor

$$\gamma = \frac{\sqrt{\sum_{I=1}^{3}(\sigma_I^-)^2}}{\sum_{I=1}^{3}|\sigma_I^-|} \qquad (41)$$

where $\sigma_I^- = -\langle -\sigma_I \rangle$ are the negative parts of principal stresses. The adjustment is done only if at least two principal stresses are negative and none of them is positive. In this way, the shape of the failure envelope becomes more realistic; see Fig. 6b. The strength under biaxial compression is now equal to the uniaxial compressive strength. According to the CEB-FIP Model Code (CEB91) it should be by 20% larger but the present version of the model does not allow an independent control of the biaxial compressive strength. For stress paths that do not generate any tensile strains, the model response is purely linear elastic. This means that nonlinear effects under highly confined compression, e.g. the so-called Hugoniot curve under hydrostatic compression, are not reproduced. Note that even though the factor γ in (41) is defined using the nominal stress σ, exactly the same value is obtained with σ replaced by the effective stress $\bar{\sigma} = \mathbb{E} : \varepsilon$, because $\sigma = (1-D)\bar{\sigma}$ and the factors $1-D$ appearing both in the numerator and the denominator of (41) cancel out. So the model remains fully explicit in the sense that stresses can be evaluated by straightforward substitution, without any iterations on the material point level.

Mazars model suffers by certain deficiencies that are typical of all isotropic damage models:

1. For a proportional loading path in the stress space the ratio between individual strain components remains constant. Consequently, the model cannot capture the experimentally observed dilatancy (volumetric expansion) at post-peak stages of the uniaxial compression test and of the shear test. Under uniaxial tension, the model predicts unlimited transverse contraction, whereas in reality the transverse strain would approach zero after the formation of a macroscopic crack.
2. When subjected to a large extension in one direction, the model completely loses stiffness not only in the direction of loading but also in the transverse directions.
3. No permanent strain is generated, i.e., unloading takes place to the origin. This could be acceptable for unloading from tension but certainly not for unloading from compression.

Deficiencies 1 and 2 motivate the development of more sophisticated models that take into account the anisotropy induced by damage. Deficiency 3 motivates the development of combined damage-plastic models, to be mentioned in chapter 4. Nevertheless, Mazars model remains quite popular in applications because it is relatively simple, easy to implement, and computationally efficient.

2 Smeared Crack Models

The concept of isotropic damage is appropriate for materials weakened by voids, but if the physical source of damage is the initiation and propagation of microcracks, isotropic stiffness degradation can be considered only as a first rough approximation. More refined damage models take into account the highly oriented nature of cracking, which is reflected by the anisotropic character of the damaged stiffness and compliance matrices.

In this section, we will look at a particular class of constitutive models developed specifically for quasibrittle materials such as concrete or rock under predominantly tensile loading. They will be referred to as smeared crack models (in the narrow sense).

There is some confusion in the literature because the expression "smeared crack" is often perceived as a counterpart of "discrete crack" and, in this sense, any softening continuum model (even if it is based on plasticity or damage) could be labeled as a "smeared crack model". However, we prefer to reserve this term for a more narrow class of models, which share some common features with but are different from plasticity and damage. Similar to plasticity (see chapter 4), they decompose the total strain into an elastic part and an inelastic part (called here the crack strain). Instead of postulating a yield condition and a flow rule, the inelastic strain due to crack opening is related directly to the traction transmitted across the crack plane.

The origins of smeared crack models for concrete fracture date back to the sixties (Rashid, 1968). Initially, the crack direction was assumed to remain fixed, and shear tractions across the crack were treated using the so-called retention factor (Suidan and Schnobrich, 1973). Later, it was proposed to allow rotation of the axes of material orthotropy (Cope et al., 1980), which stimulated the development of rotating crack models (Gupta and Akbar, 1984). The original fixed crack model was later extended to multiple non-orthogonal cracking (de Borst and Nauta, 1985).

As shown by Jirásek and Zimmermann (1998a), the rotating crack model suffers by stress locking (spurious stress transfer), which arises in finite element simulations on meshes that are not aligned with the crack directions. This phenomenon pollutes the numerical results and may lead to a misprediction of structural ductility and of the failure pattern. A remedy based on transition from the rotating crack description to a scalar damage model was proposed in Jirásek and Zimmermann (1998b).

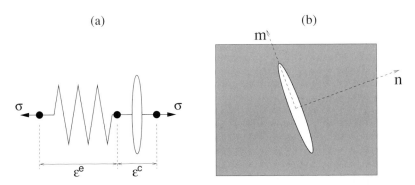

Figure 7. (a) Schematic representation of the smeared crack model as an elastic unit coupled in series to a crack unit, (b) local coordinate system aligned with the crack

2.1 One-Dimensional Smeared Crack Model

Smeared crack models decompose the total strain into two parts — one corresponds to the deformation of the uncracked material, and the other is the contribution of cracking. The response of the uncracked material can be governed by a general nonlinear material law but usually is assumed to be linear elastic. In one-dimensional setting, the strain decomposition is written as

$$\varepsilon = \varepsilon^e + \varepsilon^c \tag{42}$$

and the elastic strain ε^e is related to stress by Hooke's law

$$\sigma = E\varepsilon^e \tag{43}$$

The crack strain, ε^c, represents in a smeared manner the additional deformation due to the opening of cracks. The additive strain decomposition (42) corresponds to a rheological model in which an elastic spring is coupled in series with a unit representing the contribution of the crack, as schematically shown in Fig. 7a. Since the coupling is serial, both units transmit the same stress, σ.

Initially, the material is assumed to be in its virgin (uncracked) state, the crack strain vanishes and the overall response is linear elastic. A crack is initiated when the stress reaches the tensile strength of the material, f_t. A constitutive law governing the stress evolution after crack initiation is needed.

In early studies it was assumed that the traction transmitted by the crack drops to zero immediately after crack initiation. On the structural level, such an approach leads to results that are not objective with respect to the mesh size, as will be explained in Section 3.1. To ensure proper energy dissipation, and also to avoid unrealistic stress jumps, it is necessary to describe the loss of cohesion as a gradual process. Physically, this is justified by the fact that the formation of a macroscopic stress-free crack is in a heterogeneous material preceded by the initiation, growth and coalescence of a network of microcracks. For the purpose of modeling, we replace such a complicated system of small non-contiguous cracks by an equivalent cohesive crack, which can still transmit stress. This cohesive stress is then considered as a (usually decreasing) function of the crack strain,

$$\sigma = f^c(\varepsilon^c) \tag{44}$$

where the appropriate form of function f^c should be identified from experiments.[3]

Based on a comprehensive analysis of experimental results, Reinhardt et al. (1986) and Hordijk (1991) proposed a softening law in the form

$$\sigma = f^c(\varepsilon^c) \equiv f_t \left\{ \left[1 + \left(\frac{c_1 \varepsilon^c}{\varepsilon_f} \right)^3 \right] \exp\left(-\frac{c_2 \varepsilon^c}{\varepsilon_f} \right) - e^{-c_2} \left(1 + c_1^3 \right) \frac{\varepsilon^c}{\varepsilon_f} \right\} \tag{45}$$

where f_t is the uniaxial tensile strength, ε_f is the strain at which the crack becomes stress-free, and c_1 and c_2 are dimensionless material parameters controlling the shape of the softening curve. Their default values recommended by Hordijk (1991) are $c_1 = 3$ and $c_2 = 6.93$. The corresponding softening curve is plotted in Fig. 8 in terms of dimensionless stress σ/f_t and normalized crack strain $5.136\,\varepsilon^c/\varepsilon_f$ (the factor 5.136 leads to a unit area under the curve). Parameters f_t and ε_f fully define the Reinhardt-Hordijk law with default values of c_1 and c_2, and their change is respectively equivalent to vertical and horizontal rescaling of the normalized curve.

The Reinhardt-Hordijk law (45) gives the best fit of experimental results but is relatively complicated. Acceptable results are usually obtained with simpler relations such as the exponential law

$$\sigma = f^c(\varepsilon^c) \equiv f_t \exp\left(-\frac{\varepsilon^c}{\varepsilon_f} \right) \tag{46}$$

where ε_f is a material parameter controlling the steepness of the softening curve,

---

[3] Due to the localized character of the cracking zone, it is not possible to give an objective definition of the crack strain—it always depends on the gauge length along which the (average) strain is measured. From the physical point of view it is more meaningful to characterize the cracking material by the so-called traction-separation law, which links the cohesive stress transmitted by the crack to the crack opening, defined as the integral of all inelastic deformation across the width of the fracture process zone. This will be discussed in detail in chapter 3.

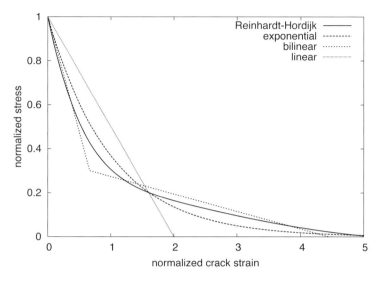

Figure 8. Normalized plot of various softening laws

or the bilinear law

$$\sigma = f^c(\varepsilon^c) \equiv \begin{cases} f_t\left(1 - \dfrac{\varepsilon^c}{\varepsilon_b}\right) + \sigma_b \dfrac{\varepsilon^c}{\varepsilon_b} & \text{if } 0 \leq \varepsilon^c \leq \varepsilon_b \\ \sigma_b \dfrac{\varepsilon_f - \varepsilon^c}{\varepsilon_f - \varepsilon_b} & \text{if } \varepsilon_b \leq \varepsilon^c \leq \varepsilon_f \\ 0 & \text{if } \varepsilon_f \leq \varepsilon^c \end{cases} \quad (47)$$

where ε_b and σ_b are the coordinates of the point at which the softening curve changes slope, and ε_f is the strain at which the cohesive stress vanishes. The normalized softening curves corresponding to the Reinhardt-Hordijk law (45), exponential law (46) and bilinear law (47) are compared in Fig. 8. The parameters have been determined such that the area under all the curves is equal to unity. It is clear that the deviations of the exponential and bilinear curves from Reinhardt's curve are relatively small. On the other hand, the linear softening curve, also plotted in Fig. 8, substantially differs from the nonlinear ones and, for concrete, can be used only as a very rough approximation.

Equations (42)–(44) fully define the one-dimensional smeared crack model (provided that the strain increases monotonically). The stress-strain curve has a linear pre-peak branch and a softening branch, and for each given stress between 0 and f_t it is easy to compute the corresponding total strain as the sum of the elastic

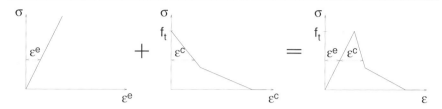

Figure 9. Total stress-strain curve obtained by summing the elastic strain and the crack strain

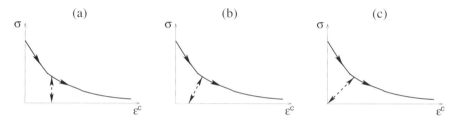

Figure 10. Various types of unloading

strain and the crack strain, as shown in Fig. 9.

Numerical simulations typically require the evaluation of stress for a given strain increment, starting from a state at which all variables are known. For this purpose, the increment of the crack strain must be determined such that the final stress in the elastic spring be equal to the final stress in the cracking unit. This condition leads to the equation

$$E(\varepsilon - \varepsilon^c) = f^c(\varepsilon^c) \tag{48}$$

describing the internal equilibrium in the rheological model in Fig. 7a. Here, ε is the given strain at the end of the step and ε^c is the unknown to be determined. For a linear (or bilinear) softening law, equation (48) is linear (or piecewise linear) and can be solved exactly. For general softening laws such as (45) or (46), the equation is nonlinear and needs to be solved iteratively. Since the functions describing softening are usually smooth and convex, the Newton method leads to very fast monotonic convergence.

For general applications, possible unloading must be taken into account and the cohesive law needs a refinement. The simplest assumption is that unloading is linear. In one extreme case, the crack strain can be considered as permanent, similar to plasticity (Fig. 10a). In another extreme case, perfect crack closure can be assumed, with vanishing crack strain upon complete stress removal (Fig. 10c).

None of these approaches is fully realistic, because crack closure certainly reduces the inelastic part of strain, but it is partially hindered by the roughness of the crack faces and by potential debris inside the cracks (Fig. 10b). Sophisticated models aiming at application to cyclic loading even consider nonlinear unloading, with stiffness recovery upon crack closure and with hysteresis effects. From the computational point of view, the equation to be solved during unloading has the same form as (48), but with a modified function f^c.

To cover the general case, including possible unloading, one should introduce a history variable κ, representing the maximum previously reached value of ε^c, and rewrite (44) as

$$\sigma = f^c(\varepsilon^c, \kappa) \tag{49}$$

The evolution of κ can formally be described by the loading-unloading conditions

$$\varepsilon^c - \kappa \leq 0, \quad \dot{\kappa} \geq 0, \quad (\varepsilon^c - \kappa)\dot{\kappa} = 0 \tag{50}$$

2.2 Multi-Dimensional Smeared Crack Models

In a general setting, equations (42)–(43) are written as

$$\varepsilon = \varepsilon^e + \varepsilon^c \tag{51}$$

$$\sigma = \mathbb{E} : \varepsilon^e \tag{52}$$

The crack strain, ε^c, represents in a smeared manner the additional deformation due to the opening of cracks. In a real material, microcracks have different sizes, shapes and orientations. They are not necessarily planar and their faces are rough. For the purpose of modeling, simplifications are needed. Therefore, we consider an equivalent computational crack which is perfectly planar and its direction is defined by a unit normal vector $\mathbf{n}$. It is usually assumed that the crack opening and sliding are affected only by the traction vector acting on the crack plane, i.e., by the first-order tensor

$$\mathbf{t}^c = \mathbf{n} \cdot \boldsymbol{\sigma} \tag{53}$$

It is useful to introduce local coordinates aligned with the crack, based on the unit vector $\mathbf{n}$ normal to the crack and on mutually orthogonal unit vectors $\mathbf{m}$ and $\mathbf{l}$ in the crack plane. With respect to such local coordinate system, the traction acting on the crack plane can be represented as

$$\mathbf{t}^c = \sigma_{nn}\mathbf{n} + \sigma_{nm}\mathbf{m} + \sigma_{nl}\mathbf{l} \tag{54}$$

where $\sigma_{nn} = \mathbf{t}^c \cdot \mathbf{n} = \mathbf{n} \cdot \boldsymbol{\sigma} \cdot \mathbf{n}$ is the normal traction and $\sigma_{nm} = \mathbf{t}^c \cdot \mathbf{m} = \mathbf{n} \cdot \boldsymbol{\sigma} \cdot \mathbf{m}$ and $\sigma_{nl} = \mathbf{t}^c \cdot \mathbf{l} = \mathbf{n} \cdot \boldsymbol{\sigma} \cdot \mathbf{l}$ are the shear tractions.

The traction components are linked to the crack opening and sliding by the generalized form of the cohesive law. It is natural to assume that opening of the crack contributes only to the normal strain ε_{nn}^c in the direction perpendicular to the crack, and sliding of the crack only to the shear strains γ_{nm}^c and γ_{nl}^c in planes perpendicular to the crack. These are the crack strain components expressed with respect to the local coordinate system with unit base vectors **n**, **m** and **l**. The corresponding strain components in global coordinates are obtained by the standard coordinate transformation. In tensorial notation, we can write

$$\varepsilon^c = \varepsilon_{nn}^c \mathbf{n} \otimes \mathbf{n} + \gamma_{nm}^c (\mathbf{n} \otimes \mathbf{m})_{\text{sym}} + \gamma_{nl}^c (\mathbf{n} \otimes \mathbf{l})_{\text{sym}} = (\mathbf{n} \otimes \mathbf{e}^c)_{\text{sym}} \tag{55}$$

where

$$\mathbf{e}^c = \varepsilon_{nn}^c \mathbf{n} + \gamma_{nm}^c \mathbf{m} + \gamma_{nl}^c \mathbf{l} \tag{56}$$

is the crack strain vector, i.e., a first-order tensor work-conjugate with $\mathbf{t}^c$.

Combining (51), (52) and (55), we obtain the stress-strain law

$$\boldsymbol{\sigma} = \mathbb{E} : (\boldsymbol{\varepsilon} - \mathbf{n} \otimes \mathbf{e}^c) \tag{57}$$

in which $\mathbf{e}^c$ plays the role of an internal variable. Note that, because of minor symmetry of $\mathbb{E}$, the symbol of symmetric part at $\mathbf{n} \otimes \mathbf{e}^c$ can be omitted.

For a virgin material, $\mathbf{e}^c = \mathbf{0}$ and the response is linear elastic. The computational crack is initiated when the stress state reaches the strength envelope, i.e., a certain limit surface in the stress space. Traditional smeared crack models control crack initiation by the Rankine criterion of maximum principal stress. Some models aiming at the description of fracture under shear and compression exploit more general criteria; see e.g. (Weihe, 1995). The initiation criterion should also specify the initial orientation of the crack. For example, the Rankine criterion postulates that, at crack initiation, the crack plane is perpendicular to the direction of maximum principal stress.

Under pure mode-I conditions, the crack propagates in its plane and the normal **n** does not change. Under general mixed-mode conditions, two conceptually different approaches are possible:

- Fixed crack models freeze the crack direction and postulate general softening laws that link all crack strain components to all components of the crack traction vector.
- Rotating crack models assume that the crack normal always remains aligned with the current direction of maximum principal strain. The shear components of crack strain, γ_{nm}^c and γ_{nl}^c, and of crack traction, σ_{nm} and σ_{nl}, are then zero, and the simple law (44) with σ and ε^c replaced by σ_{nn} and ε_{nn}^c is sufficient for the description of the softening process.

2.3 Fixed Crack Model

The fixed crack model freezes the crack direction determined at the moment of crack initiation. If the initiation criterion is based on the Rankine condition, the crack plane initially transmits a normal traction (equal to the tensile strength) but no shear tractions. Later on, the crack plane remains fixed but the principal axes can rotate. The crack plane in general transmits shear tractions that produce relative sliding of the crack faces, represented by shear components of crack strain.

In simplistic versions of the fixed crack model, the shear traction is taken as proportional to the shear crack strain, with a proportionality factor βG, where G is the shear modulus of elasticity and $\beta < 1$ is the so-called shear retention factor (Suidan and Schnobrich, 1973). This is of course not very realistic because such a cohesive crack is allowed to transmit large shear tractions even when it is widely open. If the shear retention factor is treated as a constant, it usually has to be set to a very small value, e.g. $\beta = 0.01$, to limit the spurious stress transfer that could lead to the so-called stress locking.[4] A better remedy is to make β variable, decreasing to zero as the crack opening grows (Cedolin and Poli, 1977). It is also possible to formulate the relation between $\mathbf{t}^c$ and $\mathbf{e}^c$ in the spirit of damage theory, with a scalar damage parameter depending on an equivalent crack strain that is computed from $\mathbf{e}^c$.

Whatever choice is made, the cohesive law can be written in the total form

$$\mathbf{t}^c = \boldsymbol{f}^c(\mathbf{e}^c) \tag{58}$$

or in the rate (incremental) form

$$\dot{\mathbf{t}}^c = \mathbf{E}^c \cdot \dot{\mathbf{e}}^c \tag{59}$$

where $\mathbf{E}^c = \partial \boldsymbol{f}^c / \partial \mathbf{e}^c$ is the second-order tangent crack stiffness tensor.

The traction vector $\mathbf{t}^c$, which is linked by the cohesive law (58) to the crack strain vector $\mathbf{e}^c$, must be equal to the projection of the stress tensor, which can be computed from the elastic strain. Combining equations (53) and (57) we get

$$\mathbf{t}^c = \mathbf{n} \cdot \mathbb{E} : \boldsymbol{\varepsilon} - \mathbf{n} \cdot \mathbb{E} : (\mathbf{n} \otimes \mathbf{e}^c) = \mathbf{n} \cdot \mathbb{E} : \boldsymbol{\varepsilon} - \mathbf{E} \cdot \mathbf{e}^c \tag{60}$$

where $\mathbf{E} = \mathbf{n} \cdot \mathbb{E} \cdot \mathbf{n}$ is the so-called acoustic tensor. Comparing the right-hand sides of (58) and (60), we obtain a generalized form of equation (48), which physically corresponds to the internal equilibrium condition between the tractions in the elastic unit and in the crack unit:

$$\mathbf{n} \cdot \mathbb{E} : \boldsymbol{\varepsilon} - \mathbf{E} \cdot \mathbf{e}^c = \boldsymbol{f}^c(\mathbf{e}^c) \tag{61}$$

For a given strain increment, the unknown crack strain $\mathbf{e}^c$ can be computed by solving (61), which is usually done in an iterative manner, e.g. by the Newton-Raphson method. Substitution of $\mathbf{e}^c$ into (57) then provides the corresponding stress.

---

[4] In the present context, stress locking means that spurious stresses build up around the band of cracking elements. This pollutes the numerical results and leads to an overestimated energy dissipation and nonzero residual strength of a cracked structure.

It is also useful to derive the tangent material stiffness tensor, which links the stress rate to the strain rate. From the rate form of the internal equilibrium condition (61),

$$\mathbf{n} \cdot \mathbb{E} : \dot{\varepsilon} - \mathbb{E} \cdot \dot{\mathbf{e}}^c = \mathbf{E}^c \cdot \dot{\mathbf{e}}^c \tag{62}$$

the crack strain rate

$$\dot{\mathbf{e}}^c = (\mathbf{E} + \mathbf{E}^c)^{-1} \cdot (\mathbf{n} \cdot \mathbb{E}) : \dot{\varepsilon} \tag{63}$$

is easily computed. Its substitution into the rate form of (57) then gives the rate form of the stress-strain law,

$$\dot{\boldsymbol{\sigma}} = \mathbb{E} : (\dot{\varepsilon} - \mathbf{n} \otimes \dot{\mathbf{e}}^c) = \mathbb{E} : \dot{\varepsilon} - \mathbb{E} : [\mathbf{n} \otimes (\mathbf{E} + \mathbf{E}^c)^{-1}] \cdot (\mathbf{n} \cdot \mathbb{E}) : \dot{\varepsilon} = \mathbb{E}_T : \dot{\varepsilon} \tag{64}$$

with the tangent stiffness tensor of the elastic-cracking material given by

$$\mathbb{E}_T = \mathbb{E} - (\mathbb{E} \cdot \mathbf{n}) \cdot (\mathbf{E} + \mathbf{E}^c)^{-1} \cdot (\mathbf{n} \cdot \mathbb{E}) \tag{65}$$

2.4 Rotating Crack Model

For the rotating crack model (RCM), the plane of the computational crack is allowed to rotate and is assumed to remain perpendicular to the direction of maximum principal strain. Of course, if a physical crack resides in a certain plane, the already formed crack faces cannot rotate. However, if the crack propagates under general loading, it can deviate from the original plane and become non-planar. Also, new cracks in planes that are not parallel with the initial crack plane can be initiated. All this is reflected in the model by a change of direction of the equivalent computational crack.

Alignment of the crack with the principal directions implies that the shear components of the crack strain, γ^c_{nm} and γ^c_{nl}, are zero. Equation (56) then simplifies to $\mathbf{e}^c = \varepsilon^c_{nn} \mathbf{n}$ and (57) turns into

$$\boldsymbol{\sigma} = \mathbb{E} : (\varepsilon - \mathbf{N}\varepsilon^c_{nn}) \tag{66}$$

where, for convenience, the second-order tensor $\mathbf{n} \otimes \mathbf{n}$ is denoted as $\mathbf{N}$. The normal traction on the crack plane can be evaluated by an appropriate projection of the stress tensor:

$$\sigma_{nn} = \mathbf{n} \cdot \boldsymbol{\sigma} \cdot \mathbf{n} = \mathbf{N} : \boldsymbol{\sigma} = \mathbf{N} : \mathbb{E} : \varepsilon - \mathbf{N} : \mathbb{E} : \mathbf{N}\varepsilon^c_{nn} \tag{67}$$

At the same time, this traction is linked to the normal crack strain, ε^c_{nn}, by a scalar cohesive law analogous to (44):

$$\sigma_{nn} = f^c(\varepsilon^c_{nn}) \tag{68}$$

Comparing the right-hand sides of (67) and (68), we obtain the internal equilibrium condition

$$\mathbf{N} : \mathbb{E} : \varepsilon - \mathbf{N} : \mathbb{E} : \mathbf{N}\varepsilon_{nn}^c = f^c(\varepsilon_{nn}^c) \tag{69}$$

which is analogous to (61) but is simpler, with only one scalar unknown ε_{nn}^c instead of the vectorial unknown $\mathbf{e}^c$.

If the uncracked material is isotropic, the elastic stiffness is given by $\mathbb{E} = \lambda \mathbf{1} \otimes \mathbf{1} + 2\mu \mathbb{I}_S$ where λ and μ are Lamé's constants and $\mathbb{I}_S$ is the symmetric fourth-order unit tensor. In this case, we have $\mathbf{N} : \mathbb{E} = \lambda \mathbf{1} + 2\mu \mathbf{N}$ and $\mathbf{N} : \mathbb{E} : \mathbf{N} = \lambda + 2\mu$, and (69) can be written as

$$\lambda \mathbf{1} : \varepsilon + 2\mu \mathbf{N} : \varepsilon - (\lambda + 2\mu)\varepsilon_{nn}^c = f^c(\varepsilon_{nn}^c) \tag{70}$$

where $\mathbf{1} : \varepsilon$ is the trace of the strain tensor and $\mathbf{N} : \varepsilon = \mathbf{n} \cdot \varepsilon \cdot \mathbf{n}$ is the (total) strain normal to the crack, which is in fact equal to the maximum principal strain. Similar to the one-dimensional case, the crack strain ε_{nn}^c can be obtained by the Newton method (or in closed form, if the softening law is linear or piecewise linear). Substitution of the result into (66) then provides the stress tensor.

In order to derive the tangent stiffness tensor for the rotating crack model, we first convert the internal equilibrium equation (70) to the rate form. It is important to realize that the crack normal $\mathbf{n}$ in general rotates, and so the second-order tensor $\mathbf{N} = \mathbf{n} \otimes \mathbf{n}$ is also variable in time. This needs to be taken into account when differentiating the second term in (70). The resulting rate equation thus reads

$$\lambda \mathbf{1} : \dot{\varepsilon} + 2\mu \mathbf{N} : \dot{\varepsilon} + 2\mu \dot{\mathbf{N}} : \varepsilon - (\lambda + 2\mu)\dot{\varepsilon}_{nn}^c = E^c \dot{\varepsilon}_{nn}^c \tag{71}$$

where $E^c = \mathrm{d}f^c/\mathrm{d}\varepsilon_{nn}^c$ is the softening modulus. The rates $\dot{\mathbf{n}}$ and $\dot{\mathbf{N}}$ can be expressed in terms of the current strain and the strain rate, but the derivation is somewhat tedious. As explained in the footnote,[5] $\dot{\mathbf{N}} = \mathbb{N} : \dot{\varepsilon}$ where $\mathbb{N}$ is a certain fourth-order tensor depending only on ε and $\mathbf{n}$. It can be shown that $\mathbb{N} : \varepsilon = \mathbf{O}$, and thus the the third term in (71) cancels.

---

[5] According to the fundamental assumption of the rotating crack model, the crack remains perpendicular to the direction of maximum principal strain, i.e., $\mathbf{n} = \mathbf{p}_1$ is the normalized eigenvector of ε associated with the largest eigenvalue. If the maximum principal strain ε_1 is strictly greater than the other two principal strains, ε_2 and ε_3, the expression for the rate of $\mathbf{p}_1$ can be derived by manipulating the rate form of the relations $\varepsilon \cdot \mathbf{p}_I = \varepsilon_I \mathbf{p}_I$ and $\mathbf{p}_I \cdot \mathbf{p}_J = \delta_{IJ}$, $I, J = 1, 2, 3$. The resulting formula reads

$$\dot{\mathbf{p}}_1 = \frac{\dot{\varepsilon}_{12}\mathbf{p}_2}{\varepsilon_1 - \varepsilon_2} + \frac{\dot{\varepsilon}_{13}\mathbf{p}_3}{\varepsilon_1 - \varepsilon_3} \tag{72}$$

where $\dot{\varepsilon}_{12} = \mathbf{p}_1 \cdot \dot{\varepsilon} \cdot \mathbf{p}_2$ and $\dot{\varepsilon}_{13} = \mathbf{p}_1 \cdot \dot{\varepsilon} \cdot \mathbf{p}_3$ are the components of the strain rate tensor $\dot{\varepsilon}$ with respect to the principal coordinate system. It is then straightforward to evaluate $\dot{\mathbf{N}} = \dot{\mathbf{p}}_1 \otimes \mathbf{p}_1 + \mathbf{p}_1 \otimes \dot{\mathbf{p}}_1$ and convert it to the form $\dot{\mathbf{N}} = \mathbb{N} : \dot{\varepsilon}$ where

$$\mathbb{N} = \frac{2}{\varepsilon_1 - \varepsilon_2}(\mathbf{p}_1 \otimes \mathbf{p}_2)_{\mathrm{sym}} \otimes (\mathbf{p}_1 \otimes \mathbf{p}_2)_{\mathrm{sym}} + \frac{2}{\varepsilon_1 - \varepsilon_3}(\mathbf{p}_1 \otimes \mathbf{p}_3)_{\mathrm{sym}} \otimes (\mathbf{p}_1 \otimes \mathbf{p}_3)_{\mathrm{sym}} \tag{73}$$

Based on (71) with the third term dropped, the crack strain rate corresponding to a given strain rate is easily expressed as

$$\dot{\varepsilon}_{nn}^c = \frac{\lambda \mathbf{1} : \dot{\varepsilon} + 2\mu \mathbf{N} : \dot{\varepsilon}}{\lambda + 2\mu + E^c} = \frac{\mathbf{N} : \mathbb{E}}{\lambda + 2\mu + E^c} : \dot{\varepsilon} \qquad (74)$$

Substitution of this result into the rate form of (66) leads to the direct relation between the rates of stress and strain,

$$\dot{\boldsymbol{\sigma}} = \mathbb{E} : (\dot{\varepsilon} - \dot{\mathbf{N}}\varepsilon_{nn}^c - \mathbf{N}\dot{\varepsilon}_{nn}^c) = \mathbb{E} : \dot{\varepsilon} - \mathbb{E} : \dot{\mathbf{N}} : \dot{\varepsilon}\,\varepsilon_{nn}^c - \mathbb{E} : \mathbf{N}\frac{\mathbf{N} : \mathbb{E}}{\lambda + 2\mu + E^c} : \dot{\varepsilon} \qquad (75)$$

This can be further simplified, since $\mathbb{E} : \mathbf{N} = 2\mu \mathbf{N}$ (as follows from (73)). The final formula for the tangent stiffness tensor reads

$$\mathbb{E}_T = \mathbb{E} - 2\mu \varepsilon_{nn}^c \dot{\mathbf{N}} - \frac{\mathbb{E} : \mathbf{N} \otimes \mathbf{N} : \mathbb{E}}{\lambda + 2\mu + E^c} \qquad (76)$$

The rotating crack model is probably the simplest model that can take into account cracking-induced anisotropy within a framework close to continuum damage mechanics. Nevertheless, let us point out that it has a number of drawbacks, analyzed in detail by Rots (1988) and Jirásek and Zimmermann (1998a). For instance, it is not thermodynamically consistent. This can be demonstrated by looking at formula (76) for the tangent stiffness. Its derivation is valid not only in the case of crack growth, but also during unloading (with constant positive modulus E^c), when the pre-cracked material should respond as an anisotropic linear elastic one, and it should be characterized by a constant and positive definite tangent stiffness. However, formula (76) indicates that the stiffness tensor remains constant only if tensors $\mathbf{N}$ and $\dot{\mathbf{N}}$ remain constant, i.e., if the principal axes do not rotate. As shown by Jirásek and Zimmermann (1998a), under certain circumstances the unloading stiffness can even lose positive definiteness, which leads to nonphysical instabilities.

2.5 Rotating Crack Model with Transition to Scalar Damage

In applications to failure simulation by the finite element method, it is necessary to avoid stress locking, which arises on meshes not aligned with the crack directions and is caused by the poor kinematic representation of the discontinuous displacement field around a macroscopic crack. This leads to spurious stress transfer, which pollutes the numerical results and may result into a misprediction of structural ductility and of the failure pattern.

A typical example of stress locking in a fracture simulation is presented in Fig. 11. A three-point bend specimen (Fig. 11a) is discretized by constant-strain triangular elements. The macroscopic crack is initiated at the notch tip and propagates

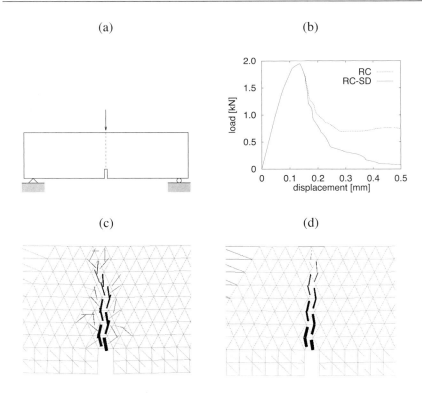

Figure 11. Three-point bending test: a) geometry and loading, b) load-displacement curves, c) crack pattern for the RC model, d) crack pattern for the RC-SD model

across the specimen. After a sufficiently large load-point displacement has been applied, the specimen should break completely, and the resisting force should vanish. However, the post-peak branch of the computed load-displacement diagram does not tend to zero but to a non-negligible residual value of the resisting force, which then remains roughly constant, or even slightly increases (dashed curve in Fig. 11b). This is indeed a paradoxical result since strains in the cracking elements are no doubt sufficiently large to make the corresponding stresses vanish. The stress-strain law is designed in such a way that the stress transferred across a crack is zero as soon as the crack opening reaches a certain critical value. For the fixed crack model with non-zero retention factor (Rots, 1988), stress locking occurs due to shear stresses transferred across the crack. The rotating crack model does not produce any shear stress on the crack faces but it still locks. The origin of this type of locking lies in the fact that the local direction of a model crack (determined from the direction of the maximum principal strain at the Gauss point) is not always aligned with the global direction of the macroscopic crack propagating across the body and represented by a band of cracking elements.

A remedy based on the modification of the constitutive law was proposed by Jirásek and Zimmermann (1998b). The basic idea is that the standard RCM is applied only during the initial stage of cracking. When the crack opens sufficiently wide the model switches to a damage-type formulation. At the moment of transition, the material has a certain secant stiffness matrix that exhibits anisotropy induced by cracking. The final stage of cracking is described by a damage model that uses this anisotropic stiffness multiplied by a scalar factor that decays to zero as cracking further progresses. Consequently, the combined model at least partially preserves the directional sensitivity of the RCM. Moreover, all stresses decay to zero as damage progresses, and so the model does not transfer spurious stresses across a widely open crack. This combined approach is called the rotating crack model with transition to scalar damage (RC-SD). Let us emphasize that the damage is scalar but the resulting stiffness is not isotropic. The failure pattern (Fig. 11d) and load-displacement diagram (solid curve in Fig. 11b) show that the RC-SD model can indeed completely release the stresses transmitted across a macroscopic crack even if the crack does not run along the element lines.

2.6 Examples of Failure Simulations

To illustrate the role of anisotropy and the influence of mesh-induced directional bias on the simulated crack trajectory, consider the double-edge-notched (DEN) specimen tested by Nooru-Mohamed (1992). The experimental setup is shown in Fig. 12a.

The nonproportional loading path 4c was chosen for the comparison. This is the most challenging test of the entire testing program, because the final failure pattern consists of two cracks with a relatively high curvature; see Fig. 12b. During the first stage, the specimen is loaded by an increasing "shear" force, P_s, until the maximum force that the specimen can carry is reached. Then, in the second stage, this force is kept constant and the specimen is stretched under displacement control in the vertical direction, which generates a vertical reaction force P.

The material models considered here include the simple isotropic damage model with the smooth Rankine or modified Mises definitions of equivalent strain (IDM-R and IDM-M), the Mazars model with two damage variables (MM), and the rotating crack model with transition to scalar damage (RCSD). The basic material parameters are chosen as $E = 29$ GPa, $\nu = 0.2$ and $f_t = 3$ MPa, and an exponential softening law is used. The ratio between the compressive and tensile strength (considered by the IDM-M and by Mazars model) is set to 10.

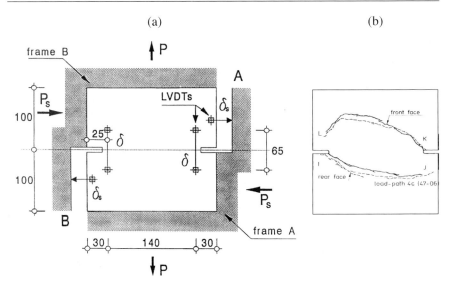

Figure 12. The DEN specimen: (a) geometry and loading, (b) observed crack pattern for loading path 4c (after Nooru-Mohamed (1992))

The failure patterns are presented in Fig. 13. All of them refer to a late stage of the loading process, when the cracking is localized into two macroscopic cracks, which are computationally represented as bands of cracking elements. The dark color corresponds to a high damage level, close to 1.

It is interesting to note that the shapes of the crack bands obtained with individual models are quite different. The simple isotropic damage model with smooth Rankine expression for equivalent strain (IDM-R) leads to a misprediction of the failure pattern (Fig. 13a). The other three models give a better prediction, with two distict curved macroscopic cracks. The simple isotropic damage model with modified Mises expression for equivalent strain (IDM-M) overestimates the curvature of the cracks (Fig. 13b) while the rotating crack model underestimates it (Fig. 13d). Mazars model globally approximates the crack trajectory quite well but locally is strongly affected by directional mesh bias (Fig. 13c).

Since the cracking bands are fully localized, their thickness depends on the size of finite elements. If the model parameters are kept the same and the mesh is refined, the thickness of the computational crack band decreases and the energy dissipated in the failure process decreases as well. The structural response thus becomes more brittle. Sensitivity of the numerical results to the size of finite elements will be analyzed in detail in the next section.

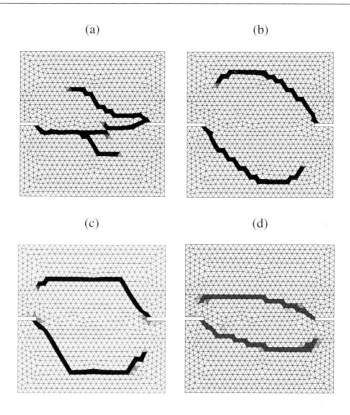

Figure 13. Crack band trajectory in the DEN test for (a) IDM-R, (b) IDM-M, (c) Mazars model, (d) RCSD

3 Strain Localization due to Softening

3.1 Strain Localization in One Dimension

The idea of modeling damaged concrete and other quasibrittle materials as strain-softening continua, which emerged in the seventies, was not immediately accepted by the entire scientific community. It was necessary to overcome a number of objections regarding the ill-posedness and lack of objectivity of such a formulation. Indeed, most of the early analyses were not truly objective and, upon mesh refinement, their results would not have converged to a meaningful solution. Instead of going into theoretical details, let us explain the nature of the problem by a simple example.

Consider a straight bar of a constant cross section A and of total length L under

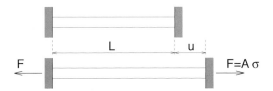

Figure 14. Bar under uniaxial tension

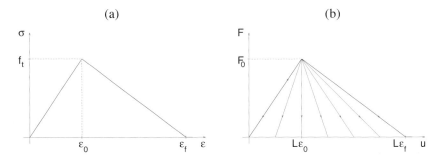

Figure 15. a) Stress-strain diagram with linear softening, b) fan of possible post-peak branches of the load-displacement diagram

uniaxial tension; see Fig. 14. The material is assumed to obey a simple stress-strain law with linear elasticity up to the peak stress, f_t, followed by linear softening; see Fig. 15a. The strain at which the transmitted stress completely disappears is denoted by ε_f. The peak stress is attained at strain $\varepsilon_0 = f_t/E$ where E is Young's modulus of elasticity. If the bar is loaded in tension by an applied displacement u at one of the supports, the response remains linear elastic up to $u_0 = L\varepsilon_0$. At this state, the force transmitted by the bar (reaction at the support) attains its maximum value, $F_0 = A f_t$. After that, the resistance of the bar starts decreasing. At each cross section, stress can decrease either at increasing strain (softening) or at decreasing strain (elastic unloading). The equation of equilibrium implies that the stress profile must remain uniform along the bar. However, at any given stress level $\bar{\sigma}$ between zero and f_t, there exist two values of strain, ε_s and ε_u, for which the constitutive equation is satisfied (Fig. 16a), and so the strain profile does not have to be uniform. In fact, any piecewise constant strain distribution that jumps between the two strain values (respectively corresponding to softening and unloading) represents a valid solution; see Fig. 16b. Let us denote by L_s the cumulative length of the softening regions and by $L_u = L - L_s$ the cumulative length of the unloading regions. When stress is completely relaxed to zero, the strain in the softening region is $\varepsilon_s = \varepsilon_f$ and the strain in the unloading region

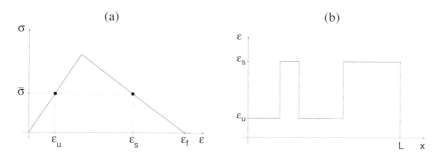

Figure 16. a) Two strain values corresponding to the same stress level, b) piecewise constant strain profile

is $\varepsilon_u = 0$; the total elongation of the bar is therefore $u_f = L_s\varepsilon_s + L_u\varepsilon_u = L_s\varepsilon_f$. Now the point is that the length L_s is undetermined, and it can be anything between zero and L. This means that the problem has infinitely many solutions, and the corresponding post-peak branches of the load-displacement diagram fill the fan shown in Fig. 15b. This fan is bounded on one side by the solution with uniform softening ($u_f = L\varepsilon_f$) and on the other side by the solution with uniform unloading ($u_f = 0$). The latter limit correctly represents the case when the bar is unloaded just before any damage takes place. All the other solutions describe a possible process in which a part of the bar is damaged and the bar loses its structural integrity. It is not clear which of these solutions is the "correct" one, i.e., the one that reflects the actual failure process.

The ambiguity is removed if imperfections are taken into account. Real material properties and sectional dimensions cannot be perfectly uniform. Suppose that the strength in a small region is slightly lower than in the remaining portion of the bar. When the applied stress reaches the reduced strength, softening starts and the stress decreases. Consequently, the material outside the weaker region must unload elastically, because its strength has not been exhausted. This leads to the conclusion that the size of the softening region is dictated by the size of the region with minimum strength. Such a region can be arbitrarily small, and so the corresponding softening branch is arbitrarily close to the elastic branch of the load-displacement diagram. Thus the standard strain-softening continuum formulation leads to a solution that has several pathological features:

1. The softening region is infinitely small.
2. The load-displacement diagram always exhibits snapback, independently of the size of the structure and of the ductility of the material.
3. The total amount of energy dissipated during the failure process is zero.

From the mathematical point of view, these annoying features are related to the loss of ellipticity of the governing differential equation. The boundary value prob-

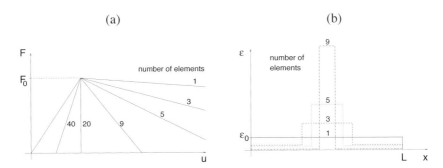

Figure 17. Effect of mesh refinement on the numerical results: a) load-displacement diagrams, b) strain profiles

lem becomes ill-posed, i.e., it does not have a unique solution with continuous dependence on the given data. From the numerical point of view, ill-posedness is manifested by pathological sensitivity of the results to the size of finite elements. For example, suppose that the bar is discretized by N_e two-node elements with linear displacement interpolation and that the weakest section is located at the vertical plane of symmetry. If the numerical algorithm properly captures the most localized solution, the softening region extends over one element, and we have $L_s = L/N_e$. The post-peak branch therefore strongly depends on the number of elements, and it approaches the initial elastic branch as the number of elements tends to infinity; see Fig. 17a (constructed for a stress-strain law with $\varepsilon_f/\varepsilon_0 = 20$). The strain profiles at $u = 2u_0$ for various mesh refinements are plotted in Fig. 17b. In the limit, the profiles tend to $2u_0\delta(x - L/2)$ where $\delta(x - L/2)$ denotes the Dirac distribution centered at $x = L/2$. The limit solution represents a displacement jump at the center, with zero strain everywhere else.

3.2 Strain Localization in Multiple Dimensions

To demonstrate that the pathological sensitivity of the numerical results to the discretization is not limited to the uniaxial tensile case, we simulate the three-point bending test of a concrete beam with and without a notch. The beam has a square cross section 100×100 mm and span 450 mm, and the notch is 5 mm thick and extends over one half of the beam depth. These dimensions correspond to the experiments performed by Kormeling and Reinhardt (1983).

Failure of the notched beam is first simulated using the finite element mesh in Fig. 18a, with minimum element size 5 mm. The elements are standard bilinear quadrilaterals with 2×2 integration points, and the elastic constants are set to $E = 20$ GPa and $\nu = 0.2$. The adopted damage law (40) corresponds to linear

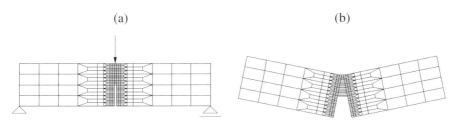

Figure 18. Notched beam under three-point bending: (a) coarsest finite element mesh with supports and load, (b) deformed mesh at complete failure (displacements exaggerated)

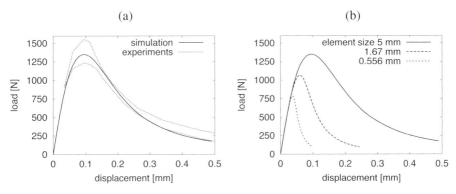

Figure 19. Load-displacement curves for notched beam: (a) simulation on coarse mesh matching experiments, (b) simulations on three different meshes

elasticity up to peak stress, followed by exponential softening. If the parameters of the damage law are set to $\varepsilon_0 = 120 \times 10^{-6}$ and $\varepsilon_f = 7 \times 10^{-3}$, the simulated peak load is within the experimental bounds and the load-displacement curve favorably compares with experimental data; see Fig. 19a. The deformed mesh at the end of the simulation (with exaggerated displacements) is plotted in Fig. 18b. It is clear that strain localizes into one vertical layer of elements starting at the notch.

If the material parameters ε_0 and ε_f are kept fixed but the mesh is refined, the results change dramatically. Fig. 19b shows the numerical load-displacement curves obtained on meshes with minimum element sizes respectively 5 mm, 1.67 mm and 0.556 mm. These element sizes are used in a narrow zone around the axis of symmetry, where the strains are expected to localize, while the other parts of the specimen are discretized by larger elements, to keep the number of unknowns and equations limited. The load-displacement diagram clearly shows that both the

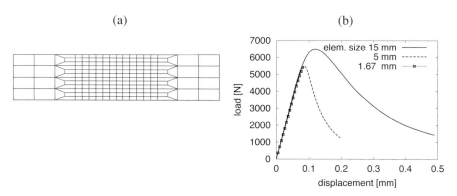

Figure 20. Simulations of unnotched beam: (a) coarsest mesh, (b) load-displacement curves obtained on three different meshes

peak load and the total dissipated energy (area under the curve) decrease as the mesh is refined. Such a spurious dependence of the results on the discretization is unacceptable.

The same material parameters have been used to simulate the three-point bending test of an unnotched beam with the same dimensions. In this case, damage is initially distributed over a region around the bottom face, and so the minimum element size is used in the central third of the beam; see Fig. 20a. The test has been run on three meshes with minimum element sizes respectively 15 mm, 5 mm and 1.67 mm. The load-displacement curves corresponding to the coarse and medium meshes differ substantially, as shown in Fig. 20b. Again, mesh refinement from 15 mm to 5 mm leads to peak load reduction and to a more brittle response. Upon further refinement, the peak load does not decrease any more, but the equilibrium iteration fails to converge at peak. The loss of convergence is due to the abrupt change of the strain increment distribution from a smoothly distributed to a highly localized one. As the mesh gets finer, the number of possible combinations of loading and unloading at individual Gauss points increases and the numerical algorithm has difficulties in finding the actual one.

The simple one-dimensional example presented in the previous section illustrated the essence of the problem with localization of inelastic strain into a process zone of an arbitrarily small width. In one dimension, localization occurs when the peak of the stress-strain diagram is reached, independently of the specific constitutive model used. Formulations based on damage mechanics, smeared cracks or softening plasticity all lead to the same type of behavior as soon as the tangent material stiffness ceases to be positive. In multiple dimensions, the analysis of the localization process is more complicated and the derivation of criteria for potential onset of localization represents a challenging mathematical problem.

The fundamental question is under which conditions the inelastic strain increments can localize in one or more narrow bands separated from the remaining part of the body by weak discontinuity surfaces. Across such surfaces, the displacement field remains continuous but the strain field can have a jump. At the onset of localization, the current strains are still continuous and the jump appears only in the strain rates. The necessary conditions for the existence of such a solution, inspired by the early works of Hadamard (1903) and Hill (1958), have been developed, among others, for plasticity by Rudnicky and Rice (1975) and Ottosen and Runesson (1991) and for damage by Rizzi et al. (1996).

3.3 Mesh-Adjusted Softening Modulus (Crack Band Approach)

Pathological sensitivity of the finite element results to the element size is not acceptable and must be avoided. The simplest remedy, frequently used in engineering applications, is based on an adjustment of the stress-strain diagram depending on the size of the element. This technique was originally proposed for softening plasticity (Pietruszczak and Mróz, 1981) but it is best explained in the context of crack models (Bažant and Oh, 1983). In a fracture process, the inelastic part of deformation is due to the opening of microcracks that later merge and form a macroscopic crack. If the crack opening is modeled as a displacement discontinuity, which is done by discrete crack models (see chapter 3), it is possible to formulate an objective traction-separation law relating the traction transmitted by the partially formed crack to the displacement jump (crack opening). The numerical results do not exhibit pathological mesh sensitivity because the physical description of the fracture process is objective—the crack opening is a well defined quantity. On the other hand, smeared crack models describe the inelastic part of deformation as the cracking strain. It is tacitly assumed that the material can still be treated as a continuum. However, as the crack opening is in reality a displacement discontinuity, it must be transformed into an equivalent strain by smearing over a certain distance. Naturally, the numerical results correspond to reality only if the width of the simulated softening region is equal to the smearing distance. These considerations lead to alternative techniques of ensuring objectivity of the model:

1. Supplement the stress-strain law by an additional material parameter specifying the actual width of the fracture process zone. Implement a localization limiter into the constitutive model, and adjust its length scale such that the numerically simulated process zone has the correct width.
2. Use the traction-separation law as the basic constitutive description. At each material point, construct the stress-strain law by transformation of the traction-separation law, taking into account the width of the numerically simulated process zone (which depends on local mesh characteristics, e.g.,

the size and type of the corresponding finite element).
The first approach is exploited by advanced localization limiters, to be discussed in Section 4. Here we focus on the second approach. The technique to be described adjusts the slope of the curve that represents the inelastic behavior depending on local mesh characteristics, i.e., it works with a mesh-adjusted softening modulus.

A discrete cohesive crack can be characterized by a traction-separation law in the general form

$$\sigma_{nn} = f^w(w_n) \tag{77}$$

where w_n is the crack opening (normal component of displacement jump) and σ_{nn} is the stress component normal to the crack (normal traction on the crack faces). For simplicity, consider a uniaxial situation. The crack traction σ_{nn} corresponds to the applied uniaxial stress σ. If the crack opening is smeared over a distance h, the resulting cracking strain is

$$\varepsilon^c = \frac{w_n}{h} = \frac{\tilde{f}^w(\sigma)}{h} \tag{78}$$

where $\tilde{f}^w$ is the inverse function of f^w. Equation (78) combined with the elastic law $\varepsilon^e = \sigma/E$ gives an inverse stress-strain relation

$$\varepsilon = \varepsilon^e + \varepsilon^c = \frac{\sigma}{E} + \frac{\tilde{f}^w(\sigma)}{h} \tag{79}$$

describing the softening branch of the stress-strain diagram; see Fig. 21.

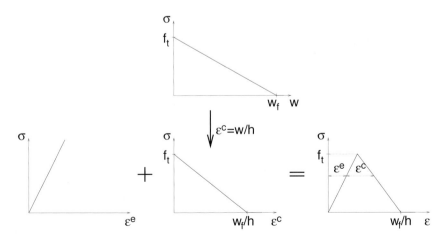

Figure 21. Stress-strain diagram derived from a traction-separation law

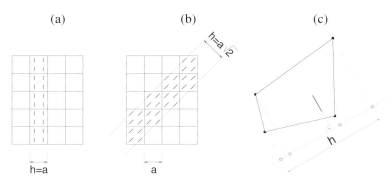

Figure 22. a) Crack band parallel to element sides, b) crack band across element diagonals, c) equivalent element size

After localization, equation (79) is valid in the softening region of length L_s. The remaining part of the bar of length $L_u = L - L_s$ is elastically unloading, and its strain is only ε^e. The total elongation of the bar is

$$u = L_u \varepsilon^e + L_s(\varepsilon^e + \varepsilon^c) = L\varepsilon^e + L_s\varepsilon^c = \frac{L\sigma}{E} + \frac{L_s}{h}\tilde{f}^w(\sigma) \qquad (80)$$

The first term, $L\sigma/E$, represents the contribution of elastic deformation. If $L_s = h$, the second term exactly corresponds to the crack opening, w_n. The load-displacement diagram is then independent of the finite element discretization, provided that we can à priori determine the size of the simulated process zone, L_s. In other words, we have to deduce the expected value of L_s from the mesh characteristics, and set h in (78) equal to that value. As shown in Section 3.1, in the uniaxial problem inelastic strain localizes in a single element, and so L_s is equal to the length of that element. In multiple dimensions, inelastic strain in general localizes in a band of elements running across the mesh and forming the so-called crack band. Usually, the band is the smallest possible pattern that still allows separation of nodes on its opposite sides. The correct value of h is affected not only by the mesh size but also by the inclination of the crack band with respect to the mesh lines. This is illustrated in Fig. 22b, which explains why the correct value of h for a band propagating along the diagonals of a regular square mesh is $\sqrt{2}$ times larger compared to the case when the band is aligned with element sides (Fig. 22a).

Based on this type of considerations and on numerical experiments, Rots (1988) proposed certain rules for the choice of the equivalent element size, h, for a number of typical situations. A more rigorous approach was developed by Oliver (1989). In practical simulations, it seems to be reasonable to compute h as the size of the element projected onto the crack normal; see Fig. 22c. This estimate can be improved by applying a correction factor proposed by Červenka et al. (1995).

Special care must be taken of higher-order elements, where the effective width of the crack band depends also on the integration scheme. For example, for a 3-node uniaxial element with quadratic displacement interpolation and 3-point Gauss integration, there exists a solution with softening localized into 2 integration points only, which corresponds to an effective width of the process zone equal to $13/18$ times the length of the element.

Very similar considerations can be done for other types of constitutive models. For example, in softening plasticity we deal with the plastic strain instead of the cracking strain but the nature of the problem remains the same. For damage models the stress is not explicitly related to the inelastic strain. This complicates the transformation of an objective traction-separation law into a size-dependent stress-strain law. In the uniaxial case the stress-strain law reads

$$\sigma = (1-D)E\varepsilon = E(\varepsilon - D\varepsilon) \tag{81}$$

where D is the damage parameter whose dependence on the (equivalent) strain is to be found. The product $D\varepsilon$ can be interpreted as the inelastic strain that should be related to the stress through (78).[6] This condition leads to the equation

$$f^w(hD\varepsilon) = (1-D)E\varepsilon \tag{82}$$

which implicitly defines the damage law $D = g(\varepsilon)$ associated with a given softening law $\sigma = f^w(w_n)$ and a given element size h. For a linear softening law

$$\sigma = f^w(w_n) = \left(1 - \frac{w_n}{w_f}\right) f_t \tag{83}$$

equation (82) is also linear (in terms of D) and can be solved analytically. The result is

$$D = g(\varepsilon) = \frac{1 - \dfrac{\varepsilon_0}{\varepsilon}}{1 - \dfrac{\varepsilon_0}{\varepsilon_f}} \qquad \text{if } \varepsilon_0 \leq \varepsilon \leq \varepsilon_f \tag{84}$$

where $\varepsilon_0 = f_t/E$ is the equivalent strain at peak stress (i.e., at the limit of elasticity), and $\varepsilon_f = w_f/h$ is the equivalent strain at the critical crack opening (i.e., when the crack becomes stress-free).

In general, equation (82) is nonlinear and cannot be solved in a closed form. For a given strain level ε, the value of $D = g(\varepsilon)$ can be obtained by solving

---

[6] It is tacitly assumed that localization starts at the onset of nonlinearity. In a general case it is necessary to use the decomposition into strain in the unloading material and additional strain in the softening material, rather than the decomposition into elastic and inelastic strain.

(82) numerically using e.g. the Newton method. For reasonable functions f^w the iterative process usually converges very fast.

The above considerations apply to the uniaxial case and to softening starting immediately at the onset of nonlinearity. If the nonlinear part of the uniaxial stress-strain diagram has a hardening branch, it has to be taken into account that localization starts at peak. Thus the pre-peak part of the diagram is taken as objective, size-independent, and the adjustment according to the element size is done only in the post-peak range. In a general multiaxial case it is not easy to detect the onset of localization and to define which geometric characteristic of the element should play the role of the size h. These problems can be bypassed if the constitutive model is regularized by a localization limiter that imposes a certain size of the localization band considered as an additional material parameter.

4 Regularized Softening Models

Fully regularized description of localized inelastic deformation can be achieved by a proper generalization of the underlying continuum theory. Generalized continua in the broad sense can be classified according to the following criteria:
1. Generalized kinematic relations (and the dual equilibrium equations).
   a) Continua with microstructure, e.g., Cosserat-type continua or strain-gradient theories.
   b) Continua with nonlocal strain, e.g., nonlocal elasticity.
2. Generalized constitutive equations.
   a) Material models with gradients of internal variables (in some cases also with gradients of thermodynamic forces).
   b) Material models with weighted spatial averages of internal variables (in some cases also of thermodynamic forces).

Here we focus on the second class of models, with enhancements on the level of the constitutive equations. Their advantage is that the kinematic and equilibrium equations remain standard, and the notions of stress and strain keep their usual meaning.

4.1 Integral-Type Nonlocal Models

Integral-type nonlocal models abandon the classical assumption of locality and admit that stress at a certain point depends not only on the state variables at that point but in general on the distribution of state variables over the whole body, or at least on their distribution in a finite neighborhood of the point under consideration. The first models of this type, proposed in the 1960s, aimed at improving the description of elastic wave dispersion in crystals. Nonlocal elasticity was further developed by Eringen, who later extended it to nonlocal elastoplasticity (Eringen, 1981, 1983). Subsequently it was found that certain nonlocal formulations can act as efficient localization limiters with a regularizing effect on problems with strain localization (Pijaudier-Cabot and Bažant, 1987). Nonlocal formulations were elaborated for a wide spectrum of models, including softening plasticity (Bažant and Lin, 1988b; Strömberg and Ristinmaa, 1996), smeared crack models (Bažant and Lin, 1988a; Jirásek and Zimmermann, 1997), or microplane models (Bažant and Ožbolt, 1990; Ožbolt and Bažant, 1996; Bažant and Luzio, 2004). For a detailed account, the reader is referred to Bažant and Jirásek (2002).

Generally speaking, the nonlocal approach consists in replacing a certain variable by its nonlocal counterpart obtained by weighted averaging over a spatial neighborhood of each point under consideration. If $f(\mathbf{x})$ is some "local" field in a domain V, the corresponding nonlocal field is defined as

$$\overline{f}(\mathbf{x}) = \int_V \alpha(\mathbf{x}, \boldsymbol{\xi}) f(\boldsymbol{\xi}) \, \mathrm{d}\boldsymbol{\xi} \tag{85}$$

where $\alpha(\mathbf{x}, \boldsymbol{\xi})$ is a given nonlocal weight function. In an infinite body, the weight function depends only on the distance between the "source" point, $\boldsymbol{\xi}$, and the "receiver" point, $\mathbf{x}$. In the vicinity of a boundary, the weight function is usually rescaled such that the nonlocal operator does not alter a uniform field. This can be achieved by setting

$$\alpha(\mathbf{x}, \boldsymbol{\xi}) = \frac{\alpha_0(\|\mathbf{x} - \boldsymbol{\xi}\|)}{\int_V \alpha_0(\|\mathbf{x} - \boldsymbol{\zeta}\|) \, \mathrm{d}\boldsymbol{\zeta}} \tag{86}$$

where $\alpha_0(r)$ is a monotonically decreasing nonnegative function of the distance $r = \|\mathbf{x} - \boldsymbol{\xi}\|$. In the one-dimensional setting, x and ξ are scalars and the domain of integration V reduces to an interval.

The weight function is often taken as the Gauss distribution function (solid curve in Fig. 23a)

$$\alpha_0(r) = \exp\left(-\frac{r^2}{2\ell^2}\right) \tag{87}$$

where ℓ is a parameter reflecting the internal length of the nonlocal continuum. Another possible choice is the truncated quartic polynomial function (dashed curve in Fig. 23a)

$$\alpha_0(r) = \left\langle 1 - \frac{r^2}{R^2} \right\rangle^2 \tag{88}$$

where R is a parameter related to the internal length. Since R corresponds to the largest distance of point $\boldsymbol{\xi}$ that affects the nonlocal average at point $\mathbf{x}$, it is called the interaction radius. The Gauss function (87) has an unbounded support, i.e., its interaction radius is $R = \infty$.

A suitable nonlocal damage formulation that restores well-posedness of the boundary value problem is obtained if damage is computed from the nonlocal

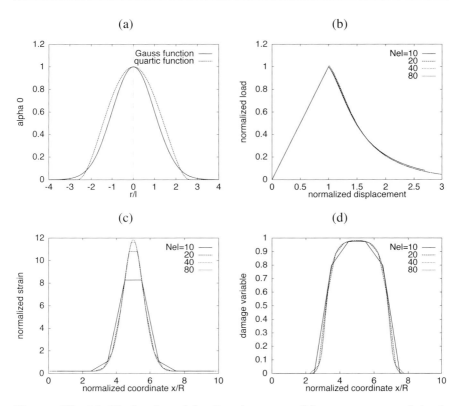

Figure 23. (a) Nonlocal weight functions α_0, (b) convergence of load-displacement diagram, (c) convergence of strain profile, (d) convergence of damage profile; Nel = number of elements

equivalent strain. In the loading function (17), the local value $\tilde{\varepsilon}$ is replaced by its weighted spatial average

$$\bar{\varepsilon}(\mathbf{x}) = \int_V \alpha(\mathbf{x}, \boldsymbol{\xi}) \tilde{\varepsilon}(\boldsymbol{\xi}) \, \mathrm{d}\boldsymbol{\xi} \qquad (89)$$

The internal variable κ is then the largest previously reached value of the nonlocal equivalent strain $\bar{\varepsilon}$. The corresponding damage variable evaluated from (11) is then substituted into the stress-strain equations (14). It is important to note that the damage variable is evaluated from the nonlocal equivalent strain $\bar{\varepsilon}$, but the strain ε that is used in (14) or (15) to compute the effective stress is considered as local. In the elastic range, the damage variable remains equal to zero, and the stress-strain relation is local.

Fig. 23b shows the load-displacement diagram for strain localization in a bar under uniaxial tension, calculated using a nonlocal damage model with the weight function (88) and with the exponential damage law (40). As the number of finite elements increases, the load-displacement curve rapidly converges to the exact solution. Convergence of strain and damage profiles generated by an applied displacement $u = 2u_0$ is documented in Fig. 23c,d. In contrast to the local model, the process zone does not shrink to a single point as the mesh is refined. Its size is controled by the interaction radius R, considered as a material parameter.

Nonlocal formulations can be developed not only for isotropic damage models but also for other constitutive models with softening, e.g. for anisotropic damage models, smeared crack models, or softening plasticity. However, one needs to be careful with the choice of the variable subjected to nonlocal averaging. For the rotating crack model, a nonlocal formulation based on the secant format of the stress-strain law was developed by Jirásek and Zimmermann (1997). They proposed to evaluate the secant stiffness from the nonlocal strain and then multiply it by the local strain to compute the stress. The stress-strain law thus remains local in the elastic range, and after the onset of cracking it serves as an efficient localization limiter.

4.2 Gradient-Enhanced Models

Explicit gradient formulation. Gradient models can be considered as the differential counterpart of integral nonlocal formulations. Instead of dealing with integrals that represent spatial interactions, we can take the microstructure into account by incorporating the influence of gradients (of the first or higher order) of internal variables into the constitutive relations. The most popular example is the gradient-dependent plasticity theory that evolved from the original ideas of Aifantis (1984). If a similar approach is used in damage mechanics, damage is assumed to be driven not only by the (local) equivalent strain $\tilde{\varepsilon}$ but also by its Laplacean, $\nabla^2 \tilde{\varepsilon}$, which represents in a generalized sense the "curvature" of the strain distribution. In the simplest case, $\tilde{\varepsilon}$ is replaced in the loading function (17) by the quantity

$$\bar{\varepsilon} = \tilde{\varepsilon} + \ell^2 \nabla^2 \tilde{\varepsilon} \tag{90}$$

where ℓ is a material parameter with the dimension of length. The damage-driving quantity $\bar{\varepsilon}$ can be considered as a specific type of nonlocal equivalent strain, which is now constructed by applying a differential operator on the local equivalent strain, rather than by applying the integral operator according to (89). Therefore, the gradient damage model based on (90) is considered as weakly nonlocal.[7]

---

[7] For strongly nonlocal models, the value of the nonlocal quantity at a certain point depends on the distribution of the corresponding local quantity in the entire body or at least in a finite neighborhood

As long as the strain distribution remains uniform (such as in a uniaxial tensile test before localization), the equivalent strain is also uniform, its Laplacean vanishes, and the model response is exactly the same as for the local formulation. After the onset of localization, the higher-order term is activated and prevents localization of damage in a set of zero measure.

It is instructive to discuss how the gradient term limits localization. Around the point that experiences the largest strain, the curvature of the strain profile is negative, and due to the Laplacean term in (90) the nonlocal equivalent strain is smaller than the local one. If the softening zone is too narrow, the negative curvature of the strain profile around its peak has a large magnitude, and the damage evolution is very slow. This slows down the strain growth in the central part of the localized zone and accelerates the growth in the adjacent regions, so that the localization zone expands. The minimum size of this zone is controled by the length parameter ℓ.

Implicit gradient formulation. Due to the presence of second derivatives of internal variables, the numerical implementation of explicit gradient models is not easy. The simplest version of the explicit gradient damage model suffers by certain deficiencies, described e.g. by Simone (2007). These problems can be overcome by the implicit gradient damage formulation (Peerlings et al., 1996), which defines the nonlocal variable indirectly as the solution of a Helmholtz-type differential equation

$$\bar{\varepsilon} - \ell^2 \nabla^2 \bar{\varepsilon} = \tilde{\varepsilon} \tag{91}$$

with the homogeneous Neumann boundary condition $\boldsymbol{n} \cdot \boldsymbol{\nabla} \bar{\varepsilon} = 0$ imposed on the entire physical boundary S of the body V.

The solution $\bar{\varepsilon}$ of the above boundary value problem can be expressed in the form of an averaging formula similar to (89) with the weight function $\alpha(\mathbf{x}, \boldsymbol{\xi})$ replaced by the Green function of the boundary value problem. For instance, for an infinite one-dimensional domain, the Green function of the Helmholtz equation (91) is given by

$$G(x,\xi) = \frac{1}{2\ell} \exp\left(-\frac{|x-\xi|}{\ell}\right) \tag{92}$$

So the implicit gradient models are equivalent to integral-type nonlocal models with special nonlocal weight functions. Despite this formal equivalence, their numerical implementation is quite different (Peerlings et al., 1996).

---

of that point, while for weakly nonlocal models it can be computed from the distribution of the local quantity in an arbitrarily small neighborhood.

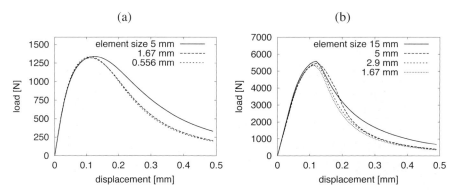

Figure 24. Load-displacement diagrams obtained with the nonlocal damage model: (a) notched beam, (b) unnotched beam

4.3 Examples of Failure Simulations

The examples in Section 3.2 demonstrate that a local damage model with softening leads to numerical results which suffer by pathological sensitivity to the discretization.

The analysis of the notched beam on three different meshes is now repeated using the nonlocal formulation of the isotropic damage model based on nonlocal equivalent strain according to formula (89). The damage law has again the exponential form (40) with parameters $\varepsilon_0 = 90 \times 10^{-6}$ and $\varepsilon_f = 7 \times 10^{-3}$. The nonlocal interaction radius that appears in the definition of the quartic nonlocal weight function (88) is set to $R = 4$ mm. The resulting load-displacement curves are plotted in Fig. 24a. The curves corresponding to the medium and fine meshes are almost coincident, which indicates that the solution converges upon mesh refinement and confirms that the nonlocal model does not suffer by pathological sensitivity to the mesh size. The load-displacement curve obtained with the coarse mesh is in the post-peak range somewhat above the converged solution. This is normal, because the element size (5 mm) is larger than the interaction radius (4 mm) and the localized process zone cannot be resolved with sufficient accuracy.

Analysis of the unnotched beam is repeated using a nonlocal model with parameters $\varepsilon_0 = 90 \times 10^{-6}$, $\varepsilon_f = 5 \times 10^{-3}$ and $R = 8$ mm. Note that in this case the element sizes are 15 mm, 5 mm, 2.9 mm and 1.67 mm, and so for $R = 4$ mm the medium meshes would still be too coarse and convergence upon mesh refinement could not be convincingly demonstrated. Of course, parameter R should be considered as a material property reflecting the internal length scale of the microstructure (e.g., the size and spacing of major heterogeneities such as largest aggregates in concrete), but in this academic example we do not try to link it to the

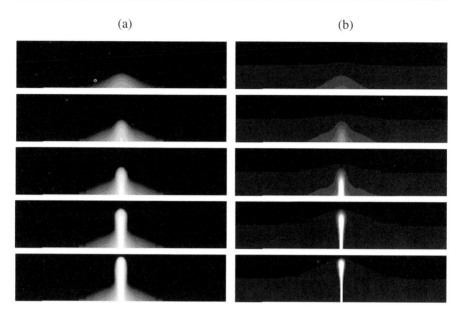

Figure 25. Evolution of the process zone in nonlocal damage simulation of unnotched beam: (a) damage, (b) strain ε_{xx}

actual material. Our aim is to show that the results are almost mesh-independent if the mesh is sufficiently fine. Indeed, the load-displacement curves plotted in Fig. 24b indicate that the two finest meshes give very similar results. The element sizes in the two coarsest meshes are too big compared to parameter R and the nonlocal interaction is not captured accurately. This is why the corresponding load-displacement curves differ from the other two, but the difference is not very dramatic.

The evolution of the process zone simulated on the fine mesh is shown in Fig. 25 in terms of damage and strain distribution at five stages of loading. Light shades of gray indicate high levels of the damage variable D and normal strain ε_{xx} (parallel to the beam axis). Black color marks regions with no damage in the left part of the figure and regions with compressive strain ($\varepsilon_{xx} < 0$) in the right part. The images in the second row correspond to a state shortly before the peak load and those in the third row to a state shortly after the peak load. Damage is irreversible and it cannot decrease, but in the post-peak range it remains constant in the predamaged region around the bottom face of the beam, except for a narrow band around the axis of symmetry that propagates to the top. The damage band keeps a certain minimum thickness while the band of increasing strains becomes progressively thinner. This is quite natural, given that the state at complete failure should represent a macroscopic stress-free crack.

Acknowledgment

The author is grateful for financial support of the Ministry of Education of the Czech Republic under Research Plan MSM 6840770003.

Bibliography

E. C. Aifantis. On the microstructural origin of certain inelastic models. *Journal of Engineering Materials and Technology, ASME*, 106:326–330, 1984.

Z. P. Bažant and M. Jirásek. Nonlocal integral formulations of plasticity and damage: Survey of progress. *Journal of Engineering Mechanics, ASCE*, 128:1119–1149, 2002.

Z. P. Bažant and F.-B. Lin. Nonlocal smeared cracking model for concrete fracture. *Journal of Engineering Mechanics, ASCE*, 114:2493–2510, 1988a.

Z. P. Bažant and F.-B. Lin. Nonlocal yield-limit degradation. *International Journal for Numerical Methods in Engineering*, 26:1805–1823, 1988b.

Z. P. Bažant and G. Di Luzio. Nonlocal microplane model with strain-softening yield limits. *International Journal of Solids and Structures*, 41:7209–7240, 2004.

Z. P. Bažant and B.-H. Oh. Crack band theory for fracture of concrete. *Materials and Structures*, 16:155–177, 1983.

Z. P. Bažant and J. Ožbolt. Nonlocal microplane model for fracture, damage, and size effect in structures. *Journal of Engineering Mechanics, ASCE*, 116:2485–2505, 1990.

L. Cedolin and S. Dei Poli. Finite element studies of shear-critical R/C beams. *Journal of Engineering Mechanics, ASCE*, 103(3):395–410, 1977.

J. L. Chaboche. Le concept de contrainte effective, appliqué à l'élasticité et à la viscoplasticité en présence d'un endommagement anisotrope. Number 295 in Col. Euromech 115, pages 737–760, Grenoble, 1979. Editions du CNRS.

CEB91. *CEB-FIP Model Code 1990, Design Code*. Comité Euro-International du Béton, Lausanne, Switzerland, 1991.

R. J. Cope, P. V. Rao, L. A. Clark, and P. Norris. Modelling of reinforced concrete behaviour for finite element analysis of bridge slabs. In C.Taylor et al, editor, *Numerical Methods for Nonlinear Problems*, volume 1, pages 457–470, Swansea, 1980. Pineridge Press.

J. P. Cordebois and F. Sidoroff. Anisotropie élastique induite par endommagement. In *Comportement mécanique des solides anisotropes*, number 295 in Colloques internationaux du CNRS, pages 761–774, Grenoble, 1979. Editions du CNRS.

R. de Borst and P. Nauta. Non-orthogonal cracks in a smeared finite element model. *Engineering Computations*, 2:35–46, 1985.

J. H. P. de Vree, W. A. M. Brekelmans, and M. A. J. van Gils. Comparison of nonlocal approaches in continuum damage mechanics. *Computers and Structures*, 55:581–588, 1995.

A. C. Eringen. On nonlocal plasticity. *International Journal of Engineering Science*, 19:1461–1474, 1981.

A. C. Eringen. Theories of nonlocal plasticity. *International Journal of Engineering Science*, 21:741–751, 1983.

A. K. Gupta and H. Akbar. Cracking in reinforced concrete analysis. *Journal of Structural Engineering, ASCE*, 110(8):1735–1746, 1984.

J. Hadamard. *Leçons sur la propagation des ondes*. Librairie Scientifique A. Hermann et Fils, Paris, 1903.

D. R. Hayhurst. Creep rupture under multi-axial state of stress. *Journal of the Mechanics and Physics of Solids*, 20:381–390, 1972.

R. Hill. A general theory of uniqueness and stability in elastic-plastic solids. *Journal of the Mechanics and Physics of Solids*, 6:236–249, 1958.

D. A. Hordijk. *Local approach to fatigue of concrete*. PhD thesis, Delft University of Technology, Delft, The Netherlands, 1991.

M. Jirásek and T. Zimmermann. Rotating crack model with transition to scalar damage: I. Local formulation, II. Nonlocal formulation and adaptivity. LSC Internal Report 97/01, Swiss Federal Institute of Technology, Lausanne, Switzerland, 1997.

M. Jirásek and Th. Zimmermann. Analysis of rotating crack model. *Journal of Engineering Mechanics, ASCE*, 124:842–851, 1998a.

M. Jirásek and Th. Zimmermann. Rotating crack model with transition to scalar damage. *Journal of Engineering Mechanics, ASCE*, 124:277–284, 1998b.

L. M. Kachanov. Time of the rupture process under creep conditions. *Izvestija Akademii Nauk SSSR, Otdelenie Techniceskich Nauk*, 8:26–31, 1958.

H. A. Kormeling and H. W. Reinhardt. Determination of the fracture energy of normal concrete and epoxy modified concrete. Technical Report 5-83-18, Stevin Lab, Delft University of Technology, 1983.

D. Krajcinovic and G. U. Fonseka. The continuous damage theory of brittle materials. *Journal of Applied Mechanics, ASME*, 48:809–824, 1981.

H. Kupfer, H. K. Hilsdorf, and H. Rüsch. Behavior of concrete under biaxial stresses. *Journal of the American Concrete Institute*, 66:656–666, 1969.

F. A. Leckie and D. R. Hayhurst. Creep rupture of structures. *Proceedings of the Royal Society A*, 340:323–347, 1974.

J. Lemaitre. Evaluation of dissipation, damage in metals submitted to dynamic loading. In *Proc. 1st International Conference on Mechanical Behavior of Materials*, 1971.

J. Lemaitre and R. Desmorat. *Engineering Damage Mechanics*. Springer, Berlin, Heidelberg, New York, 2005.

J. Mazars. Application de la mécanique de l'endommagement au comportement non linéaire et à la rupture du béton de structure. Thèse de Doctorat d'Etat, Université Paris VI., France, 1984.

J. Mazars. A description of micro and macroscale damage of concrete structures. *International Journal of Fracture*, 25:729–737, 1986.

M. B. Nooru-Mohamed. *Mixed-mode fracture of concrete: An experimental approach*. PhD thesis, Delft University of Technology, The Netherlands, 1992.

J. Oliver. A consistent characteristic length for smeared cracking models. *International Journal for Numerical Methods in Engineering*, 28:461–474, 1989.

N. Ottosen and K. Runesson. Properties of discontinuous bifurcation solutions in elasto-plasticity. *International Journal of Solids and Structures*, 27:401–421, 1991.

J. Ožbolt and Z. P. Bažant. Numerical smeared fracture analysis: Nonlocal microcrack interaction approach. *International Journal for Numerical Methods in Engineering*, 39:635–661, 1996.

R. H. J. Peerlings, R. de Borst, W. A. M. Brekelmans, and J. H. P. de Vree. Gradient-enhanced damage for quasi-brittle materials. *International Journal for Numerical Methods in Engineering*, 39:3391–3403, 1996.

S. Pietruszczak and Z. Mróz. Finite element analysis of deformation of strain-softening materials. *International Journal for Numerical Methods in Engineering*, 17:327–334, 1981.

G. Pijaudier-Cabot and Z. P. Bažant. Nonlocal damage theory. *Journal of Engineering Mechanics, ASCE*, 113:1512–1533, 1987.

Y. N. Rabotnov. Creep rupture. In *Proc. 12th International Congress of Applied Mechanics*, Stanford, 1968.

Y. R. Rashid. Ultimate strength analysis of prestressed concrete pressure vessels. *Nuclear Engineering and Design*, 7:334–344, 1968.

H. W. Reinhardt, H. A. W. Cornelissen, and D. A. Hordijk. Tensile tests and failure analysis of concrete. *Journal of Structural Engineering, ASCE*, 112: 2462–2477, 1986.

E. Rizzi, I. Carol, and K. Willam. Localization analysis of elastic degradation with application to scalar damage. *Journal of Engineering Mechanics, ASCE*, 121: 541–554, 1996.

J. G. Rots. *Computational modeling of concrete fracture*. PhD thesis, Delft University of Technology, Delft, The Netherlands, 1988.

J. W. Rudnicky and J. R. Rice. Conditions for the localization of deformation in pressure-sensitive dilatant materials. *Journal of the Mechanics and Physics of Solids*, 23:371–394, 1975.

C. Saouridis. *Identification et numérisation objectives des comportements adoucissants: Une approche multiéchelle de l'endommagement du béton*. PhD thesis, Université Paris VI., 1988.

A. Simone. Explicit and implicit gradient-enhanced damage models. *Revue européenne de génie civil*, 11:1023–1044, 2007.

L. Strömberg and M. Ristinmaa. FE-formulation of a nonlocal plasticity theory. *Computer Methods in Applied Mechanics and Engineering*, 136:127–144, 1996.

M. Suidan and W. C. Schnobrich. Finite element analysis of reinforced concrete. *Journal of the Structural Division, ASCE*, 99(10):2109–2122, 1973.

A. A. Vakulenko and M. L. Kachanov. Continuum theory of medium with cracks (in Russian). *Mekhanika Tverdogo Tela*, (4):159–166, 1971.

V. Červenka, R. Pukl, J. Ožbolt, and R. Eligehausen. Mesh sensitivity in smeared finite element analysis of concrete fracture. In F. H. Wittmann, editor, *Fracture Mechanics of Concrete Structures*, pages 1387–1396. Aedificatio Publishers, Freiburg, Germany, 1995.

S. Weihe. *Modelle der fiktiven Rißbildung zur Berechnung der Initiierung und Ausbreitung von Rissen*. PhD thesis, Universität Stuttgart, 1995.

Cracking and Fracture of Concrete at Meso-level using Zero-thickness Interface Elements

Ignacio Carol*, Andrés Idiart**, Carlos López* and Antonio Caballero***

* School of Civil Engineering (ETSECCPB), Technical University of Catalonia (UPC)
Barcelona, Spain
** OXAND S.A., 49, Av. F. Roosevelt, 77210, Avon, France
*** BBR VT International, 8603, Schwerzenbach (ZH), Switzerland

1. Introduction

1.1. Historical Aspects, General Considerations

Since the pioneering *numerical concrete* of Roelfstra *et al.* (1985), mesomechanical analysis has been emerging as a powerful approach to predict the complex behavior of heterogeneous materials starting from relatively simple assumptions on the behavior of the constituents. The price to pay, however, is high computational cost associated to modeling explicitly the first-level heterogeneities of the material, in the case of concrete normally the larger aggregate pieces floating in a mortar matrix. Since that first attempt, a certain number of such models have been already proposed, first in 2D, more recently in 3D, some based on particle representations of the "DEM"-type or associated truss or beam structures between particle centers (Vonk 1992; Cusatis *et al.* 2003a, 2003b, 2006; Bolander and Saito 1998; Bolander *et al.* 2000), some based on the regular lattice concept (Schlangen and van Mier 1993; Lilliu and van Mier 2003), and some on the classical continuum-type (Stankowski 1990; Wang and Huet 1993).

In all those models, a key factor is how to represent cracking. In the first and second type (particles, lattices), cracking may be identified *a posteriori* as adjacent adequately aligned failed beam or truss elements. This leads apparently meaningful cracks, but makes it difficult to verify concepts such as fracture energy, size effect, etc. In the third type (continuum), cracks may be represented either via smeared (*e.g.* Stankowski 1990) or discrete approach (*e.g.* Wang and Huet 1993). The smeared approach leads to some well-known problems such as inobjectivity with mesh size or limited deformation modes of the standard continuum elements, which have led to a variety of techniques for regularization, crack tracking, etc. but no universal method in sight yet for handling a general cracking problem.

The discrete approach avoids the above difficulties, although it generates others. Except for the simplified case of crack path known *a priori* (*e.g.* 3-point bending beam) the general implementation of the discrete approach with travelling crack tips requires intensive remeshing (*e.g.* Ingraffea and Saouma 1985) and exhibits unresolved challenges, being limited in practice to one or two (or very few) simultaneous cracks. An alternative approach, already mentioned in Rots (1988), consists of considering each line of the mesh as a potential crack, insert zero-thickness interface elements equipped with a fracture-based constitutive law, and let them open, close and reopen depending on local stress evolution. This is the approach described in this chapter. In subsequent sections, the work for concrete developed by the authors during the last decade, first in 2D, then in 3D and also incorporating other effects such as time-dependent, diffusion-driven and durability-related phenomena, is described.

1.2. Meso-structural Geometries, Discretization

The approach originally developed for the analysis of concrete specimens in 2D, described in detail in López (1999) and Carol *et al.* (2001) is only briefly summarized here. Following previous work by Stankowski (1990), the particle geometry was generated using Delaunai/Voronoi polygons which were then shrunk, and the space left between particles was considered filled by a matrix formed by mortar plus the smaller aggregates. Both phases were discretized using standard triangular continuum elements. A sample mesh of this type is depicted in Figure 1, with separate representation of the mortar matrix (Figure 1a) and aggregate particles (Figure 1b). The possibility of cracking is introduced via zero-thickness interface elements with double nodes (Figure 1d), which are inserted along all the lines of the mesh which are considered as potential crack lines. In the examples of the figure, these are all aggregate-matrix contacts plus a number of lines connecting aggregates through the matrix, which are laid out following the proposal in Vonk (1992) (Figure 1c).

1.3. Fracture-based Interface Constitutive Law with Aging Effect

The constitutive model used for interface elements has its origins in the time-independent formulation proposed in the early nineties (Carol and Prat 1990), which was later modified and improved in Carol *et al.* (1997), López (1999), Carol *et al.* (2001), Caballero et al. (2006a), Caballero *et al.* (2008). A more detailed description of the constitutive law with aging can be found in Idiart

(2009), López et al. (2007). In the following, the fundamentals of the model are summarized, together with some verification examples.

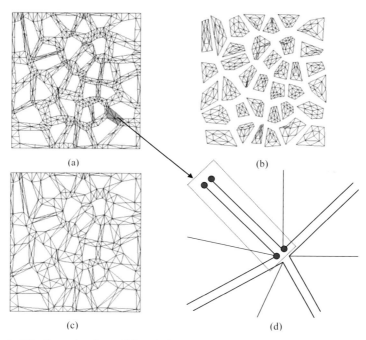

Figure 1. FE discretization of the 6x6 arrangement: a) matrix, b) aggregates, c) interfaces inserted, and d) details of discretization.

The model is formulated in terms of the normal and tangential stress components in the mid-plane of the joint element $\boldsymbol{\sigma} = [\sigma_N, \sigma_T]^t$ and the corresponding relative displacements $\boldsymbol{u} = [u_N, u_T]^t$ (^{t} = transposed). The formulation has the structure of work-softening elasto-plasticity, in which plastic relative displacements can be identified with crack openings. Non-linear Fracture Mechanics concepts are also introduced in order to define the softening behaviour, by using as a history variable the work dissipated in the fracture process (denoted as W^{cr}). The main features of the plastic model are represented in Figure 2.

The initial loading (failure) surface $F = 0$ is given as a three-parameter hyperbola (tensile strength χ, asymptotic "cohesion" c, and asymptotic friction angle $\tan\phi$, Figure 2a).

$$F = \sigma_T^2 - (c - \sigma_N \cdot \tan\phi)^2 + (c - \chi \cdot \tan\phi)^2 \tag{1}$$

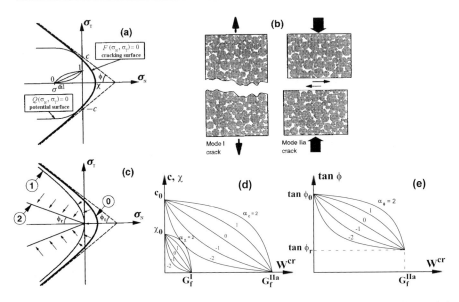

Figure 2. Cracking laws: (a) hyperbolic cracking surface F and plastic potential Q; (b) fundamental modes of fracture; (c) evolution of cracking surface; (d) softening laws for χ and c; (e) softening law for $\tan\phi$.

Classic Mode I fracture occurs in pure tension, and a second Mode IIa is defined under shear and high compression, with no dilatancy allowed (Figure 2b). The fracture energies G_f^I and G_f^{IIa} are two model parameters. After initial cracking, χ, c and $\tan\phi$ decrease (Figures 2d and 2e), and the loading surface shrinks, degenerating in the limit case into a pair of straight lines representing pure friction (Figure 2c). The process is driven by the energy spent in fracture process, W^{cr}, the increments of which are taken equal to the increments of plastic work, less frictional work in compression. Total exhaustion of tensile strength ($\chi = 0$) is reached for $W^{cr} = G_f^I$ (curve "1" in Figure 2c), and residual friction (c = 0 and $\tan\phi = \tan\phi_r$) is reached for $W^{cr} = G_f^{IIa}$ (curve "2" in Figure 2c). Additional "shape" parameters α_χ, α_c and α_ϕ allow for different shapes of the softening laws (linear decay for $\alpha_\chi = \alpha_c = \alpha_\phi = 0$, see Figures 2d and 2e).

A non-associated plastic potential Q is adopted (i.e. $Q \neq F$) (Figure 2a) in order to progressively eliminate dilatancy for high levels of compression (dilatancy vanishes progressively for $\sigma_N \to \sigma_{dil}$). Dilatancy is also decreased during the fracture process, so that it vanishes for $W^{cr} = G_f^{IIa}$. The dilatancy decay functions also include shape parameters α_σ^{dil} and α_c^{dil} (also linear decay for zero values of shape parameters).

Although physically questionable, the model requires some elastic stiffness necessary for the formulation of the interfaces in the context of the displacement-based FE method. The elastic stiffness matrix is assumed diagonal with constant K_N and K_T values, which can be regarded as penalty coefficients necessary to calculate the interface stresses (interface elements in the elastic regime, before cracking, should not add any elastic compliance). In practice this means that their values should be as high as possible without introducing numerical problems.

The aging effect is also considered in the model through the evolution of the main parameters of the fracture surface (χ, c) with time, as well as the fracture energies G_f^I and G_f^{IIa} (the latter assumed to be proportional to G_f^I). To this end, the following monotonically increasing function of time is introduced:

$$f(t) = A \cdot f(t_0) \cdot \left[1 - \exp\left(-k \cdot (t/t_0)^p\right)\right], \quad \text{with} \quad A = (1-\exp(-k))^{-1} \quad (2)$$

where t is the age of the material, t_0 a reference age (usually 28 days), and $f(t_0)$ the parameter value at t_0. Two shape parameters are introduced: p (Figure 3a), and k (Figure 3b). As a result, if the aging effect is considered, the initial fracture surface will expand in time (e.g. from curve "0" to curve "1" in Figure 4), and this effect will compete with the mechanical degradation of the interface as the crack opens and slides. This interaction is achieved by modifying the history variable, which is no longer W^{cr} directly, but a new dimensionless quantity ξ, defined incrementally as

$$d\xi = \frac{dW^{cr}}{G_F^I(t)} \quad (3)$$

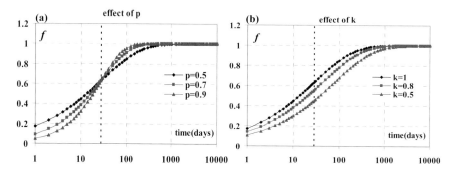

Figure 3. Evolution of the parameters (relative to the asymptotic value of f): (a) effect of parameter p on the s-shape of the curve, and (b) effect of parameter k on the value of f at time t_0 (equal to 28 days, and marked with a dashed line).

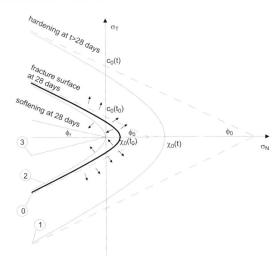

Figure 4. Cracking surface considering the aging effect: evolution with time and degradation due to energy spent fracture.

The numerical implementation is based on an implicit backward Euler integration scheme with a consistent tangent matrix operator, as described in Caballero et al. (2008).

Three verification examples of the constitutive model without aging effect are followed by some examples with aging. The first numerical test is pure tension. The σ-u curves obtained for various values of the fracture energy G_f^I are represented in Figure 5 (other relevant parameters are K_N = 1000 MPa/mm, tensile strength χ_0 = 3 MPa and all shape parameters equal to zero). Note that, even with a zero shape parameter (linear softening function in terms of W^{cr}), the resulting softening curve in terms of crack opening is of the exponential type, with total area under each curve equal to the prescribed value of G_f^I.

The second example is a numerical shear test. First, normal compression of a prescribed value is applied. Then, the shear relative displacement is increased progressively with constant normal stress, until a residual state is reached. The parameters used are $K_N = K_T$ = 25000 MPa/mm, $\tan\phi_0 = \tan\phi_r$ = 0.8785, χ_0 = 3 MPa, c_0 = 4.5 MPa, G_f^I = 0.03 N/mm, G_f^{IIa} = 0.06 N/mm, σ^{dil} = 30 MPa, α_σ^{dil} = -2 and all other shape parameters equal zero. The results are depicted in Figures 6a and 6b. In Figure 6a, shear stresses are represented against shear relative displacement for various values of normal compression. Note that, after the peak, all curves tend to a residual value of the shear stress, which corresponds to basic friction times the normal stress. In Figure 6b, dilatancy is represented by normal against shear relative displacements.

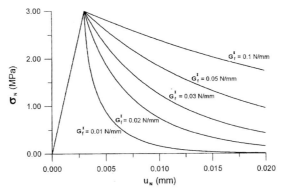

Figure 5. Pure tension: normal stress vs. relative displacement for different values of fracture energy.

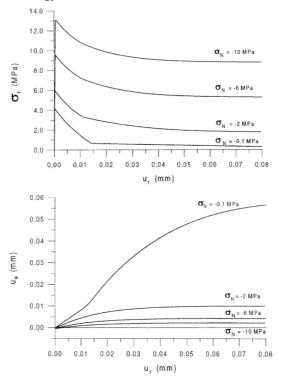

Figure 6. Shear under constant compression: (a) shear stress vs. shear relative displacement for various values of σ_N and (b) evolution of dilatancy for various values of σ_N.

The third constitutive example corresponds to an experimental test by Hassanzadeh (1990). Normal and shear relative displacements are prescribed to an interface in two steps. First, only normal (opening) is applied till the peak stress is reached. From this point, shear displacement is applied simultaneously to the normal in a fixed proportion characterized by the angle $\tan\theta = u_N/u_T$.

Tests were run for various θ, starting with the limit case $\theta = 90°$ which actually corresponds to continuing the test in pure tension. The parameter values used are $K_N = K_T = 200$ MPa/m, $\tan\phi_0 = \tan\phi_r = 0.9$, $\chi_0 = 2.8$ MPa, $c_0 = 7$ MPa, $G_f^I = 0.1$ N/mm, $G_f^{IIa} = 1.0$ N/mm, $\sigma^{dil} = 56$ MPa, $\alpha_\chi = 0$, $\alpha_c = 1.5$, $\alpha_\sigma^{dil} = -2.7$, $\alpha_c^{dil} = 3$. The results obtained in the second part of the test are represented in Figs. 7a and 7b.

In Figure 7a, normal stress is represented versus normal relative displacements. While for $\theta = 90°$ (pure tension), the usual exponential-type of decay is obtained, the imposition of a certain proportion of shear relative displacement causes the stresses to drop faster, change sign into compression, reach a peak and then vanish asymptotically. This is due to development of shear dilatancy that would exceed the prescribed normal opening rate.

In Figure 7b, the shear stresses corresponding to the same tests are plotted against shear relative displacements. In both figures, numerical results (continuous lines) are represented together with experimental dots, showing how the proposed model not only gives logical numerical results, but is also capable of fitting experimental data obtained during non-trivial fully coupled normal/shear loading scenarios. Additional details of the constitutive formulation and verification examples can be found in Carol *et al.* 1997, López (1999), Carol *et al.* (2001).

The first example with aging is a shear test under constant normal compressive stresses. The test consists of applying first a compressive stress (-1MPa), and then prescribe a gradually increasing shear relative displacement while compression remains constant. The application of the whole loading (both the normal load and shear relative displacements) is considered to take place "instantaneously" and the calculation is repeated for ages 28 days and 280 days.

The results are shown in Figure 8 in terms of shear stresses vs. shear relative displacements. The corresponding fracture surfaces at the different states, identified with numbers, are also represented in the figure. The curves exhibit an elastic stiffness up to the peak (points 1 or 2), followed by a descending branch (softening) and a residual shear stress (point 3). As expected, the curve for 280 days (point 2) shows higher peak values than for 28 days (point 1).

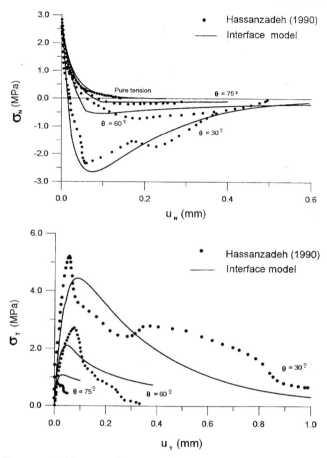

Figure 7. Hassanzadeh's tests: (a) normal stress vs. relative displacement and (b) shear stress vs. relative displacement.

The combined effect of aging and degradation is illustrated with a variation of the previous shear test. At 28 days the normal stress is applied and the shear test is initiated as before, but then, at some intermediate point in the descending branch, the application of the shear displacements is stopped and held constant until the age of 280 days, at which point the test is resumed. The results obtained in this "interrupted shear test" may be also seen in Figure 8. The first part of the shear test before the interruption obviously coincides with the curve at 28 days (point 1 and descending branch thereafter), until the interruption point 4. Upon resuming the shear displacement, though, the loading surface has been expanding due to aging, and the stress must first increase elastically until the loading surface is regained, point 5. Only then the softening is resumed until residual state. Note finally that shear stress and softening curves followed after the interruption lay between those for 28 and 280 days, closer to the latter if the interruption was earlier in the softening branch, and to the former if later.

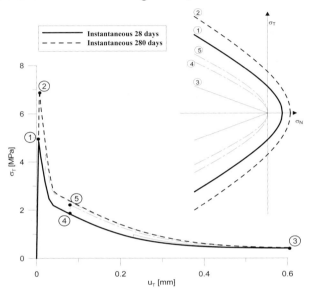

Figure 8. Constitutive behavior for the case of a shear test under constant compression with different loading histories, as a function of shear stresses and relative displacements.

1.4. Aging Viscoelastic Model for the Matrix-phase

In order to simulate the time-dependent deformations in concrete at the meso-level, a model for basic creep (i.e. creep under uniform moisture conditions) needs to be introduced for the matrix phase. Aggregates are assumed to remain linear elastic and time-independent. For the mortar matrix, the model implemented consists of an aging Maxwell-chain, which is equivalent to a Dirichlet series expansion of the relaxation function $R(t,t')$, dual to the usual creep compliance function $J(t,t')$, in which t' represents the age at loading. It is based on previous work by Bazant and coworkers (Bazant and Panula, 1978; Bazant, 1982).

Because the matrix exhibits a time-dependent mechanism while the aggregates do not, the parameters of the Maxwell Chain for the matrix have to be determined via "inverse analysis", i.e. their values have to somehow be guessed in order to produce the desired overall viscoelastic behavior corresponding to the concrete (López et al., 2001). In the present case, this adjustment is made manually so that the overall creep corresponds to the creep compliance function J(t,t') given in the concrete design code of Spain (EHE, 1998).

2. Original Results in 2D under Mechanical Loading

In the first developing phase of this approach (López, 1999; López et al., 2008a, 2008b), cases analyzed ranged from uniaxial tension, uniaxial compression and biaxial tension-compression, to Brazilian and other more complex loading cases and effects, all in 2D. All of them gave realistic results in terms of macroscopic (average) stress-strain curves as well as microcrack pattern and evolution.

The typical results obtained for a specimen under uniaxial tension are represented in Figure 9. Figure 9a compares the overall stress-displacement curve obtained with the model with experimental results in (Hordijk, 1992). The resulting crack pattern at point C of the loading history is presented in Figure 9b, in which line thickness represents the norm of the opening/sliding displacement, in red if the crack is active and in blue if unloaded.

Figure 10 shows similar results for uniaxial compression. Note the correct lateral expansion evolution (left), which leads to the right volumetric expansion after the peak. Note also the relatively dense cracking obtained (right), in the end mostly arrested (in blue) except along three active macrocracks (in red).

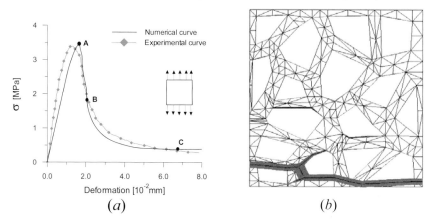

Figure 9. Typical results in uniaxial tension: (a) average stress-deformation curve, compared to experimental results (Hordijk, 1992), and (b) crack pattern obtained in uniaxial tension along y-axis.

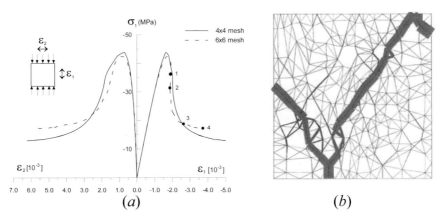

Figure 10. Results in uniaxial compression: (a) average stress-strain curves for the 4x4 and 6x6 concrete specimens, and (b) crack patterns in terms of the norm of plastic displacement vector corresponding to the 6x6 mesh.

Figure 11 includes the results for the Brazilian test, in which the specimen is loaded vertically on small platens in the middle of top and bottom faces, and the crack mouth opening displacement (CMOD) is measured between two points at mid-height of specimen, on each side of the vertical symmetry line along which the tensile crack will eventually develop. Note the effect of the platen width, which makes the overall response to change from peak-softening (narrow platen) to second peak (wider platen). The crack pattern starts with the vertical tensile crack in the center of the specimen, followed by the wedge-type shear-compression cracks under the platens. The overall response is the result of superposition of the two mechanisms that sequentially dominate the first peak (brittle tensile crack) and then the second peak (ductile shear compression with friction).

The effect of specimen height and the verification of fracture energy concept was verified with three specimens of similar width but increasing length (Ciancio et al. 2002; López et al. 2004): These specimens were analyzed under uniaxial tension and uniaxial compression, in order to reproduce the experimental results reported in van Mier (1997), in which the various overall stress-strain curves may look quite different after the peak, while they are much more coincident in terms of stress-post peak displacement. The figure 12 shows the resulting crack patterns for the compression analysis of the three specimens, in which in spite of the different length, similar total length of macrocrack develops, only zigzagging more across specimen due to the geometry constraint.

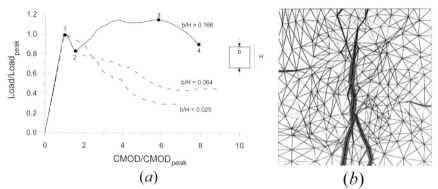

Figure 11. Brazilian test: (a) relative load-CMOD diagrams for three b/H ratios, and (b) crack patterns for b/H=0.166 (magnitude of plastic relative displacements at the interfaces).

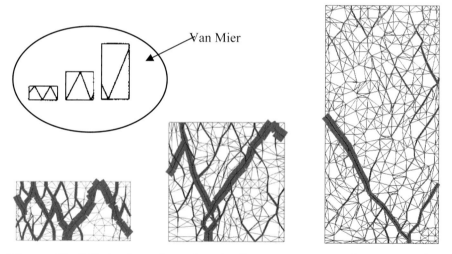

Figure 12. Effect of specimen length: fracture process under compression loading.

Another application of the model at that stage was high-strength or light-weight aggregate concrete, in which mortar has comparable or higher mechanical properties than the aggregates: In this case, adequate discretization and systematic use of interface elements were also employed within the aggregates, with interface properties adequately chosen. In this way, depending on properties and loading, cracks could also develop through aggregates, leading in that case to higher overall brittleness (López 1999; López et al. 2008b).

Basic creep behavior of concrete was also analyzed. The matrix phase (mortar plus small aggregates) was assumed visco-elastic, while particles (larger aggregates) were assumed simply elastic. Under constant stress, this would lead to overall time-dependent deformations (creep), which were however lower than those the matrix itself would experience under similar external load. This was due to internal stress redistribution consisting roughly of a stress decrease in the matrix compensated by a stresses increase at the aggregates and aggregate-matrix interfaces. For certain load levels, the latter effect led to the increase of cracking with time, causing non-linear creep and even failure under sustained load, all of which resulting in qualitatively correct Rüsch (1960) curves (López et al. 2001).

More details about these and other application examples of mechanical behavior in 2D can be found in López (1999); López et al. (2000); Carol et al. (2001) and López et al. (2008a, 2008b). Applications of the same meso-mechanical model with zero-thickness fracture-based interfaces have been also developed for other materials, more specifically for porous sandstone rock and for cellular materials such as trabecular bone tissue. In the former case the motivating problem is (undesirable) sand production in oil reservoirs (Garolera et al. 2005), and in the second inelastic behavior of that porous biological material, with special emphasis on anisotropy (Roa et al. 2000; Carol et al. 2001; Roa 2004).

3. Extension to 3D

In a second stage of development the meso-mechanical model was extended to 3D. Although between 2D and 3D the concepts do not change much, this extension has required very significant efforts and new non-trivial developments in all fronts of activity, *i.e.* geometry and mesh generation, constitutive model, computational efficiency, graphic representation, etc. All these developments, together with some numerical results are summarized in this section. More details can be found in Caballero (2005); Caballero et al. (2006a, 2006b); Caballero et al. (2007), Caballero et al (2008).

3.1. Geometry and Meshes

The procedure for particle geometry and mesh generation in 3D is schematized in Figure 13. The departing point is a regular distribution of points (Figure 13c), arranged according to the "Body Centered Cube" (BCC) distribution (Figure 13a), which leads to basic 14-side polyhedra (Figure 13b) and exhibits advantages in front of alternative possibilities (Okabe et al. 1992).

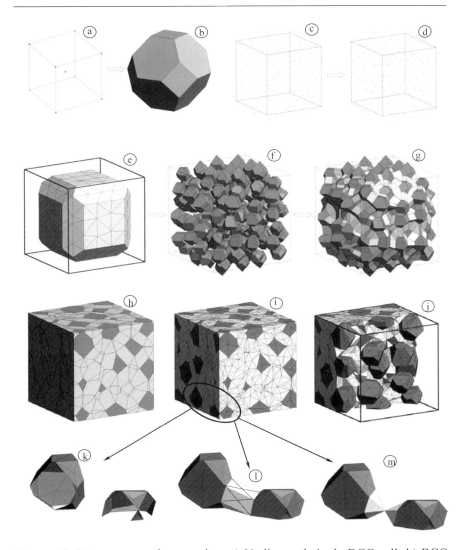

Figure 13. Meso-structural generation: a) Undistorted single BCC cell; b) BCC polyhedron obtained from the undistorted BCC distribution of points; c) regular distribution of points (BCC distribution); d) distorted distribution of points; e) Delaunay mesh obtained from distorted distribution of points in the volume interior; f) Resulting Voronoï polyhedra (aggregates); g) filling up space in between the polyhedra (mortar); h) cutting process; i and j) mesh generation and interface element insertion; k) aggregate-mortar interfaces; l and m) mortar-mortar interfaces.

The position of the initial points is randomly perturbed (Figure 13d), and on them a Delaunay mesh of tetrahedra is generated (Figure 13e). The dual Voronoï polygons are then shrunk (Figure 13f), the space in between filled with matrix blocks (Figure 13g) and the specimen sides "cut" to obtain flat surfaces (Figure 13h). A first correction step of resulting ill-conditioned blocks (too flat or needle-type) is followed by the insertion of interface planes along the aggregate-mortar interface (Figure 13k) and also across the mortar itself (Figure 13l and 13m), in such way that the main potential fracture surfaces around and between aggregates are represented.

A second check and correction of potentially ill-conditioned blocks is performed, before the FE meshing of the continuum elements, which simply consists in a subdivision of each tetrahedral block in a fixed number of tetrahedral finite elements. Finally, the interface elements are introduced along all potential fracture planes already present in the FE mesh, which basically consists of an orderly duplication of the nodes and subsequent changes in element nodal connectivities. This process increases the number of nodes considerably while the number of continuum elements remains unchanged.

3.2. Extension of the Constitutive Model to 3D

The interface constitutive formulation in 3D (Caballero 2005, Caballero et al, 2008) is an extension of the previous 2D model. Interface behavior is formulated in terms of one normal and two tangential traction components on the plane of the joint, $\sigma = [\sigma_N, \sigma_{T1}, \sigma_{T2}]^t$ and the corresponding relative displacements $u = [u_N, u_{T1}, u_{T2}]^t$. The initial loading (failure) surface $F = 0$, is given as a three-parameter hyperboloid with the same parameters tensile strength χ, asymptotic "cohesion" c, and asymptotic friction angle $\tan\phi$:

$$F = \sigma_{T1}^2 + \sigma_{T2}^2 - (c - \sigma_N \tan\phi)^2 + (c - \chi \tan\phi)^2 \qquad (4)$$

As novelties with respect to its 2D counterpart (1), the 3D extension was developed to be more robust and ensure faster overall (structural level) convergence in the calculations. This was achieved through incremental changes in the formulation such as smooth transition functions for the softening and dilatancy laws, as well as a more advanced numerical implementation using consistent tangent procedures together with sub-incrementation. All this is described in detail, together with some implementation examples, in a recent publication (Caballero et al. 2008).

3.3. Computational and Graphics Aspects

The new 3D formulation has been implemented into the existing, in-house developed FE code, DRAC (Prat *et al.* 1993). The FE code itself has also needed considerable improvement in order to solve the resulting large systems in reasonable times. New iterative solvers with partial factorization have been implemented. A Line-Search strategy combined with quadratic Arc-Length constitutes the iterative procedure for non-linear calculations at the structural level. Graphic representation in 3D needed at the pre- and post-processing stages has been handled with GiD (Ribó and Riera 1997), for which adequate interfaces have been developed with FE code DRAC.

3.4. Results

The 3D examples consist of cubical specimens of 4x4x4 cm^3 subject to three different loading cases. The specimen contains 14 aggregate pieces with a volume fraction of 25%, and the resulting FE mesh contains 6254 nodes, 5755 tetrahedra and 3991 interface elements. The material parameters used are: E=70 GPa (aggregate), E=25 GPa (mortar) and ν=0.20 (both) for the continuum elements; for the aggregate-mortar interfaces: $K_N=K_T=10^9$ MPa/m, $\tan\phi_0 = 0.6$, $\tan\phi_r = 0.2$, $\chi_0= 2$ MPa, $c_0=7.0$ MPa, $G_f^I=0.03$ N/mm, $G_f^{IIa}=10G_f^I$; for the mortar-mortar interfaces the same parameters except for $\chi_0=4$ MPa, $c_0=14.0$ MPa, $G_f^I=0.06$ N/mm.

Uniaxial loading is applied via prescribed displacements along the desired axis for the corresponding face nodes, while lateral displacements are left free, except for the minimum number of restrictions to avoid rigid body motions. The results of applying uniaxial tension along *x*, *y* and *z*-axis are represented, in terms of average stress-strain curves, in Figure 14. The results of the uniaxial compression loading, also along *x*, *y* and *z*, are represented in Figure 15a (axial stress *vs.* axial strain) and Figure 15b (axial stress *vs.* volumetric strain). Note the realistic shape of all the curves, especially the post-peak dilatancy in compression.

The same loading cases for the *z*-axis only, are also illustrated in Figures 16 and 17 in terms of microcrack evolution and deformations. In those figures, the sequence of cracking, for the loading levels marked as A to E in figures 14 and 15, are represented. Figures 16 and 17 are organized in a similar way, with 3 columns of 5 plots each. The left column includes those interface elements which are in plastic loading, while the middle column includes those that, after initial cracking, currently are in elastic unloading. Finally, the right column depicts the deformed mesh at each load level. Note the realistic representation of micro-cracking in Figures 16 and 17, which is widely distributed before peak, and then localizes spontaneously in only a few macro-cracks while the rest unload. More details about the results of these loading cases may be found in Caballero *et al.* (2006b).

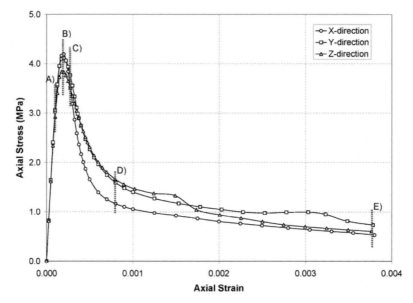

Figure 14. Stress-strain curves for the uniaxial tension tests in *x*, *y* and *z*.

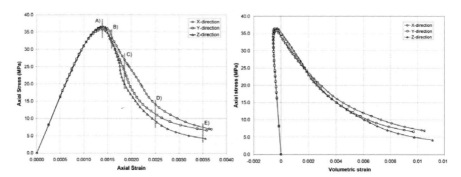

Figure 15. Stress-strain curves for the uniaxial compression tests in *x*, *y* and *z*, and stress-volumetric strain diagram curves for the uniaxial compression test in *x*, *y* and *z*.

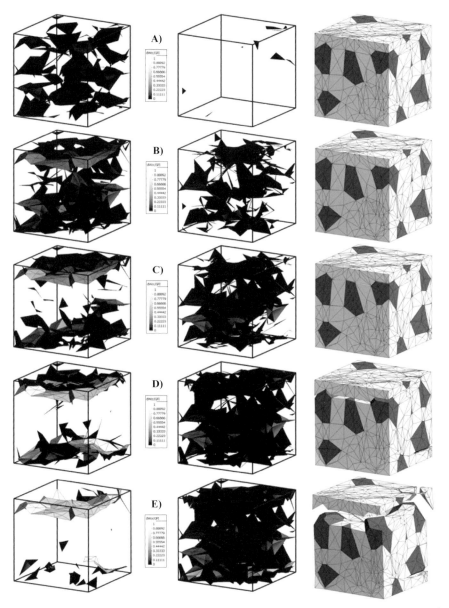

Figure 16. Results of tension test in the z direction. History parameter W^{cr}/G_F^I on opening interfaces (left column), closing interfaces (middle column), and deformed shape (right column), at loading states A, B, C, D and E.

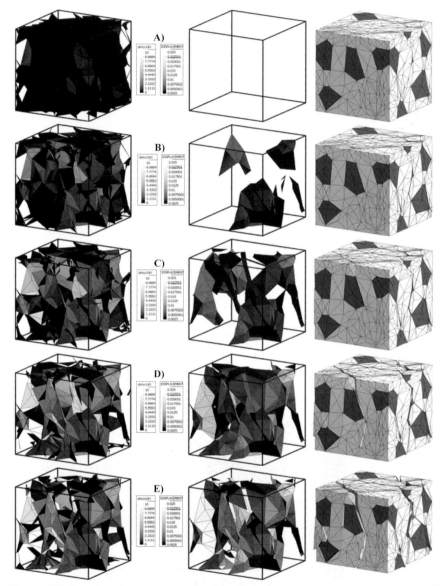

Figure 17. Results of compression test in the z direction. History parameter W^{cr}/G_F^I on opening interfaces (left column), norm of relative displacement on opening interfaces (middle column) and deformed shape of specimen (right column), at loading states A to E.

Cracking and Fracture of Concrete at Meso-level

The third loading case presented is biaxial loading, which is applied simultaneously along axes x and y onto the same 14-aggregate specimen, as represented in Figure 18 (view from a bottom perspective). In order to maintain fixed proportions between the x and y loads, in this case the load platens are also discretized with stiff elastic elements and interface elements in between platens and specimen.

The peak loads obtained for increasing loading and fixed σ_x/σ_y ratio, are represented in Figure 19 in terms the angle θ of the path in the normalized σ_x-σ_y space, showing good qualitative agreement with the classical biaxial envelope (Kupfer *et al.* 1969).

The results are also represented in terms of cracking in Figure 20. In that figure, the side view of specimen from z, y and x directions are depicted in the left, central and right columns, while the two rows correspond to the cases of 60 and 225 degrees (Figure 19), (that is, to the cases of biaxial tension with $\sigma_x = 2\sigma_y$ and equi-biaxial compression).

Note that in the first case, cracks run mainly parallel to the z-axis (one can "see through" the left diagram), that is they are mostly in-plane cracks, while in the second they are mainly out-of- plane cracks (perpendicular to z, i.e. one can "see through" the right and center diagrams but not trough the left one). Note that this type of effect can only be captured with a 3D model.

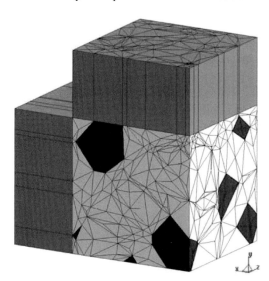

Figure 18. Finite element mesh for concrete and rigid platens used in the biaxial numerical tests.

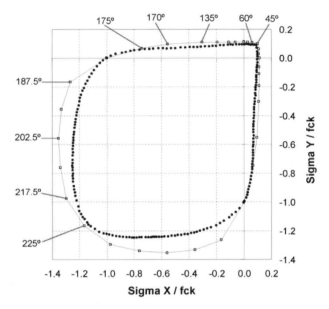

Figure 19. Comparison between experimental (marked using dots) and numerical failure envelope for biaxial loading (continuous line).

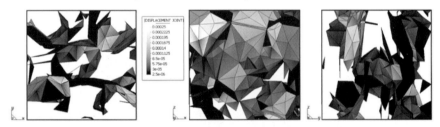

Figure 20a. θ = 60°. Cracking state in terms of relative displacement norm. Planes XY (a), XZ (b) and YZ (c).

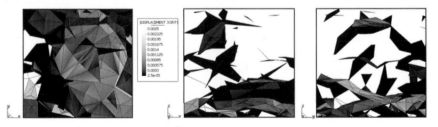

Figure 20b. θ = 225°. Cracking state in terms of relative displacement norm. Planes XY (a), XZ (b) and YZ (c).

4. Deterioration due to Diffusion-driven Environmental Phenomena: Drying Shrinkage and External Sulphate Attack

In recent years, the 2D meso-mechanical model of Section 2 has been extended in the direction of incorporating new more complex aspects of concrete behavior such as coupled environmental-related degradation phenomena. One of the key ingredients in that sense is moisture diffusion through the pore system, which is the main driving force inducing mechanical effects such as drying shrinkage or drying creep, and plays also a significant role in more complex conditions such as high temperature behavior and other durability-related phenomena. Also relevant is the diffusion of potentially aggressive ions through the cement paste matrix, which may lead to the formation of expansive reaction products, as in the external sulfate attack case, or to the dissolution of cement hydrates, as in the case of calcium leaching. In this section, the hygro-mechanical modeling of drying shrinkage and chemo-mechanical analysis of external sulfate attack are addressed. The fundamentals of the moisture diffusion model, as well as the diffusion-reaction model for sulphate ions are briefly presented, and the main findings are summarized. In the proposed model, hygro and chemo-mechanical coupling is achieved through a staggered approach. One code (DRACFLOW) performs the nonlinear moisture diffusion(-reaction) analysis, and the results in terms of volumetric strains at the local level serve as input to the second code (DRAC), solving the mechanical problem. The updated displacement field (nodal variables) obtained in the latter, from which new crack openings are derived, will in turn alter the diffusion(-reaction) analysis. Thus, this loop must be successively repeated within each time step until a certain tolerance is satisfied, before proceeding with the following time interval.

4.1. Hygro-mechanical Modeling of Drying Shrinkage

Drying shrinkage of concrete is governed by the drying through the pore system of the cement paste. Moisture migration is driven by a non-linear diffusion process in which, from an initial saturated state, moisture moves towards the concrete surface exposed to a lower relative humidity (RH) environment. Shrinkage strains are then related to the water loss (Bazant, 1988; Granger *et al.*, 1997), different at each material point. Deformation compatibility leads in general to cross-sectional or structural stresses and, in some cases, cracking. Strains measured at the drying surface depend not only on the external (and internal) restraints to deformation, but also on the non-uniform moisture distribution through a cross-section of a concrete member. Drying-induced cracks may accelerate the drying process, since they represent a preferential way out for moisture diffusion. This consideration is usually neglected in most of the models proposed in the literature. This is mainly due to the complexity of a correct numerical implementation and the difficulties encountered for validating a theoretical model. Drying through the cracks is explicitly taken into account in the present model, as will be outlined in the sequel.

It should be mentioned that there have been some previous attempts to model the main features of drying shrinkage of concrete with the help of mesostructural models (Tsubaki *et al.*, 1992; Sadouki and van Mier, 1997; Sadouki and Wittmann, 2001; Schlangen *et al.*, 2007; Grassl *et al.*, 2010). Most of that previous work, however, considered a reduced number of particles, focused only on either the moisture diffusion or the mechanical analysis of shrinkage, or exhibited some limited representation of the mechanical behavior and cracking.

Since the early work of Bazant and Najjar (1972), it is generally accepted that moisture movement in concrete follows a non-linear diffusion-type equation which may be advantageously written in terms of the local RH (H) as

$$\frac{dw_e}{dt} = \frac{dw_e}{dH} \cdot \frac{dH}{dt} = C(H) \cdot \frac{dH}{dt} = -\text{div}\left(D_{\text{eff}}(H)\text{grad}(H)\right) \tag{5}$$

where w_e [g/cm^3] is the evaporable water content, $C(H)$ is the moisture capacity matrix [g/cm^3], calculated as the derivative of the desorption isotherm with respect to H. D_{eff} [cm^2/s] is the effective diffusion coefficient, which strongly depends on H itself. In this work, following a previous study within the research group (Roncero, 1999), this dependency has been represented as

$$D_{\text{eff}}(H) = D_0 + (D_1 - D_0)f(\beta, H), \text{ with } f(\beta, H) = \frac{e^{-\beta} \cdot H}{1 + (e^{-\beta} - 1) \cdot H} \tag{6}$$

in which D_0 and D_1 are the values for $H=0$ and $H=1$ respectively, and f is a hyperbolic transition function which depends also on a shape parameter β. The dependence of the model diffusivity on H is shown in Figure 21a for different values of the shape factor. A value of 3.8 was shown to fit experimental data on mortar samples (Roncero, 1999).

Cracks may affect the overall diffusivity since they represent potential preferential channels for moisture migration. In the case that the mechanical analysis determines crack formation or propagation, the diffusion process will in turn be affected. To analyze this problem, the same FE mesh is used for the diffusion and mechanical calculations, which simplifies the coupled H-M scheme. This has required the development of interface elements with double nodes also for the diffusion analysis (Segura and Carol, 2004). Such elements incorporate longitudinal (K_L [cm^2/s]) as well as transversal (K_T [cm^2/s]) diffusivities. In the absence of specific information, transversal diffusivity is assigned a high value, representing the case of no jump in RH across the crack. For the longitudinal diffusion an expression similar to (6) is used once the crack has opened, in which the value for saturated flow K_I is in this case given by the so-called "cubic law" and K_0 is set as a small fraction of K_I:

$$K_L(H,u) = K_0(u) + [K_I(u) - K_0(u)] f(\beta_K, H), \text{ and } K_I(u) = \eta \cdot u^3 \quad (7)$$

In (7), u [cm] is the crack opening, η [1/(cm.s)] is a parameter relating crack width with diffusivity, to be determined, β_K is a model parameter, and f is the function given in (6), with β replaced by β_K.

The model also includes a desorption isotherm, which relates the RH to the evaporable water content in the pores w_e, which is needed first for calculating the specimen weight loss, but also for predicting the "shrinkage at a point" (not to be confused with specimen or cross-section shrinkage, which depends also on geometry, etc). In this case, the isotherm proposed in Norling (1997) is adopted. If the formulation accounting for a nonlinear moisture capacity $C(H)$ is considered, the desorption isotherm will have an effect on the H field, entering the nonlinear problem ($C(H)$ is calculated as the derivative of the desorption isotherm). This relation is written in (8) and (9) and is shown in Figure 21b, together with its derivative, for typical values of the model parameters.

$$w_e(H)/c = a_1 \cdot (1 - e^{-a_3 \cdot H}) + a_2 \cdot (e^{a_3 \cdot H} - 1) \quad (8)$$

$$a_1 = \frac{0.15 \cdot \alpha_{hydr}}{1 - e^{-a_3}}, \quad a_2 = \frac{w_0/c - 0.33 \cdot \alpha_{hydr}}{e^{a_3} - 1}, \quad a_3 = -(w_0/c) \cdot f_1 + f_2 \quad (9)$$

In (8) and (9), c is the cement content [g/cm^3], a_1, a_2 and a_3 are functions of the degree of hydration (α_{hydr}) and the initial w/c ratio (w_0/c), and f_1 and f_2 represent two shape factors. A sensitivity study of the different parameters on the overall weight losses can be found elsewhere (Roncero, 1999).

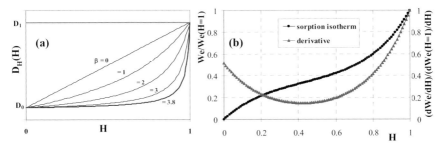

Figure 21. (a) Diffusivity vs. relative humidity and (b) desorption isotherm used.

In the literature, the local volume change ("shrinkage at a point") due to drying has been typically formulated in two ways: as a function of the water loss (g/liter) (Granger et al., 1997) or as a function of the change in relative humidity (Bazant and Najjar, 1972; Sadouki and Wittmann, 2001). Also, the difficulties encountered for experimentally determining those dependencies at a local level, have traditionally lead most researchers to simply assume a constant shrinkage coefficient α_{shr}, relating volumetric strains to either weight losses or RH variations.

However, a linear relationship may sometimes yield only rough approximations. To overcome this problem, inverse analysis was used in Alvaredo (1995) to estimate different shrinkage coefficients at various RH levels, and van Zijl (1999) proposed to treat α_{shr} as a linear or hyperbolic function of RH. In the present work, shrinkage at the point has been assumed to be related linearly to the local water loss per unit volume in the first set of simulations (López et al., 2005a). Further calibration of model parameters with experimental results has shown the advantages of considering a nonlinear relation between strains and weight losses in the simulations (Idiart et al., 2010a). The dependency adopted in this paper is based on the work in van Zijl (1999), only that in our case strains are related to weight losses. The following general expression for the shrinkage coefficient is used:

$$\alpha_{shr}(w_e) = \frac{n \cdot \varepsilon_{shr}^{\infty}}{\left(w_e^{env}\right)^n - \left(w_e^0\right)^n} \overline{w}_e^{n-1}, \text{ with } n=1,2,3 \qquad (10)$$

In (10), $\varepsilon_{shr}^{\infty}$ represents the final shrinkage strain corresponding to the environmental RH (H^{env}), w_e^{env} and w_e^0 are the moisture contents corresponding to H^{env} and initial internal H, respectively (from the desorption isotherm), and $\overline{w}_e$ represents the average moisture content within the considered time interval. Note that if n=1, a constant α_{shr} is retrieved. The results of this model when drying from saturation to a zero RH value have been qualitatively compared with experimental results on thin concrete slices in Idiart et al. (2011a), suggesting that the case of a constant α_{shr} may be more suited for high-performance concrete, whereas ordinary Portland cement behavior is better captured with a quadratic α_{shr}.

In order to evaluate the coupled behavior of the model, a series of simulations has been performed over the same mesostructural FE mesh represented in Figure 22 (14x14cm^2 concrete specimen with 6x6 aggregates arrangement and 26% volume fraction). As initial condition, $H=1$ throughout the sample is assumed. At $t_0=28$ days, $H=0.5$ is imposed on left and right faces (no moisture flow is allowed through top and bottom faces). The specimen is assumed to be simply supported.

The parameters for the diffusion model through the matrix have been adopted from previous work (Roncero, 1999), fitted to experimental results in OPC mortar specimens. Mechanical parameters are set so that a conventional concrete ($f_c\sim$40MPa) is represented. Aging viscoelasticity and aging interface elements have been considered for the matrix behavior, while aggregates behave linear elastically and remain saturated throughout the simulation. Calculations have been repeated for uncoupled and coupled situations (López et al., 2005a,b). The results at three different drying states are presented for the uncoupled and coupled cases in Figure 22 in terms of H field, and in Figure 23 in terms of weight loss evolution. The effect of coupling consists of a slightly higher degree of drying, quantified by lower H at the interior of the sample and higher weight losses than in the uncoupled case, although it is not very pronounced in this case.

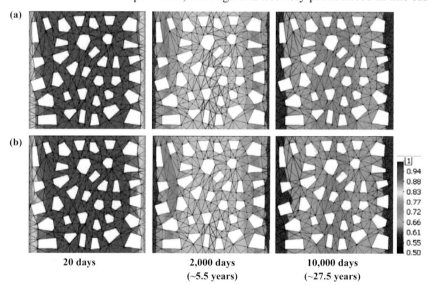

Figure 22. Distribution of RH for a 14x14cm^2 concrete specimen at 20, 2,000 and 10,000 days of drying: (a) uncoupled and (b) coupled cases.

Figure 24 depicts a sequence of the evolution in time of the energy spent in fracture processes for the coupled case with aging viscoelasticity. Initially, cracks perpendicular to the two drying surfaces develop (loaded cracks in red). As the drying front penetrates the specimen, surface cracks unload (arrested cracks in blue). At an advanced state of drying, microcracks are also observed in the interior of the sample, on the aggregate-matrix interfaces but most importantly radiating from the aggregates. This is due to the restraining effect caused by embedded inclusions in a shrinking matrix. These results qualitatively agree with experimental observations (Bisschop and van Mier, 2002), and have motivated a deeper study on the effect of aggregates on the drying-induced microcracking, which may be found in Idiart *et al.* (2007), Idiart *et al.* (2011a, 2011b).

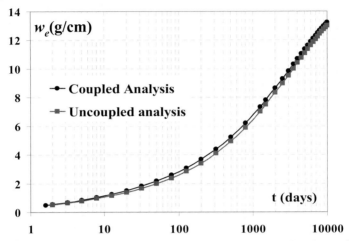

Figure 23. Evolution in time of total w_e [g/cm] for uncoupled and coupled cases.

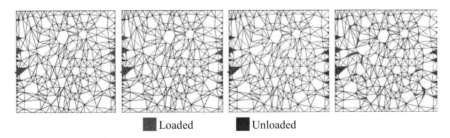

Figure 24. Fracture energy spent in interface elements for 20, 200, 2,000 and 10,000 days, left to right (coupled case with aging viscoelasticity for the matrix).

In order to evaluate the effect of aging viscoelasticity of the matrix phase, the coupled case was repeated with linear elasticity in the matrix (mechanical properties corresponding to 28 days). Figure 25 compares the deformed meshes of both cases at ~5.5 years of drying, showing that cracking is much more pronounced if linear elasticity is considered (Idiart, 2009). This can be explained by 2 reasons: (1) due to the absence of the aging-related increase of strength (since shrinkage develops with time, highest stresses will appear at a late stage, leading to more cracking if the strength remains constant than if it is increased over time); (2) the absence of visco-elasticity in the matrix will also lead to higher cracking (the effect of visco-elastic stress relaxation is not present in a simply elastic matrix, which generates higher stresses and results in cracking).

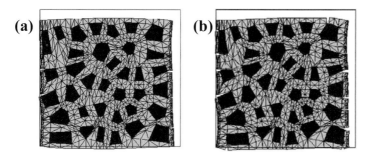

Figure 25. Deformed meshes at 2,000 days of drying (scale factor=150): coupled cases with (a) aging viscoelastic, and (b) linear elastic matrix (the blue box represents the dimensions of the undeformed mesh).

As a second verification example, classical drying shrinkage experiments on cylindrical concrete specimens (Granger *et al.*, 1997) have been analyzed with the present model. A normal strength concrete, denoted as Penly concrete in (Granger et al., 1997) has been selected for the simulations. Specimens were subjected to H=50% (top and bottom faces were sealed). This experiments are particularly relevant for evaluating a coupled hygro-mechanical model due to the fact that the overall relation between longitudinal strains and total weight losses was measured, allowing to link the mechanical behavior (longitudinal strain) with the drying process (weight loss). The main drawback for the present 2D mesostructural model is the cylindrical shape of the specimens, since in an axisymmetric analysis the aggregates would represent bodies of revolution within a more flexible matrix. Nonetheless, the lack of similar experimental results obtained on prismatic specimens has forced the selection of these experiments. In order to mitigate this undesired effect, a 'semi-axisymmetric' moisture diffusion analysis has been performed (the rotation axis is positioned at D/2, D=diameter of cylinder, and the angle covered is limited to a π), coupled with a 2D plane stress analysis for the mechanical problem, over the same FE mesh. In this way, the cylindrical shape is kept for the moisture diffusion problem, which is an essential feature, while avoiding the undesired above-mentioned effects.

The employed FE mesh (16x50cm^2, height is equal to the strain measurement basis used in experiments) has 26% aggregate volume fraction with max.-min. sizes of 25.9-10mm, respectively (Idiart et al., 2010a) (Figure 26). In order to simulate the central part of a larger specimen (specimens were 100cm long), the end faces are forced to remain planar throughout the simulation, with free horizontal displacements. Material parameters finally adopted are summarized in Table 1.

Table 1. Adopted parameters for the diffusion and mechanical models yielding the best fit to experimental results.

Material parameters adopted			
Diffusion analysis (matrix)		Mechanical analysis (continuum)	
Initial humidity	100%	E_{matrix}	aging Maxwell chain
D_0 (cm^2/day)	5x10^{-5}	E_{aggr} (MPa)	70000
D_1 (cm^2/day)	2x10^{-1}	v_{matrix} ; v_{aggr}	0.2 ; 0.2
β (–)	3.0	Mechanical analysis (interface elements)*	
C (g. cem/cm^3 mat)	0.473	χ (MPa)	2.0 ; 4.0
$\alpha_{hydration}$	0.90	c (MPa)	7.0 ; 14.0
w_0/c	0.50	tan ϕ	0.6 ; 0.6
$f1$; $f2$	5.0 ; 8.0	tan ϕ_r	0.2 ; 0.2
α_{shr}	0.01	G_f^I	0.03 ; 0.06
Diffusion analysis (interface elements)		G_f^{IIa}	10 . G_f^I
η (1/day)	100.0x10^6	σ_{dil} (MPa)	40
K_0/K_1	0.01	p_χ, p_c, p_{GF}	0.4, 0.5, 0.8
β_k (–)	0.0	k_χ, k_c, k_{GF}	1.0, 1.0, 1.0

First, the evolution of weight loss as a function of time is compared to experimental results in Figure 26. It can be observed that the effect of coupling is small and may be neglected in this case, and that the numerical results agree well with the experimental ones. However, when the same weight losses are represented against the longitudinal strains, as shown in Figure 27, the numerical results obtained using the constant shrinkage coefficient α_{shr} (lowest curve in yellow with circles) exhibit some departure from the experimental values. Although the skin microcracking effect is well captured, reflected by small longitudinal strains for large weight losses at the beginning of drying, two experimental features seem not well represented by the simulations: the numerical curve seems to be shifted horizontally with respect to the experimental one, and the second curved part of the experimental curve (strains grow more slowly than weight loss), is not well captured by the model. A deeper study of the results has led to 2 main potential reasons for these differences. On one side, the consideration of a constant α_{shr} could be the cause of the lack of a second curved part in the diagrams. For this reason, simulations have been repeated considering a linear and a parabolic dependence of α_{shr} on the weight loss, otherwise keeping all other material parameters the same.

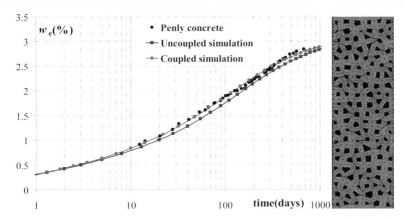

Figure 26. Comparison of coupled and uncoupled analyses with experimental results from (Granger et al., 1997): weight loss (% of the sample weight) vs. time. FE mesh used (right).

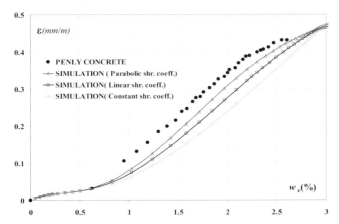

Figure 27. Comparison of numerical and experimental results from (Granger et al., 1997) in terms of weight losses and longitudinal strains [mm/m]: influence of the shrinkage coefficient (constant, linear and parabolic functions) on the uncoupled analysis.

Figure 27 also presents these results for the uncoupled cases. A considerable improvement of the global behavior can be observed, especially for advanced drying states (second curved part), which is now satisfactorily captured. However, a small gap still exists between experimental and simulation results. This may be due to a second factor having an influence on the results: the elasto-plastic nature of the model used for the interface elements, which implies that the crack closure effect is not taken into account (relative displacements are irreversible on unloading with the initial stiffness). If crack closure were not prevented, the longitudinal strains would probably increase (crack closure may be regarded as extra shrinkage strains), and the numerical curve would be horizontally shifted to the left. Further investigation would require a new constitutive interface model with secant unloading, which is at present one of the focuses of on-going work. More details of this study on drying shrinkage at the meso-level can be found in Idiart (2009), Idiart *et al.* (2010a).

4.2. Chemo-mechanical Analysis of External Sulfate Attack

Durability of concrete has become a very relevant issue in structural design over the last decades. Significant progress has been achieved regarding our understanding of the chemistry behind the different degradation processes. However, our knowledge of the interaction of some these processes and the resulting mechanical behavior is still limited. One practically relevant degradation process is the external sulfate attack, characterized by the ingress of sulfate ions from the surrounding medium, finally leading to expansion, cracking, spalling, and eventually to the complete disintegration of the material (Skalny *et al.*, 2002; Lee *et al.*,2008; Santhanam *et al.*, 2002; Schmidt *et al.*, 2009).

It is generally accepted that the cause of this expansion is the formation of ettringite from the sulfate ions and the different aluminate phases in the hardened cement paste. Even though this problem has been studied for a long time, the mechanisms by which sulfate ingress leads to overall expansion are not yet totally understood. Recent advances in the experimental (Schmidt et al., 2009; Chabrelie, 2010) and also in the modeling fields (Bary, 2008; Flatt and Scherer, 2008) have shown encouraging results towards this goal. There is a renewed interest in rationally describing this deterioration process, mainly related to the need of assessing the durability of underground nuclear waste containments and tunnel linings. The presence of cracks may play an important role also in this case, facilitating the ingress of the aggressive solution, and thus accelerating the degradation process.

A considerable effort has been devoted in recent years to develop chemo-mechanical models to predict the behavior of concrete under external sulfate attack (Schmidt-Döhl and Rostásy, 1999; Tixier and Mobasher, 2003; Bary, 2008; Basista and Weglewski, 2008). The common feature in all of these models is that they attempt to quantify the mechanical consequences of the degradation processes on the basis of a chemo-transport calculation of the main species involved and the relevant reaction products formed. However, only some of them perform a full force-deformation mechanical analysis, and none of them seems to explicitly account for the effects of opening cracks in facilitating ion diffusion and accelerating the overall degradation process, which may be a relevant feature in real cases.

Description of the chemo-transport model

The relevant chemical reactions that potentially take place during the exposure to a sodium sulfate solution are well-known (Tixier and Mobasher, 2003; Skalny *et al.*, 2002). First, portlandite (CH) and calcium silicate hydrates (CSH) react with the ingressing sulfates to form gypsum ($C\bar{S}H_2$). Decalcification of CSH, leading to a degradation of mechanical properties of cement paste, has been left out of the present work, as is only likely to happen at high sulfate concentrations.

In a second step, ettringite forms from monosulfoaluminate ($C_4A\bar{S}H_{12}$), unreacted tricalcium aluminate grains (C_3A), tetracalcium aluminate hydrate (C_4AH_{13}), and alumino-ferrite phase (C_4AF) (Tixier and Mobasher, 2003; Idiart *et al.*, 2011b), although this last reaction seems to take place at a much lower rate than the preceding ones. The effect of the precipitation of gypsum (for high sulfate concentrations) on the observed overall expansion is still not clear (Santhanam *et al.*, 2001; Tian and Cohen, 2000) and will not be further considered in this work. In the present model, following (Tixier and Mobasher, 2003; Basista and Weglewski, 2008), it is assumed that ingressing sulfates first react with CH to form gypsum. Thereafter, sulfates (in the form of gypsum) react with the different aluminate phases of the cement paste to form ettringite.

The preceding chemical reactions take place according to the sulfates and calcium aluminates availability in time and space. This is achieved numerically by the use of a diffusion equation for the ingressing sulfate ions with a second-order reaction term for the precipitation of ettringite, following (Tixier and Mobasher, 2003). In order to simplify the numerical treatment, it was proposed in that work to lump all chemical reactions leading to ettringite precipitation in one single expression:

$$C_{eq} + q\bar{S} \rightarrow C_6A\bar{S}_3H_{32} \tag{11}$$

where $C_{eq} = \gamma_1 C_4AH_{13} + \gamma_2 C_4A\bar{S}H_{12} + \gamma_3 C_3A + \gamma_4 C_4AF$ in [mol/m^3 of material] represents the equivalent lumped reacting calcium aluminates, $q = 3\gamma_1 + 2\gamma_2 + 3\gamma_3 + 4\gamma_4$ represents the stoichiometric weighted coefficient of the sulfate phase, and γ_i are defined as the proportion of each aluminate phase (C_i[mol/m^3 of material]) to the total aluminate content. In this way, the system of equations regarding the diffusion-reaction process for sulfate concentration, U[mol/m^3 of material], and calcium aluminate depletion yields

$$\text{(a) } \frac{\partial U}{\partial t} = \frac{\partial}{\partial x}\left(D_U \frac{\partial U}{\partial x}\right) - k.U.C_{eq}, \text{ and (b) } \frac{\partial C_{eq}}{\partial t} = -k\frac{U.C_{eq}}{q} \qquad (12)$$

where $D_U[m^2/s]$ is the diffusion coefficient for sulfate ions through the porous network, $k[m^3/(mol.s)]$ is the lumped rate of take-up of sulfates, and t[s] and x[m] are the time and space coordinates, respectively. This formulation implies the use of a single reaction rate coefficient (kinetics) for all the reactions. The implementation of this diffusion-reaction model has been verified with some simplified cases having an analytical solution in Idiart (2009).

In this study, it is proposed to treat reactions separately to consider different kinetics for each reaction individually. The details of the formulation in this case may be found elsewhere (Idiart, 2009; Idiart et al., 2011b), as additional equations for the various calcium aluminate phases are introduced. The simplified version with a lumped reaction is still applicable in case that the kinetics of the individual reactions is not known.

Additionally, transport of ions through the cracks is explicitly considered in the model. Given a local orthogonal coordinate system (x,y) within a crack, the mass conservation for an incompressible fluid filling the discontinuity, with a sulfate concentration U, and neglecting the variation in crack width (u[m]) within an infinitesimal time increment, may be written as (Idiart et al., 2010c)

$$\text{(a) } -\frac{d\hat{J}_1}{dx} + q^- + q^+ = 0, \text{ with (b) } \hat{J}_1 = u \cdot J_1 = -u \cdot D_U^{cr} \frac{dU}{dx} = \hat{D}_U^{cr} \frac{dU}{dx} \qquad (13)$$

where $\hat{J}_1$ is the longitudinal local transport rate and q^- and q^+ are the leak-off terms from the surrounding medium. The total longitudinal flow along the discontinuity is derived from the first Fick's law (the diffusion coefficient through the cracks, $\hat{D}_U^{cr}[m^3/s]$, is to be determined). Plugging in (13b) into (13a) a diffusion equation in terms of ionic concentration gradient is obtained.

One of the advantages of the present mesostructural model (equipped with a discrete cracking approach) is that the effective diffusion through the matrix can be explicitly split into the behavior through the uncracked porous matrix, and the diffusion through the cracks. In this work, the diffusion coefficient of the porous media is assumed to be dependent on the pore filling effect (the diffusivity decreases as pores are filled with reaction products). In this way, the model accounts for the decrease in diffusivity due to the pore filling effect and an increase of the overall diffusivity due to cracking and spalling phenomena (it is assumed that ettringite does not precipitate within the cracks).

Different models have been proposed in the literature to introduce the pore filling effect (Garboczi and Bentz, 1992; Samson and Marchand, 2007). In the present study, a hyperbolic function depending on the initial porosity (Φ_{ini}) has been adopted, which yields comparable trends to the one proposed in Samson and Marchand (2007), and is given by

$$D_U(\Phi_{cap}) = D_0 + (D_1 - D_0) f(\beta_D, \Phi_{cap}) \qquad (14)$$

where function f is given in (6b), replacing H with Φ_{cap}/Φ_{ini}, Φ_{cap} is the updated porosity value, which considers porosity changes due to precipitation of species. Initial capillary porosity is calculated through a classical Powers' model, and the updated porosity value considers the decrease due to the precipitation of ettringite in the pores through expression (15).

$$\Phi_{cap} = \Phi_{ini} - \alpha_s C_{eq}^{react}, \text{ if } \alpha_s C_{eq}^{react} < \Phi_{ini} \text{ (and 0 otherwise)} \qquad (15)$$

where the term $\alpha_s \cdot C_{eq}^{react}$ accounts for the volume of ettringite that forms in the pore space.

As previously outlined, the effect of cracks on the ion transport is explicitly considered in the model via the introduction of interface elements. In the cases where spalling occurs, a drastic change in the boundary conditions of the chemo-transport analysis is expected to occur, accelerating the degradation process. To introduce this effect, a relation between crack width and its diffusivity must be considered. Recent experiments have been carried out in Djerbi et al. (2008) to determine the chloride diffusivity through cracked concrete. They found a linear relationship between diffusivity and crack opening until a crack width of approx. 100μm, after which it remains constant and approximately equal to the diffusivity in free solution. This relation is multiplied by the crack width for its implementation in the framework presented above, in order to relate the total transport through a crack with the concentration gradient. As a result, a quadratic law of the diffusivity in terms of crack width is obtained until the threshold crack opening is reached, after which the diffusivity increases linearly with crack width (Idiart et al., 2011b). A comparison of the model response when assuming different crack width - diffusion coefficient relations may be found in Idiart (2009).

The origin and mechanisms of the observed overall expansion of concrete during external sulfate attack are, up to now, still controversial. It is fairly well accepted that ettringite is the main, if not the only, cause for expansion, but the mechanism by which ettringite formation leads to expansion and cracking of concrete is yet not very clear (Brown and Taylor, 1998). Most of the models proposed in the literature consider ettringite as the only responsible for expansion. Regarding the mechanism, mainly two directions have been followed. On one hand, some researchers consider the precipitation of crystals in a supersaturated pore solution, leading to a crystallization pressure exerted on the pore walls (Flatt and Scherer, 2008; Bary, 2008). However, this approach has not yet been validated with simulations of expansion over time of exposure to a sulfate solution. In the second approach, it is assumed that the volumetric strain is a result of the volume change associated with the chemical reactions involving calcium aluminate phases and the ingressing sulfates. For the individual reactions involved, the volumetric change due to the difference in molar volumes between reactants and reaction products can be calculated using stoichiometric constants (Clifton and Pommersheim, 1994; Tixier and Mobasher, 2003).

In this work, ettringite is assumed to be the only reaction product governing expansions and volumetric strains are obtained following the second approach. The volumetric change of the individual reactions is calculated as (Tixier and Mobasher, 2003):

$$\frac{\Delta V_i}{V_i} = \frac{m^{ettr} - m^i + a_i \cdot m^{gypsum}}{m^i + a_i \cdot m^{gypsum}} = \frac{m^{ettr}}{m^i + a_i \cdot m^{gypsum}} - 1 \qquad (16)$$

in which m^i [m^3/mol] is the molar volume of each species and a_i is the stoichiometric coefficient involved in each reaction (see Idiart, 2011b) for more details. The volumetric strain (ε_v) is then calculated as follows

$$\varepsilon_v = \sum_{i=1}^{n} C_i^{react} \frac{\Delta V_i}{V_i} \overline{m}^i - f \cdot \Phi_{ini}, \text{ with } \overline{m}^i = m^i + a_i \cdot m^{gypsum} \text{ [m}^3\text{/mol]} \qquad (17)$$

In (17), C_i^{react} (amount of aluminates that have reacted) is the difference between the initial concentration of a given aluminate phase and the respective unreacted amount. The second term in the RHS of (17) is introduced to account for the fact that part of the ettringite formed will not translate into expansions. Most of the models consider that a fraction f of the capillary porosity has to be filled before any expansion is observed (Clifton and Pommersheim, 1994; Tixier and Mobasher, 2003; Basista and Weglewski, 2009). Typical values of parameter f are in the range 0.05-0.40 (Tixier and Mobasher, 2003). The scarcity and low range of the values needed to fit experimental data suggest this assumption may be rather arbitrary (Brown and Taylor, 1998; Taylor et al., 2001). Despite this consideration, it has been decided in this work to maintain the formulation presented, mainly because the model, while being simpler, yields very reasonable approximations, as will be shown in the following section.

Main results of the model

As a first step, the coupled behavior of the model when cracking occurs has been evaluated through an academic example consisting of a small concrete sample immersed in a sulfate solution. Coupled and uncoupled simulations have been conducted with a FE mesh representing 6x6cm^2 and 4x4 aggregate arrangements, with 26% vol. fraction and 15mm max. aggregate size. All faces of the specimen are exposed to the sodium sulfate solution with a fixed concentration of 35.2[mol/m^3]. For simplicity, the lumped-reaction version of the model has been used. An initial concentration of 9.13 wt.% of C$_3$A in the cement is considered, corresponding to a CEM I 52.5R cement. Knowing the cement content of the matrix phase, representing mortar plus smaller aggregates, an initial C$_3$A content may be derived (Idiart *et al.*, 2011b).

Material parameters considered for the diffusion-reaction analysis in the two cases are as follows: D_I=1.96x10^{-12}[m^2/s], k=0.92x10^{-9}[m^3/(mol.s)], q=3, f=0.05, w/c=0.5, α=0.9, D_0/D_I = 0.05, β_D=1.5 and α_s=1.133x10^{-4}[m^3/mol]. The effect of cracks on the diffusion of sulfate ions is considered in the coupled simulations assuming as a first step a much more rapid diffusion through the crack with respect to the uncracked continuum (worst case scenario). Linear elasticity is assumed in the continuum (E$_{aggr}$=70GPa, E$_{matrix}$=25GPa, ν = 0,2 for both phases). For the aggregate-matrix interface elements: χ_0=2MPa, c$_0$=7MPa, tanϕ_0=0.7, tan ϕ_r=0.4, G$_F^I$=0.03N/mm, G$_F^{IIa}$=10.G$_F^I$, σ_{dil}=40MPa; for the matrix-matrix joint elements: same parameters, except χ_0=4MPa, c$_0$=14MPa, G$_F^I$=0.06N/mm.

Figure 28 depicts the results of the evolution of ettringite [kg/m^3] for the (a) uncoupled and (b) coupled cases, for three different advanced exposure times between 1,522 and 2,572 days. It may be observed that the front of ettringite formation, due to the penetration of the sulfates, progressively advances towards the center of the specimen, much faster in the coupled case, once cracking and spalling develop. In turn, Figure 29 illustrates the evolution of the crack pattern in terms of the energy spent in fracture, for four different earlier exposure times between 672 and 1,022 days, for which coupled and uncoupled analyses yield similar results (opening cracks highlighted in red; arrested cracks in blue). Cracks begin at the specimen corners, along the aggregate-matrix interfaces (Figure 29a), and then evolve with increasing expansion, get connected through matrix cracks and eventually produce the spalling of a mortar layer, at 1,000 days of exposure (approx.), as observed in the figure in the form of cracks parallel to the exposed surfaces over the entire perimeter. Figure 30 shows the evolution of the fracture process for the same later exposure times as Figure 29, for (a) uncoupled and (b) coupled analyses. A correlation between the ettringite concentration in Figure 28 and crack patterns is clearly observed. As soon as the corner cracks connect with the exposed surfaces, the coupled simulation accentuates the influence of these cracks as preferential penetration channels, leading to an increase of ettringite formation and thus to a higher level of expansions. In turn, a higher degree of internal microcracking between and around the aggregates is reached in the coupled case, as compared to the uncoupled one. Note that in the coupled analysis the internal cracking is much more pronounced when passing from 1,522 to 2,022 days of exposure, and that from 2,022 to 2,572 days micro-cracking is propagated around all the aggregates in the sample, representing the total disaggregation between matrix and aggregate, while cracks through the matrix connecting contiguous aggregates, unload. The simulation captures the crack patterns and the spalling phenomena observed in lab experiments, with the resulting reduction of cross-section area of the sample, and even its almost complete disintegration. This can be seen qualitatively in Figure 31, comparing the deformed mesh from the simulation to the final state of a real concrete sample (Al-Amoudi, 2002).

Recent experiments of external sodium sulfate attack on OPC concrete prisms of $10x10x40cm^3$ (Wee, 2000) have also been simulated with the present model. The results of these simulations may be found elsewhere (Idiart, 2009; Idiart et al., 2011b), together with a more detailed description of the model and other applications.

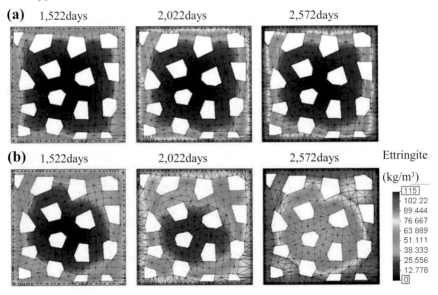

Figure 28. Ettringite concentration [kg/m^3] for three different exposure times (1,522, 2,022 and 2,572 days): (a) uncoupled, and (b) coupled analyses.

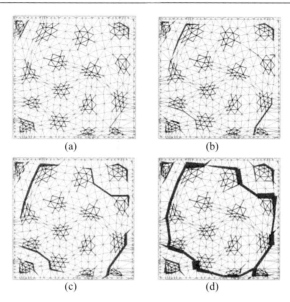

Figure 29. Crack pattern evolution in terms of energy spent in fracture, for the following exposure times: (a) 672, (b) 722, (c) 772, and (d) 1,022 days.

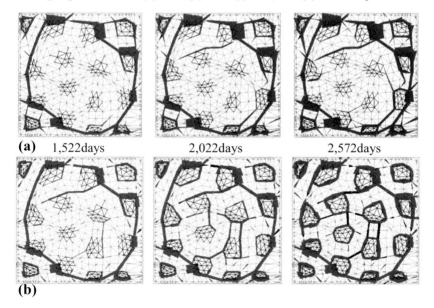

Figure 30. Crack pattern evolution in terms of energy spent in fracture for the (a) uncoupled and (b) coupled cases at same exposure times of Figure 28.

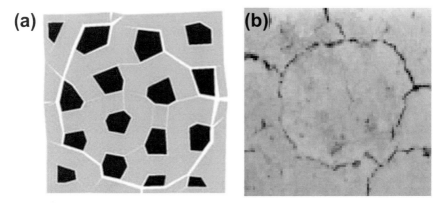

Figure 31. Qualitative numerical-experimental comparison: (a) deformed mesh, and (b) crack pattern in a concrete specimen (Al-Amoudi, 2002).

5. Concluding remarks and on-going developments

In spite of relatively rudimentary geometries and specimens analyzed so far, the approach described for meso-mechanical analysis of concrete and other heterogeneous quasi-brittle materials, has led to very satisfactory results at meso and macro levels of observation. In particular one can highlight the realistic micro-crack patterns and evolution obtained, with localization happening spontaneously only as the result of micro-crack interaction, with their opening and closing depending on local stress-strain conditions and the principles of mechanics exclusively (*i.e.* without the need of introducing any extra *ad-hoc* procedure such as tracking algorithms, etc.). Also remarkable are the overall curves in terms of average stress and strain of specimen obtained, which really capture the main well known features observed in experiments such as uniaxial compression dilatancy in the post-peak, biaxial failure envelope, creep, shrinkage, etc. Overall, this may be interpreted in the sense that the model really incorporates the essential aspects of the material behavior and, therefore, even if each ingredient is not too sophisticated, the most important observed features of concrete are already reproduced. The extension of the model to diffusion-driven environmental coupled phenomena such as drying shrinkage and external sulphate attack, also turn out very satisfactory. It also shows that the discrete representation of cracking using zero-thickness interface element, not only leads to a sound mechanical behaviour, but it also makes it possible to introduce explicitly the localized effect of cracks on diffusion, and clarify the true influence of coupling for each problem, as it turns weak for drying shrinkage but totally dominant for external sulphate attack.

It is also fair to acknowledge some limitations/drawbacks of the approach described, such as the high number of degrees of freedom due to node multiplication at interface intersections, with the subsequent computational cost which imposes limitations on the size of the problem to be analyzed. Also, crack locations are restricted to the initial mesh lines/surfaces, which may certainly represent a source of bias on the potential crack trajectories. Finally, the current procedure for geometry generation technique (Voronoi/Delaunay) leads to systems of particles of roughly similar sizes, which might be undesirable in some cases. Current efforts are aimed at the exploitation of the models developed, at their extension to 3D and to include new phenomena, and at the improvement of related numerical techniques to reduce/eliminate current drawbacks, in particular to avoid systematic node duplication due to the use of interface elements from the beginning of the analysis. On-going work along this line includes new methods for evaluation of stress-tractions across mesh lines emanating from a nodal point in meshes without interface elements, and cracking conditions at such nodal points, among others (Ciancio *et al.* 2006, 2007).

Acknowledgements

This research was supported by grants MAT2003-02481, BIA2006-12717 funded by MEC (Madrid), and 80015/A04 funded by MFOM (Madrid), and is currently sponsored by grant BIA2009-10491 from MICINN (Madrid). The second, third and fourth authors wish to thank MEC for the doctoral fellowships and "Ramón y Cajal" research position received during the period in which their contributions to this work were developed.

References

O.S.B. Al-Amoudi. Attack on plain and blended cements exposed to aggressive sulfate environments. *Cem. Concr. Comp.*, 24:305-316, 2002.

A. Alvaredo. Crack formation under hygral or thermal gradients. *Fracture Mech. Concrete Structures (FraMCoS2)*, Zürich, Switzerland, ed. F. Wittmann, vol. 2:1423-1441, 1995.

B. Bary. Simplified coupled chemo-mechanical modeling of cement pastes behavior subjected to combined leaching and external sulfate attack. *Int. J. Numer. Anal. Meth. Geomech.* 32:1791-1816, 2008.

M. Basista, W. Weglewski. Micromechanical modelling of sulphate corrosion in concrete: influence of ettringite forming reaction. *Theor. Appl. Mech.*, 35:29-52, 2008.

Z.P Bazant, L.J. Najjar, L.J. Nonlinear water diffusion in nonsaturated concrete. *Cement and Concrete Research*, vol. 1:461-473, 1972.

ZP. Bazant, L. Panula. Practical prediction of time-dependent deformations of concrete. Part II: basic creep. *Mat. Struct.*, vol. 11:317-328, 1978.

ZP. Bazant. Input of creep and shrinkage characteristics for a structural analysis program. *Mat. Struct.*, 15:283-290, 1982.

Z.P. Bazant. Mathematical model of creep and shrinkage of concrete, ed. ZP Bazant, John Wiley & Sons Ltd, 1988.

J. Bisschop, J.M.G. van Mier. Effect of aggregates on drying shrinkage microcracking in cement-based composites. *Materials and Structures*, vol. 35: 453-461, 2002.

J.E. Bolander, S. Saito. Fracture analysis using spring networks with random geometry. *Engineering Fracture Mechanics*, vol. 61:569-591, 1998.

J.E. Bolander, G.S. Hong G.S., K. Yoshitake. Structural concrete analysis using rigid-body-spring networks. *J. Comp. Aided Civil and Infrastructure Engng.*, vol. 15: 120-133, 2000.

P. Brown, H. Taylor. The role of ettringite in external sulfate attack. In: Marchand & Skalny eds., *Mat. Science of Concrete: Sulfate Attack Mech.*, Amer. Cer. Soc. Press: 73-98, 1998.

A. Caballero. *3D meso-mechanical numerical analysis of concrete using interface elements*. PhD thesis, ETSECCPB-UPC, E-08034 Barcelona, 2005.

A. Caballero, C.M. López, I. Carol. 3D meso-structural analysis of concrete specimens under uniaxial tension. *Computer Methods in Applied Mechanics and Engineering*, vol. 195, n° 52:7182-7195, 2006a.

A. Caballero, I. Carol, C.M. López. A meso-level approach to the 3D numerical analysis of cracking and fracture of concrete materials. *Fatigue and Fracture of Engineering Materials and Structures*, vol. 29:979-991, 2006b.

A. Caballero, I. Carol, C.M. López. 3D mesomechanical analysis of concrete specimens under biaxial loading. *Fatigue and Fracture of Engineering Materials and Structures*, vol. 30:877-886, 2007.

A. Caballero, K.J. Willam, I. Carol. Consistent tangent formulation for 3D interface modeling of cracking/fracture in quasi-brittle materials. *Comp. Meth. Appl. Mech. Eng.*, 197:2804-2822, 2008.

I. Carol, P. Prat. A statically constrained microplane model for smeared analysis of concrete cracking. *Computer aided analysis and design of concrete struct.*, eds. Bicanic, Mang, Pineridge Press, Austria: 919-930, 1990.

I. Carol, P. Prat, C.M. López. A normal/shear cracking model. Application to discrete crack analysis. *ASCE Journal of Engng. Mechanics*, vol. 123, n° 8: 765-773, 1997.

I. Carol, C.M. López, O. Roa. Micromechanical analysis of quasi-brittle materials using fracture-based interface elements. *Int. Journal for Numerical Methods in Engineering*, vol. 52:193-215, 2001.

A. Chabrelie. *Mechanisms of degradation of concrete by external sulfate ions under laboratory and field conditions*. PhD thesis, EPFL, Lausanne, Switzerland, 2010.

D. Ciancio, C.M. López, I. Carol, M. Cuomo. Numerical simulation of fracture localized phenomena in concrete specimens.In *3rd Joint Conference of Italian Group of Comp. Mech. and Ibero-latin American Assotiation of Comp. Meth. in Engng (Gimc 2002)*, 2002.

D. Ciancio, I. Carol, M. Cuomo. On inter-element forces in the FEM-displacement formulation, and implications for stress recovery. *Int. J. Numer. Meth. Engng*, vol. 66:502–528, 2006.

D. Ciancio, I. Carol, M. Cuomo. Crack opening conditions at 'corner nodes' in FE analysis with cracking along mesh lines. *Engng Fracture Mechanics*, vol. 74:1963-1982, 2007.

J.R. Clifton, J.M. Pommersheim. Sulfate attack of cementitious materials: volumetric relations and expansions. NISTIR 5390, NIST, 1994.

G. Cusatis, Z.P. Bazant, L. Cedolin. Confinement-shear lattice model for concrete damage in tension and compression: I. Theory. *ASCE J. Engrg. Mech.*, vol. 129, n° 12:1439-1448, 2003a.

G. Cusatis, Z.P. Bazant, L. Cedolin. Confinement-shear lattice model for concrete damage in tension and compression: II. Computation and validation. *ASCE J. Engrg. Mech.*, vol. 129, n° 12:1449-1458, 2003b.

G. Cusatis, Z.P. Bazant, L. Cedolin. Confinement-shear lattice CSL model for fracture propagation in concrete. *Comput. Methods Appl. Mech. Engrg.* 195: 7154-7171, 2006.

A. Djerbi, S. Bonnet, A. Khelidj, V. Baroghel-Bouny Influence of traversing crack on chloride diffusion into concrete. *Cem. Concr. Res.*, 2008, 38:877-883, 2008.

EHE (1998), Instrucción de Hormigón Estructural EHE. Min. de Fomento, Madrid (in Spanish).

R.J. Flatt, G.W. Scherer. Thermodynamics of crystallization stresses in DEF. *Cem. Concr. Res.*, 38:325-336, 2008.

E.J. Garboczi, D.P. Bentz. Computer simulation of the diffusivity of cement-based materials. *Journal of Materials Science*, 27: 2083-2092, 1992.

D. Garolera, C.M. López, I. Carol, P. Papanastasiou. Micromechanical analysis of the rock sanding problem. *Journal of the Mechanical Behaviour of Materials*, vol. 13, n° 1-2: 45-53, 2005.

L. Granger, J.M. Torrenti, P. Acker. Thoughts about drying shrinkage: scale effects and modelling. *Mat. Struct.*, 30:96-105, 1997.

P. Grassl, H.S. Wong, N.R. Buenfeld, Influence of aggregate size and volume fraction on shrinkage induced micro-cracking of concrete and mortar, *Cem. Concr. Res.* 40, 85-93, 2010.

M. Hassanzadeh. Determination of fracture zone properties in mixed mode I and II. *Engineering Fracture Mechanics,* 35(4/5):845-853, 1990.

DA. Hordijk. Tensile and tensile fatigue behaviour of concrete; experiments, modeling and analyses. *HERON,* vol. 37, Stevin Laboratory and TNO Research, Delft, The Netherlands, 1992.

AE. Idiart. *Coupled analysis of degradation processes in concrete specimens at the meso-level.* PhD thesis, UPC, Barcelona, Spain, 2009.

A.E. Idiart, C.M. López, I. Carol. H-M mesostructural analysis of the effect of aggregates on the drying shrinkage microcracking of concrete. *Proc. COMPLAS IX, eds. Oñate & Owen,* Barcelona, Spain: 466-469, 2007.

A.E. Idiart, J. Bisschop, A. Caballero, P. Lura. A numerical and experimental study of aggregate-induced shrinkage cracking in cementitious composites, 2010. (Submitted for publication).

A.E. Idiart, C.M. López, I. Carol. Modeling of drying shrinkage of concrete specimens at the meso-level, *Materials and Structures*, Volume 44, No. 2, 2011a. doi: 415-435, DOI: 10.1617/s11527-010-9636-2

A.E. Idiart, C.M. López, I. Carol. Chemo-mechanical analysis of concrete cracking and degradation due to external sulfate attack: a meso-scale model, *Cement and Concrete Composites*, Vol. 33, pp. 411–423, 2011b. doi:10.1016/j.cemconcomp.2010.12.001

A.R. Ingraffea, V.E. Saouma. Numerical modelling of discrete crack propagation in reinforced and plain concrete. In G. Sih and A. Di Tomasso, (Eds.), *Fracture Mechanics of concrete:* 171-225. Martinus Nijhoff, Dordrecht, The Netherlands, 1985.

H. Kupfer, H. Hilsdorf, H. Rüsch. Behavior of concrete under biaxial stresses. *J. American Concrete Inst.*, vol. 66: 656–666, 1969.

S. Lee, R. Hooton, H. Jung, D. Park, C. Choi. Effect of limestone filler on the deterioration of mortars and pastes exposed to sulfate solutions at ambient temperature. *Cem. Concr. Res.*, 38:68-76, 2008.

G. Lilliu, J.G.M. van Mier. 3D lattice type fracture model for concrete. *Engineering Fracture Mechanics*, vol. 70:927-941, 2003.

C.M. López. *Microstructural analysis of concrete fracture using interface elements. Application to various concretes* (in Spanish), PhD Thesis, ETSECCPB-UPC, E-08034 Barcelona, 1999.

CM. López, I. Carol, A. Aguado. Microstructural analysis of concrete fracture using interface elements. *ECCOMAS2000*, Barcelona, CIMNE, 2000.

CM. López, I. Carol, J. Murcia. Mesostructural modeling of basic creep at various stress levels. In *Creep, Shrinkage and Durability Mechanics of Concrete and other Quasi-Brittle materials*, Elsevier Publishers. Oxford, pp 101-106, 2001.

C.M. López, D. Ciancio, I. Carol. Effect of interface dilatancy and specimen size in meso-mechanical analysis of concrete. In *Métodos Computacionais em Engenharia (Proc. of the Int. Conf. in Lisbon 31/May 2/Jun)*, C.A. Mota Soares *et al.* (Eds), APMTAC–SEMNI, in CD-ROM, ISBN 972-49-2008-9, 2004.

C.M. López, J.M. Segura, A.E. Idiart, I. Carol. Mesomechanical Modeling of Drying Shrinkage Using Interface Elements. *Creep, Shrinkage and Durability of Concrete and Concrete Structures* (Concreep 7), G. Pijaudier-Cabot, B. Gérard, P. Acker (Eds.), Nantes, 107-112, 2005a.

CM. López, AE. Idiart, I. Carol. Mesomechanical analysis of concrete deterioration including time dependence. In *Computational Plasticity (COMPLAS VIII)*, D.R.J., Owen, E. Oñate, B. Suarez (Eds.), CIMNE, Barcelona:1059-1062, 2005b.

C.M. López, A.E. Idiart, I. Carol. Desarrollo y aplicaciones mesomecánicas en hormigón de una ley constitutiva de junta con envejecimiento. *XXIV* GEF, Burgos, Spain (in Spanish), 2007.

CM. López, I. Carol, A. Aguado. Meso-structural study of concrete fracture using interface elements. I: numerical model and tensile behavior. *Mat. Struct.*, vol 41:583-599, 2008a.

CM. López, I. Carol, A. Aguado. Meso-structural study of concrete fracture using interface elements. II: compression, biaxial and Brazilian test. *Mat. Struct.*, 41:601-620, 2008b.

K. Norling. *Moisture Conditions in High Performance Concrete*. PhD Thesis, Chalmers Univ. Tech., Goteborg, Sweden, 1997.

A. Okabe, B. Boots, K. Sugihara. *Spatial tessellations: concepts and applications of Voronoi diagrams*, John Wiley & Sons, 1992.

P. Prat, A. Gens, I. Carol, A. Ledesma, J. Gili. DRAC: A computer software for the analysis of rock mechanics problems. In *Applic. of computer methods in rock mech.*, 2, H. Liu (Ed.), Xian, China, Shaanxi Science and Tech. Press: 1361-1368, 1993.

R. Ribó, M. Riera. Gid, the personal pre and post processor. *Technical report*, CIMNE-UPC-Barcelona, 1997.

O. Roa, C.M. López, P. Pisoni, M. Pini, I. Carol, R. Contro. Microstructural analysis of cancellous bone taking into account geometrically non-linear effects. *ECCOMAS 2000, CIMNE-UPC*, Barcelona, 2000.

O. Roa. *Análisis microestructural del comportamiento mecánico del hueso trabecular* (in Spanish), PhD Thesis, ETSECCPB-UPC, E-08034 Barcelona, 2004.

P.E. Roelfstra, H. Sadouki, F.H. Wittmann. Le béton numérique. *Materials and Structures*, vol. 18:309-317, 1985.

J. Roncero. *Effect of superplasticizers on the behavior of concrete in the fresh and hardened states: implications for high performance concretes.* PhD thesis, UPC, Barcelona, Spain, 1999.

H. Rüsch. Research toward a general flexural theory for structural concrete. ACI Journal, 57:1–28, 1960.

J.G. Rots. *Computational modelling of concrete fracture*, PhD thesis, Delft University of Technology, Delft, 1988.

H. Sadouki, JGM van Mier. Meso-level analysis of moisture flow in cement composites using a lattice-type approach. *Mat. Struct.*, 30:579-587, 1997.

H. Sadouki, F.H. Wittmann. Damage in a composite material under combined mechanical and hygral load. *Lecture Notes in Physics, Cont. & Discont. Mod. of Cohesive-frictional Mat.*, Springer-Verlag Berlin Heidelberg, 568: 293-307, 2001.

E. Samson, J. Marchand. Modeling the transport of ions in unsaturated cement-based materials. *Computers and Structures*, 2007, 85:1740-1756, 2007.

M. Santhanam, M.D. Cohen, J. Olek. Sulfate attack research - whither now? *Cem. Concr. Res.*, 31:845-851, 2001.

M. Santhanam, M.D. Cohen, J. Olek. Mechanism of sulfate attack: a fresh look. Part 1: Summary of experimental results. *Cem. Concr. Res.*, 32:325-332, 2002.

E. Schlangen, J.G.M. van Mier. Lattice model for simulating fracture of concrete. In F.H. Wittmann (Ed.) *Numerical Models in Fracture Mechanics of Concrete*, Balkema, Rotterdam:195-205, 1993.

E. Schlangen, E.A. Koenders, K. van Breugel. Influence of internal dilation on the fracture behaviour of multi-phase materials. *Engng. Fract. Mech.*, 74:18-33, 2007.

T. Schmidt, B. Lothenbach, M. Romer, J. Neuenschwander, K. Scrivener. Physical and microstructural aspects of sulfate attack on ordinary and limestone blended Portland cements. *Cem. Concr. Res.*, 39:1111-1121, 2009.

F. Schmidt-Döhl, F. Rostásy. A model for the calculation of combined chemical reactions and transport processes and its application to the corrosion of mineral-building materials Parts I and II. *Cem. Conc. Res.*, 29:1039-1053, 1999.

J.M. Segura, I. Carol. On zero-thickness interface elements for diffusion problems. *Int. J. Num. Anal. Meth. Geomech.*, 28(9):947-962, 2004.

J. Skalny, J. Marchand, I. Odler. *Sulfate Attack on Concrete.* Spon Press, 2002.

T. Stankowski. *Numerical simulation of progressive failure in particle composites*, PhD Thesis, Dept. CEAE, University of Colorado, Boulder, CO 80309-0428, 1990.

H. Taylor, C. Famy, K. Scrivener. Delayed ettringite formation. *Cem. Concr. Res.*, 31:683-693, 2001.

B. Tian, M.D. Cohen. Does gypsum formation during sulfate attack on concrete lead to expansion? *Cem. Concr. Res.*, 30:117-123, 2000.

R. Tixier, B. Mobasher. Modeling of damage in cement-based materials subjected to external sulfate attack. Parts I and II. *ASCE J. Mat. Civil Engng.*, 15:305-322, 2003.

T. Tsubaki, M. Das, K. Shitaba. Cracking and damage in concrete due to nonuniform shrinkage. In Z. P. Bazant (Ed.), *Fracture Mechanics of Concrete Structures (FraMCoS1)*, Colorado, USA: 971-976, 1992.

J.G.M. van Mier. *Fracture Processes of Concrete*. CRC Press, 1997.

G.P. van Zijl. *Computational modelling of masonry creep and shrinkage*, PhD thesis, Delft University of Technology, Delft, 1999.

R. Vonk. *Softening of concrete loaded in compression*, PhD thesis, Technische Universiteit Eindhoven, Postbus 513, 5600 MB Eindhoven, 1992.

J. Wang, C. Huet. A numerical model for studying the influences of pre-existing microcracks and granular character on the fracture of concrete materials and structures. In C.Huet (Ed.), *Micromechanics of Concrete and Cementitious Composites*. Presses Politechniques et Universitaires Romandes, Lausanne, Suiza, pp. 229, 1993.

T.H. Wee, A.K. Suryavanshi, S.F. Wong, A.K. Anisur Rahman. Sulfate resistance of concrete containing mineral admixtures. *ACI Mat. J.*, 97(5):536-549, 2000.

Crack Models with Embedded Discontinuities

Alfredo E. Huespe[†] and Javier Oliver[‡]

[†] CIMEC-INTEC-UNL-CONICET, Santa Fe, Argentina.
[‡] Escola Técnica Superior d'Engyners de Camins, Canals i Ports, Technical University of Catalonia, Barcelona, Spain.

1 Fracture Problem Approaches Based on Continuum Constitutive Relations

This chapter presents a methodological approach for modeling concrete crack problems based on continuum constitutive relations and strong discontinuity kinematics. Fundamental aspects of this approach are presented in the initial Sections 1-2. The topics and ideas discussed in those points follow closely the work of Oliver (2002). A Finite Element technique with embedded strong discontinuities, particularly adapted for this methodology, is shown in Section 3. In the final Sections 4-6, some algorithmic aspects and several applications of the approach are addressed.

1.1 Motivation: Idealization of the Fracture Process Zone (FPZ) in Quasi-Brittle Material

In some ceramic materials like concrete, the microscopic observation of the zone where the crack propagation process takes place, displays a series of phenomena that can be schematically described through Figure 1, and explained as follows (Bazant (1986)).

Let us consider a body subjected to a loading process with a propagating crack. As the fracture advances, in front of the crack tip it is observed the formation, and growth, of a densely distributed micro-crack pattern that induces a non-linear

---

[1]The authors thank the following colleagues: O. Manzoli, M.D. Pulido, P.J. Sánchez, S. Blanco, D.L. Linero, G. Días, E. Samaniego and E. Chavez. They have provided an important and very esteemed collaboration to develop the CSDA methodology during several years. Also, the authors acknowledge the financial support from the agencies of Argentina: ANPCyT and CONICET through grant PICT 2006-01232 and PIP 112-200901-00341; as also, the Spanish Ministry of Science and Innovation and the Catalan Government Research Department, through grants BIA2008-00411 and 2009 SGR 1510.

material response with a notable dissipated energy fraction. In this initial stage of the fracture process, the material remains macroscopically stable and even when the strains are becoming larger in a small bounded region, at this stage there are not a marked concentration of strains. This zone is displayed in gray in Figure 1.

With an additional loading increase, the interaction effect between different micro-cracks as they grow, gives place to an unstable process: some micro-cracks coalesce to form one dominant macro-crack which governs the following stage of the fracture process, while other arrest. The material point crossed by the macro-crack, does not lose immediately the loading carrying capacity. An averaged manifestation of this micro structural loss of stability is a macroscopic behavior displaying strain-softening, which induces unloading, or micro-cracks arrest, of those point surrounding the dominant macro-crack. Also, it leads to a high concentration, or localization, of strains in a band of small thickness interpreted as the dominant macro-crack domain. In Figure 1, we sketch this fracture process zone as the localized strain cohesive zone.

The formation of new stress-free surfaces, i.e. after the complete stress relaxation in the dominant macro-crack surfaces, is only reached after an additional increase of loads. In the same Figure 1, we sketch this part of the fracture process as the non-cohesive zone.

From the point of view of the mathematical description, the fracture process zone can be idealized through the following progressive stages:

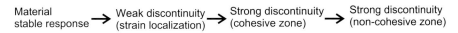

i) an initial zone where the material remains stable, however it loses the linear response. The strains are gradually increasing, but no strain localization effects are observable;

ii) a transition zone where the material response, at the macroscopic scale, becomes unstable. The strains become large and localize into a finite band-width, forming a weak discontinuity zone, i.e. the velocity field remains continuous but a jump in the spatial distribution of strain rates is observed;

iii) the width of the strain localization zone collapses to zero (into a surface) while cohesive forces subsist. Thus, the velocity field can be idealized as having a jump, i.e. a strong discontinuity, across the crack surface;

iv) cohesive forces become null and a stress-free zone arises on the surfaces of the crack.

With this idealization of the FPZ, and motivated by the fact that the classical phenomenological continuum damage models proposed by Kachanov in the 60's describe reasonably well the first stage of the idealized FPZ, where a dense distribution of micro-cracks are active and progressing; we wonder if: "being the

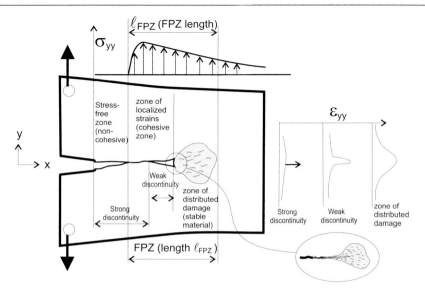

Figure 1. Phenomenological description of a quasi-brittle fracture process. Stress σ_{yy} is displayed along the fracture process zone. Also, distribution of strains ε_{yy} is depicted in three different stages of the process evolution.

fracture problem an essentially discontinuous phenomenon, is it possible to describe a response of the complete process by means of continuum mechanics?". As also: "can we use a unified framework for modeling the transition from the continuum to the strong discontinuity, including the stress-free crack, regime?

Throughout this chapter it is presented a unified continuum-based model, with a continuum kinematics and continuum constitutive relations, which describes acceptably well quasi-brittle fracture problems in all its stages, including the transition from the stable material response to the stress-free strong discontinuity regime.

The classical Fracture Mechanics Approach and the Fracture Energy Concept. Evaluation and simulation of fracture problems have been historically dealt by Fracture Mechanics. The best well known subdiscipline of this subject is the Linear Elastic Fracture Mechanics (LEFM), which has been developed for nearly one century. It provides valuable and simple procedures for predicting the consequences of cracks in structures, such as the evaluation of failure loads when flaws or cracks are present in the structure. The application range of this subdiscipline includes all those problems displaying a small inelastic deformation zone close to the crack tip.

However, as Bazant has pointed out in several contributions, since 1960 it is well known that LEFM does not provide a correct prediction for concrete fracture. Following Bazant, the reason because it fails, is due to the rather large FPZ size of concrete (ℓ_{FPZ} in Figure 1) if compared with a characteristic length of the structure, D. From here, the name of "quasi-brittle" is used to characterize concrete fracture in contrast with brittle-fracture which can be well estimated by the LEFM approach.

Energy-based procedures were one of the initial approaches utilized in Fracture Mechanics. The **Fracture Energy concept**, G_f, defined as: "the energy that is required to extend the crack surface a unit of area", comes from this approach. Considering the idealized fracture process we have explained above, the total dissipated energy into the FPZ is the necessary energy, G_f, required for generating new stress-free crack surfaces of unit area. Therefore, G_f could be related with the external load energy once the dissipated energy, outside the FPZ, is known. However, in general, the discrimination of the zone where the fracture process is taking place from those zones where the bulk inelastic process evolves, and is not having a direct role with the separation work, is not clearly done.

The wide use given to G_f, as a parameter characterizing fracture mechanics problems, relies on a key assumption: G_f is considered to ba a material constant. This is a strong assumption, which is not totally valid in general. Evidently, it means that G_f, for example, does not depend on the loading path that every point of the FPZ follows during the fracturing process. However, it is an acceptable good approach for some type of fracture problems; particularly it has been widely adopted in modeling concrete tensile-dominant fracture problems.

Additionally, and more importantly for the present purposes, is that by adopting tghe fracture energy concept and assuming that it is a constant parameter for a given material, a characteristic length is induced in the material model. In fact, a standard formula in Fracture Mechanics (see Bazant and Planas (1998), and Hellan (1984)), provides a material length parameter, ℓ_{ch}, defined by:

$$\ell_{ch} = \frac{G_f E}{f_t^2} \tag{1}$$

where E is the Young's modulus and f_t a representative ultimate stress (tensile strength in concrete or yield stress in plastic materials), that can be related with the FPZ size, $\ell_{FPZ}, (\ell_{FPZ} \approx \ell_{ch})$, and similar to Irwin's approximation of the plastic zone size related with ductile fracture (Hellan, 1984).

1.2 Stability and Uniqueness of the Mechanical Boundary Value Problem (BVP)

As it is explained in the idealized model of the FPZ, material stability and strain localization mechanisms play a fundamental role in the formation and propagation of cracks. Therefore, we summarize the material stability concept applied to continuum material models, and briefly analyze the mathematical condition that they should satisfy in order that a solution with strain localization is possible.

Stability conditions. A general stability condition for structural mechanical systems, expressed as a BVP, with constant traction t^* on the boundary Γ_σ, Figure 2-a, can be expressed through the positivity of the second order work (see Hill (1958), and Figure 2-b):

$$\frac{1}{2}\int_\Omega \dot{\boldsymbol{\sigma}} : \dot{\boldsymbol{\varepsilon}}\, d\Omega > 0 \qquad \forall\, \dot{\boldsymbol{\varepsilon}} \text{ admissible} \qquad (2)$$

where $\dot{\boldsymbol{\sigma}}$ and $\dot{\boldsymbol{\varepsilon}}$ are related through the constitutive equation.

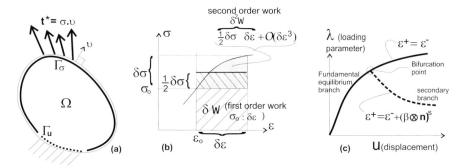

Figure 2. (a) The mechanical BVP; (b) second order work; (c) equilibrium path displaying a bifurcation point.

Therefore, when a material model satisfies for every point that:

$$\dot{\boldsymbol{\sigma}} : \dot{\boldsymbol{\varepsilon}} > 0 \qquad (3)$$

the stability condition (2) is assured. Condition (3) is known as the H-stability condition. We remark that there are alternative stability criteria.

Those materials verifying (3) must have a stress-strain curve (σ, ε) with positive slope for the complete range of strains. This condition precludes any softening response. Thus, it is an excessively strong requirement for those continuum constitutive models we are interested in.

Uniqueness conditions. Uniqueness of the related BVP solution can be guaranteed under general conditions, (Hill, 1958). However, in the present work, we are interested in analyzing a more restricted type of solution uniqueness.

Considering a body Ω subjected to a loading process with a smooth strain field; we analyze the possible existence of solutions displaying bifurcation points, see Figure 2-c, with secondary branches being characterized by weak discontinuity solutions. It means that the second solution displays a continuous displacement field, but strains and stresses are discontinuous across a given surface S.

The analysis follows the classical methodology proposed by Rice (1976): the Maxwell's compatibility relations define the structure of strain tensors displaying jumps or discontinuities. Then, a perturbed strain with this structure is introduced in the BVP governing equations, from where, the conditions on the material model admitting a non-unique solution can be found.

Maxwell's compatibility conditions. Let us consider a continuous scalar function ϕ, defined in the body Ω with a discontinuous gradient across a surface S, as shown in Figure 3-a. The surface has an orthogonal unit vector n and divides the domain Ω in two parts that we denote as Ω^+ and Ω^-, as shown in Figure 3-a. The vector n points toward Ω^+.

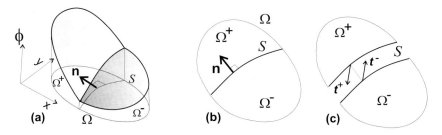

Figure 3. (a) Scalar function ϕ with a weak discontinuity across the surface S; (b-c) body displaying a weak discontinuity across the surface S.

The jump of the gradient of ϕ, i.e. the gradient difference on each side of S: $\nabla\phi^+ - \nabla\phi^-$, where $(\cdot)^+$ and $(\cdot)^-$ denote the values on each side of S, is:

$$\nabla\phi^+ - \nabla\phi^- = [\![\nabla\phi]\!] = \xi n \qquad (4)$$

being ξ a proportionality scalar factor. The projection of $[\![\nabla\phi]\!]$ on S is zero because the continuity of ϕ along S.

If a vectorial continuous function, u, has a discontinuous gradient on S, the gradient of each components of u satisfies equation (4). Thus, the jump of the gradient is:

$$\nabla u^+ - \nabla u^- = [\![\nabla u]\!] = \beta \otimes n \qquad (5)$$

where, β is a vector. In the particular situation that u is a continuous displacement field with a discontinuous gradient, the strain jump on S is:

$$[\![\varepsilon]\!] = [\![\nabla u]\!]^{sym} = \varepsilon^+ - \varepsilon^- = (\beta \otimes n)^{sym} \qquad (6)$$

the notation $(\cdot)^{sym}$ means the symmetric part of $(\cdot)$.

Bifurcation analysis. Let us consider on the body Ω, subjected to a loading process, the existence of a weak discontinuity solution, with a strain jump across the surface S, as shown in Figure 3-b. The continuity of the (rate of) traction vector on S is a condition required by the body equilibrium, see Figure3-b:c:

$$(\dot{t}^+ + \dot{t}^-) = \dot{\sigma}^+ \cdot n - \dot{\sigma}^- \cdot n = [\![\dot{\sigma}]\!] \cdot n = 0 \qquad (7)$$

Replacing the stress rate $\dot{\sigma}$ by the constitutive relation: $\dot{\sigma} = \mathbb{D}_T : \dot{\varepsilon}$, where $\mathbb{D}_T$ is the tangent tensor of the constitutive relation; we can rewrite equation (7) as follows:

$$\dot{\sigma}^+ \cdot n - \dot{\sigma}^- \cdot n = [(\mathbb{D}_T)^+ : \dot{\varepsilon}^+ - (\mathbb{D}_T)^- : \dot{\varepsilon}^-] \cdot n = 0 \qquad (8)$$

The smooth fundamental solution $\dot{\varepsilon}^+ = \dot{\varepsilon}^-$ verifies equation (8). However, we search for a second solution with strains having a jump across S. Thus, the strain field of the second solution follows equation (6): $\dot{\varepsilon}^+ = \dot{\varepsilon}^- + (\dot{\beta} \otimes n)^{sym}$, which, replaced in equation (8), gives:

$$[(\mathbb{D}_T)^+ : (\dot{\varepsilon}^- + (\dot{\beta} \otimes n)^{sym}) - (\mathbb{D}_T)^- : \dot{\varepsilon}^-] \cdot n = 0 \qquad (9)$$

and assuming that $(\mathbb{D}_T)^+ = (\mathbb{D}_T)^- = \mathbb{D}_T$ (this assumption should be carefully considered), then:

$$[\mathbb{D}_T : (\dot{\beta} \otimes n)^{sym}] \cdot n = \underbrace{[n \cdot \mathbb{D}_T \cdot n]}_{Q} \dot{\beta} = 0 \qquad (10)$$

where the tensor Q is the so called acoustic tensor, or localization tensor. Existence of a solution $[\![\dot{\varepsilon}]\!]$ different from zero, means that $\dot{\beta} \neq 0$ for some vector n. Which is only possible if:

$$\det Q(n) = 0 \qquad (11)$$

Contrarily, the condition

$$\det Q(n) > 0 \quad \forall\, n \qquad (12)$$

called strong ellipticity condition, precludes any type of discontinuous bifurcation.
Remark: for a rate-independent material, H-stability ($\dot{\varepsilon} : \mathbb{D}_T : \dot{\varepsilon} > 0\ \forall \dot{\varepsilon}$) implies strong ellipticity.

1.3 Example of a Material Model Subjected to Stability Loss and Bifurcation: Isotropic continuum damage model for concrete

Table 1 describes a specific isotropic continuum damage model, with a scalar internal variable D, the damage varaible, describing the elastic stiffness degradation due to micro cracking: $D = 0$, for the undamaged material, and $D = 1$ for the fully damaged material (for more generic continuum damage models, see for example Lemaitre and Desmorat (2005)).

In Table 1, and following Oliver (2000), the damage variable D depends on an internal strain-like variable r and its stress-like conjugate internal variable, q which depends on r. The free energy, denoted W, depends on the strain tensor ε and the damage variable D. The elastic strain energy for the undamaged material is denoted W_0, and $\mathbb{E}$ is the Hooke's elastic tensor: μ and λ are the Lamè's parameters and $\mathbb{I}$ and $\mathbb{1}$ are the fourth and second order identity tensors respectively.

In equation (17), $\bar{\sigma}$ is the effective stress. Its positive counterpart is then defined as:

$$\bar{\sigma}^+ = \langle \bar{\sigma}_i \rangle \, \boldsymbol{p}_i \otimes \boldsymbol{p}_i \tag{13}$$

where $\langle \bar{\sigma}_i \rangle$ stands for the positive part (McAuley brackets) of the i-th principal effective stress $\bar{\sigma}_i$ ($\langle \bar{\sigma}_i \rangle = \bar{\sigma}_i$ for $\bar{\sigma}_i > 0$ and $\langle \bar{\sigma}_i \rangle = 0$ for $\bar{\sigma}_i \leq 0$) and $\boldsymbol{p}_i$ stands for the i-th principal stress direction.

Equation (18) defines the damage function f and the initial elastic domain as: $f < 0$. This domain is unbounded for compressive stress states ($\bar{\sigma}^+ = \boldsymbol{0}$). Therefore, damage evolution is only possible with tensile stress states, as it is usually observed in concrete crack phenomenon. Alternatively, the definition of τ_ε can be performed with $\bar{\sigma}$, such as: $\tau_\varepsilon = \sqrt{\bar{\sigma} : \varepsilon}$, in which case, the elastic damage, in the stress space, is a symmetric domain with respect to the origin (null stress) point.

The stresses $\boldsymbol{\sigma}$ and the stress-like variable q are determined from equations (17) and (21). Equation (21) defines the softening law in terms of the softening parameter H. In equation (19), f_t and E are, respectively, the tensile strength and the Young's modulus. Finally, equation (22) is the rate constitutive law in terms of the tangent modulus $\mathbb{D}_T$, which is defined in equation (23). And (20) defines the loading-unloading conditions of the damage model.

Bifurcation Analysis for the Isotropic Damage Model. We analyze the existence of discontinuous-strain bifurcation modes, induced by this damage model. Without loss of generality we study the damage model defined by the symmetric elastic domain, with $\tau_\varepsilon = \sqrt{\bar{\sigma} : \varepsilon}$.

Equation (11) must be evaluated with the acoustic tensor $\boldsymbol{Q}(\boldsymbol{n}, H) = \boldsymbol{n} \cdot \mathbb{D}_T(H) \cdot \boldsymbol{n}$, where $\mathbb{D}_T$ is defined in (23). In the unloading case, $\mathbb{D}_T$ becomes proportional to the elastic tensor, with a proportionality factor that is positive.

Table 1. Isotropic continuum damage model.

Free energy

$$W(\boldsymbol{\varepsilon}, D) = (1-D)W_0 \quad ; \quad W_0 = \frac{1}{2}\boldsymbol{\varepsilon} : \mathbb{E} : \boldsymbol{\varepsilon} \quad (14)$$

$$\mathbb{E} = 2\mu \mathbb{I} + \lambda(\mathbb{1} \otimes \mathbb{1}) \quad (15)$$

Damage variable

$$D(r) = 1 - \frac{q(r)}{r} \quad ; \quad q \geq 0 \quad ; \quad r \geq 0 \quad (16)$$

Constitutive equation

$$\boldsymbol{\sigma} = (1-D)\mathbb{E} : \boldsymbol{\varepsilon} = \frac{q}{r}\underbrace{\mathbb{E} : \boldsymbol{\varepsilon}}_{\bar{\boldsymbol{\sigma}}} = \frac{q}{r}\bar{\boldsymbol{\sigma}} \quad (17)$$

Damage function

$$f(\boldsymbol{\varepsilon}, r) = \tau_\varepsilon(\boldsymbol{\varepsilon}) - r \quad ; \quad \tau_\varepsilon(\boldsymbol{\varepsilon}) = \sqrt{\bar{\boldsymbol{\sigma}}^+ : \boldsymbol{\varepsilon}} \quad (18)$$

Initial conditions (time t=0)

$$r|_{t=0} = r_0 = \frac{f_t}{\sqrt{E}} \quad ; \quad q|_{t=0} = r_0 \quad (19)$$

Loading-unloading conditions

$$\dot{r} \geq 0 \quad ; \quad f \leq 0 \quad ; \quad \dot{r}f = 0 \quad (20)$$

Softening law

$$\dot{q} = H\dot{r} \quad ; \quad H \leq 0 \quad ; \quad (q \in [0, r_0]) \quad (21)$$

$$H = \begin{cases} H = constant \;:\; (linear\ softening) \\ H = -H_0 \exp(-\frac{r_0}{G_f}(r-r_0)) \;:\; (exponential\ softening) \end{cases}$$

Incremental constitutive law

$$\dot{\boldsymbol{\sigma}} = \mathbb{D}_T : \dot{\boldsymbol{\varepsilon}} \quad (22)$$

Tangent tensor of the constitutive relation

$$\mathbb{D}_T = \frac{\partial \boldsymbol{\sigma}}{\partial \boldsymbol{\varepsilon}} = \begin{cases} \mathbb{D}_T = (1-D)\mathbb{E} = \frac{q}{r}\mathbb{E} \quad (for\ unloading) \\ \mathbb{D}_T = \frac{q}{r}\mathbb{E} - \frac{q-Hr}{r^3}\bar{\boldsymbol{\sigma}}^+ \otimes \bar{\boldsymbol{\sigma}} \quad (for\ loading) \end{cases} \quad (23)$$

Therefore, discontinuous bifurcation is precluded.

In the loading case, and due to the function τ_ε chosen to perform this analysis, the term $\bar{\sigma}^+$ in (23) is replaced by $\bar{\sigma}$. Thus, bifurcation is possible if:

$$\det \boldsymbol{Q} = \det \left[\boldsymbol{n} \left(\frac{q}{r} \mathbb{E} - \frac{q - Hr}{r^3} (\bar{\sigma} \otimes \bar{\sigma}) \right) \boldsymbol{n} \right] =$$
$$= (\frac{q}{r})^3 \det \boldsymbol{Q}^e \left[1 - (\frac{q - Hr}{qr^2}) \left((\bar{\sigma} \cdot \boldsymbol{n}) \cdot (\boldsymbol{Q}^e)^{-1} \cdot (\bar{\sigma} \cdot \boldsymbol{n}) \right) \right] = 0. \quad (24)$$

By definition, $r > 0$ and $q > 0$, as well as:

$$\det \boldsymbol{Q}^e = \det(\boldsymbol{n} \cdot \mathbb{E} \cdot \boldsymbol{n}) = \mu^2 (2\mu + \lambda) > 0 \quad (25)$$

Therefore, equation (24) can be rewritten as:

$$\left[1 - (\frac{q - Hr}{qr^2}) \left([\bar{\sigma} \cdot \boldsymbol{n}] \cdot (\boldsymbol{Q}^e)^{-1} \cdot [\bar{\sigma} \cdot \boldsymbol{n}] \right) \right] = [1 - \frac{Z(\boldsymbol{n})}{\xi(H)}] = 0 \quad (26)$$

Equation (26) implicitly defines the value of the softening modulus H as a function of the directions $\boldsymbol{n}$. During a loading process, the critical softening H_{crit} and normal $\boldsymbol{n}_{crit}$, are identified as those values that first verify equation (26). They can be evaluated by means of a closed formula used in Oliver and Huespe (2004). Additional details of this procedure can be obtained there.

2 Material Failure Analysis Using the Continuum-Strong Discontinuity Approach (CSDA)

2.1 Motivation

The formation of weak discontinuities, characterized by continuous displacements but discontinuous strains across a surface S, Figure 4-a, could give rise to concentration of strains whenever two discontinuity surfaces, S and S', bound a band of finite width, such as that shown in Figure 4-b. The concept of strong discontinuity is recovered when the width of the localization band tends to zero, and the value of the strain jump tends to infinity. Thus, the strong discontinuity problem can be regarded as a limit case of the strain localization (weak discontinuity) one.

The previous Section presented the necessary conditions that a mechanical problem must verify, in order to show bifurcation points with secondary equilibrium branches characterized by non-smooth, discontinuous, strain solutions.

In the present Section, additional details of the mechanical problem displaying a non-smooth kinematics, with strong discontinuities, are addressed. Particularly,

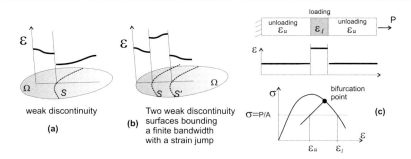

Figure 4. Formation of a finite width region bounding a strain jump (a-b) as a result of a bifurcation point. (c) The 1D problem with a weak discontinuity.

we show a methodological framework that makes possible the consistency between strong discontinuity kinematics and continuum constitutive models. In the interface of a strong discontinuity, where displacement jumps arise, this methodology provides a discrete traction-separation law similar to that proposed by classical cohesive models. The so derived traction-separation law is compatible and consistent with the continuum constitutive relation of the bulk, while, the classical discrete approach defines cohesive forces which are independent of the bulk constitutive relation.

Conventional one-dimensional (1D) damage model In Table 2, the damage model presented in the previous section is particularized for the 1D case, which is used to solve a 1D bar problem. With this solution, we show the inconsistent and nonphysical response provided by the conventional approach when it is used for modeling structural problems characterized by materials with strain softening. Symbols and equations in Table 2 are those explained in Table 1.

With the initial conditions given by (29), it results $D(t = 0) = 0$ and the damage criterion is verified for $\sigma = f_t$, where f_t is the material tensile strength.

During a monotonic loading process, with $\dot{\varepsilon} \geq 0$ and $\dot{r} > 0$, and after reaching the condition $f = 0$, the identity $\dot{r} = \sqrt{E}\dot{\varepsilon}$ follows from the loading condition: $\dot{f} = 0$. Then, the rate equation (33) is obtained from this identity which gives a constitutive modulus $D_T = HE$.

A plot of the stress-strain response can be observed in Figure 5-a, while, the softening law $q(r)$, is plotted in Figure 5-b.

The 1D bar problem Let us consider the 1D bar problem of Figure 5-c, with a material described by the damage model of Table 2. After the limit load is reached, in the softening regime, one possible solution is obtained by considering that a

Table 2. One-dimensional continuum damage model.

Free energy and stress-strain relation

$$W(\varepsilon, r) = \frac{1}{2}\frac{q(r)}{r}E\,\varepsilon^2 \quad ; \quad \sigma = \frac{q(r)}{r}E\,\varepsilon \tag{27}$$

Damage function

$$f(\varepsilon, r) = \tau_\varepsilon(\varepsilon) - r \quad ; \quad \tau_\varepsilon(\varepsilon) = \sqrt{E\,\varepsilon^2} \tag{28}$$

Initial conditions

$$r|_{t=0} = r_0 = \frac{f_t}{\sqrt{E}} \quad ; \quad q|_{t=0} = r_0 = \frac{f_t}{\sqrt{E}} \tag{29}$$

Loading-unloading conditions

$$\dot{r} \geq 0 \quad ; \quad f \leq 0 \quad ; \quad \dot{r}f = 0 \tag{30}$$

Softening law

$$\dot{q} = H\dot{r} \quad ; \quad H = \text{constant} < 0 \tag{31}$$

Incremental constitutive law

$$\dot{\sigma} = \frac{q}{r}E\dot{\varepsilon} \quad ; \quad \text{unloading case} \tag{32}$$

$$\dot{\sigma} = \frac{q}{r}E\dot{\varepsilon} + \overbrace{q_{,r}\frac{r-q}{r^2}}^{H}E\varepsilon\dot{r} = HE\dot{\varepsilon} \quad ; \quad \text{loading case} \tag{33}$$

fraction β, $(0 \leq \beta \leq 1)$, of the bar length ℓ, is in loading, while the remaining part, with a length $(1 - \beta)\ell$, is elastically unloading.

Denoting the variables in the loading zone with subindex *l* and those of the unloading zone with subindex *u*; the relation between the load P and the total displacement at the end of the bar Δu_T, can be obtained through the following set of equations:

i) *Compatibility equation*

$$\Delta u_T = \Delta u_l + \Delta u_u = (\beta)\ell\varepsilon_l + (1-\beta)\ell\varepsilon_u \tag{34}$$

where Δu_l and Δu_u are the length increment of the bar zones corresponding to loading and unloading, respectively.

ii) *Equilibrium equation and traction boundary condition*

$$\sigma_l = \sigma_u = \frac{P}{A} \tag{35}$$

iii) *Constitutive relation, see Figure 5-d*

$$\sigma_l = f_t + EH(\varepsilon_l - \frac{f_t}{E}) \quad ; \quad \sigma_u = E\,\varepsilon_u \qquad (36)$$

After replacing equations (35-36) in equation (34), it can be found:

$$P = \frac{AE}{[1 - (1 - \frac{1}{H})\beta]\ell}\left[\Delta u_T + \frac{\beta\ell}{EH}(1 - H)f_t\right] \qquad (37)$$

or, in rates:

$$\dot{P} = k\Delta\dot{u}_T \quad ; \quad k = \frac{AE}{[1 - (1 - \frac{1}{H})\beta]\ell} \qquad (38)$$

The set of values $\beta \in [0,1]$ represents admissible solutions to the mathematical problem. Therefore, in the softening regime, the bar stiffness k depends on the length β of the inelastic zone, see Figure 5-e.

Balance of energy in the bar problem. Considering that the reduced dissipation $\mathcal{D}$ of an inelastic process is the difference between the stress power and the internal energy rate, see Simo and Hughes (1998), the damage model dissipation is given by:

$$\mathcal{D} = \sigma\dot{\varepsilon} - \dot{W} = \sigma\dot{\varepsilon} - [\frac{1}{2}\underbrace{\overbrace{H\,r - q}^{\partial_r q}}_{\partial_r W}\underbrace{(E\varepsilon^2)}_{}\dot{r} + \underbrace{(\frac{q}{r}E\varepsilon)}_{\partial_\varepsilon W = \sigma}\dot{\varepsilon}]$$

$$= \frac{1}{2}(q - Hr)\dot{r} = \frac{1}{2}(1 - H)r_o\dot{r} \qquad (39)$$

where loading conditions are considered ($f = 0$). Also, considering that from equation (28): $r^2 = E\varepsilon^2$, a linear (constant) softening model: $H < 0$, such as that shown in Figure 5-b, with $q = q_o + H(r - r_o)$ and by the initial conditions: $q_o = r_o$; the dissipated energy until reaching the complete material degradation ($r = r_d$ in Figure 5-b) can be determined as follows:

$$Y = \int_{t=0}^{t=\infty}\mathcal{D}\,dt = \int_{r=r_o}^{r=r_d}\frac{1}{2}(1-H)r_o\,dr$$

$$= \frac{1}{2}(1-H)r_o\underbrace{(r_d - r_o)}_{\frac{-r_o}{H}} = \frac{1}{2}(1 - \frac{1}{H})r_o^2 \qquad (40)$$

which, once multiplied by the volume of the damaged zone, gives the total structural dissipated energy as:

$$Y_\Omega = \frac{1}{2}(1 - \frac{1}{H})r_o^2\beta A\ell \qquad (41)$$

On the other side, we evaluate the external work produced by the force P, as the area below the equilibrium curve in Figure 5-e. It is equal to:

$$\text{Area} = \int_0^{\Delta u_{end}} P du = \frac{1}{2}(1 - \frac{1}{H})r_o^2 \beta A \ell \qquad (42)$$

which, as required by the energy balance, coincides with the structural dissipated energy at rupture Y_Ω. It is noted that the total structural dissipated energy (rupture energy) tends to zero as the solution tends to the strong discontinuity limit, with $\beta \to 0$, which is a nonphysical structural response.

Remark: From equation (42), it can be observed that the rupture work depends on the damaged zone length: $\beta \ell$. This length is not provided by the constitutive relation or by any other equation of the model. Thus, it is arbitrary. Multiplicity of solutions and the nonphysical response in the strong discontinuity limit are characteristics of an ill-posed mathematical problem. In fact, BVP with conventional material models having strain softening provide ill-posed mathematical problems. Well-posed property is recovered if the constitutive relation introduces a characteristic length, such as the width of the strain localization zone. In this case, we will refer to them as **regularized constitutive models**.

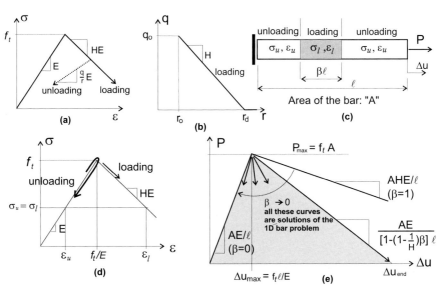

Figure 5. (a-b) 1D damage model. (c-d) Bar problem; (e) load vs. displacement solution of the bar problem.

Remark: As a consequence, numerical simulations of this type of problems present serious deficiencies whenever non-regularized constitutive relations are used. It is noted that, in discrete models derived from finite element formulations, a characteristic length is induced by the finite element mesh size. Thus, in some well defined cases, this discretization property could be used as a valid regularization procedure.

Alternatively, it is possible to assume that the total rupture work of the bar, per unit of area A, is a material parameter. According to the concept introduced in the previous section, we call this parameter the fracture energy: G_f. Its adoption is an important methodological assumption that can be interpreted in two different ways:

i) considering that the softening modulus, H, is a given parameter of the material continuum damage model. Then, from equation (41) and the definition of G_f: $G_f = Y_\Omega / A$, a characteristic length is induced as follows:

$$\beta \ell = \frac{2 G_f}{r_o^2} \frac{H}{H-1} \qquad (43)$$

and represent the size of the strain localization zone, i.e. the width of the strain localization zone is defined once H and G_f are given;

ii) alternatively, it can be assumed that the rupture process zone is concentrated in a very thin band ($\beta \ell \to 0$). Then, for capturing a correct rupture work, it is necessary that the continuum softening parameter $H \to 0$ such that, the ratio $\frac{H}{\beta \ell}$ is bounded and takes the value:

$$\lim_{H \to 0; \beta \ell \to 0} \frac{H}{\beta \ell} = \bar{H} = -\frac{1}{2} \frac{r_o^2}{G_f} = -\frac{1}{2} \frac{f_t^2}{G_f E} \qquad (44)$$

In the last identity, from (29), it is replaced $r_o^2 = f_t^2 / E$. We call $\bar{H}$ the intrinsic softening modulus which has a dimension of Length^{-1}. Replacing, in the constitutive equations, the modulus H by $\bar{H}$, it introduces automatically a length dimension in the problem. Therefore, considering a strong discontinuity solution, the material parameter H is determined by G_f.

2.2 The 1D Continuum-Strong Discontinuity Approach (CSDA)

We analyze the strong discontinuity solution of the bar problem that happens when strains localize in the rupture surface S ($\beta \ell \to 0$) and the displacement field displays a finite jump across this surface. In this case, we search for a continuum approach that describes the rupture process.

In order to proceed with this approach, some additional ingredients must be considered:

i) the assumption that the material constitutive relation of the continuum model describes the mechanical response of the dissipative phenomena taking place in the discontinuity zone S;

ii) it is noted that, as the width of the strain localization band goes to zero, the strains become unbounded. Thus, a regularization procedure must be introduced such that the kinematic singular terms, due to the unbounded growth of strains, become compatible with the continuum material model.
The fracture problem approach adopting these assumptions has been called the Continuum-Strong Discontinuity Approach (CSDA) (Oliver, 1996a, Oliver et al., 1999, Oliver, 2000 and Oliver et al., 2002).

Kinematics Regularization. Let us consider the bar rupture problem of Figure 6-a, with a damage zone having a width that goes to zero. Then, in the limit, this zone tends to S and the displacement field $u(x)$ displays a discontinuity across S. By denoting the displacement jump across this surface: $[\![u]\!] = u^+ - u^-$, were u^+ is the displacement on Ω evaluated on the right part of S (Ω^+), and u^- evaluated on the left part of S (Ω^-), we can write the displacement field as follows:

$$u(x) = \bar{u}(x) + \mathcal{H}(x)[\![u]\!] \quad ; \quad \mathcal{H}(x) = \begin{cases} 0 & \forall x \in \Omega^- \\ 1 & \forall x \in \Omega^+ \end{cases} \quad (45)$$

where $\bar{u}$ is a smooth field and $\mathcal{H}(x)$ the Heaviside's step function.

Introducing the generalized derivative concept: $\partial_x \mathcal{H}(x) = \delta_s(x)$, where $\delta_s(x)$ is the Dirac's delta function on S, the strain field ε, compatible with the displacement (45), is given by:

$$\varepsilon(x) = \partial_x \bar{u}(x) + \partial_x \mathcal{H}(x)[\![u]\!] = \underbrace{\bar{\varepsilon}(x)}_{regular\ term} + \underbrace{\delta_s(x)}_{singular\ term} [\![u]\!] \quad (46)$$

For the subsequent mathematical treatment, we will approach the Dirac's function by a regularized sequence of functions. Let Ω_s^k be the bar finite thickness zone, of width k, which includes the discontinuity surface S and such that $\lim_{k \to 0} \Omega_s^k = S$, see Figure 6-b. Then, defining:

$$\delta_s^k(x) = \frac{1}{k}\mu(x) \quad ; \quad \mu(x) = \begin{cases} 0 & \forall x \in \Omega \backslash \Omega_s^k \\ 1 & \forall x \in \Omega_s^k \end{cases} \quad (47)$$

the Dirac's Delta function can be approached by the k-regularized sequence

$$\lim_{k \to 0} \delta_s^k(x) = \delta_s(x) \quad (48)$$

Replacing (48) in (46), we obtain a k-regularized sequence of strains. One term of this sequence can be interpreted as a weak discontinuity solution, where, through a finite bandwidth k, the strain has a jump, see Figure 6-b. Strong discontinuities are recovered when $k \to 0$. Thus, with this regularized kinematics, we characterize two families of discontinuities:

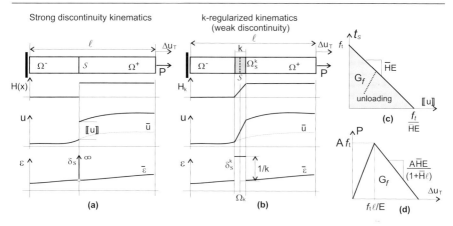

Figure 6. Strong discontinuity approach for the bar problem.

$$\varepsilon(x) = \bar{\varepsilon}(x) + \frac{1}{h}\mu(x)[\![u]\!] \quad (49)$$

Discontinuous kinematics

$h \neq 0$: weak discontinuity

$h = k \to 0$: strong discontinuity

Constitutive Model Regularization: Traction-Separation Law as a Projection of the Continuum Constitutive Relation. Problems involving strong discontinuities can be analyzed by assuming that Ω^+ and Ω^- are two independent bodies interacting mechanically through cohesive forces applied along the surface S, such as it has been assumed in the classical discrete cohesive models. In this type of mechanical approach, the continuum is described by means of a constitutive relation that is independent of the traction-separation law defined in the fracture process zone. Some issues of the discrete cohesive methodology can be questioned, as it happens for example when cohesive forces depend on the crack tip stress state, typically the triaxiality ratio. In this case, the cohesive discrete law must depend on the stress state, and thus, the traction-separation law parameters must be stress dependent, see for example the work of Siegmund and Brocks (2000) where an application of this approach is used in order to simulate ductile fracture problems.

As an alternative to discrete cohesive approaches, the CSDA provides cohesive laws that are consistent with the continuum constitutive relation. The traction-separation law is defined via a continuum model projection, in the discontinuity surface, that is obtained after introducing the discontinuous kinematics in the continuum constitutive relation. It is necessary to impose some constraints such that the continuum model provides bounded stresses when unbounded strains arise in Ω_s^k. These constraints are determined as follows:

i) The equilibrium condition requires that:

$$\sigma_s^k = \sigma_{\Omega \backslash \Omega_s^k} = \frac{P}{A} \quad (50)$$

where σ_s^k is understood as the stress in Ω_s^k and $\sigma_{\Omega \backslash \Omega_s^k}$ is the stress outside Ω_s^k.

ii) Substitution of the discontinuous kinematics, equation (49), in the constitutive model of Table 2, gives:

$$\sigma_s = \lim_{k \to 0} \sigma_s^k = \lim_{k \to 0} \frac{q_s}{r_s} E(\bar{\varepsilon} + \frac{1}{k}[\![u]\!]) = \lim_{k \to 0} \frac{1}{kr_s} \underbrace{[q_s E([\![u]\!])]}_{bounded} \quad (51)$$

where the bounded character of: $q_s \in [0, f_t/\sqrt{E}]$, E and $[\![u]\!]$ is emphasized. Equation (50) constraints the k–sequence of stresses, σ_s^k, to be bounded. Therefore, equation (51) requires that

$$\lim_{k \to 0} kr_s = \bar{\alpha} \quad (52)$$

must be bounded.

iii) The regularization of the constitutive model redefines the softening modulus H, following the concept introduced in equation (44), through the intrinsic softening modulus $\bar{H}$, which is given by:

$$\bar{H} = \frac{H}{k} = -\frac{1}{2} \frac{f_t^2}{G_f E} \quad (53)$$

and replacing H by $\bar{H}$ in the softening law, equation (31), the rate of the internal variable $\dot{q}_s$ is defined by:

$$\dot{q}_s = H\dot{r}_s = \lim_{k \to 0} \left(\frac{H}{k}\right) k\dot{r}_s = \bar{H}\dot{\bar{\alpha}} \quad (54)$$

which provides a bounded term $\dot{q}_s$ during the complete damage process, even when $\dot{r}_s$ is not bounded.

Thus, the stress σ_s (traction t_s), equation (51), can be rewritten as:

$$(\sigma_s =)t_s = \frac{q_s}{\bar{\alpha}}(E[\![u]\!]) \quad (55)$$

Remark: a discrete traction-separation model (t_s, $[\![u]\!]$), see Table 3, consistent with the damage continuum model, is obtained from equations (53-55) jointly with the regularized damage function, see equation (28):

$$f_s = \lim_{k \to 0} k(\sqrt{(E\varepsilon^2)} - r) = \sqrt{(E[\![u]\!]^2)} - \bar{\alpha} \qquad (56)$$

and the loading-unloading conditions: $f_s \leq 0$; $\dot{\bar{\alpha}} \geq 0$; $\dot{\bar{\alpha}} f_s = 0$.

Table 3. Discrete one-dimensional traction-separation law resulting from the continuum damage model projection (t_b is the activation time of the strong discontinuity mode).

Traction-separation law
$t_s = \dfrac{q_s}{\bar{\alpha}} E [\![u]\!]$ (57)
Discrete Damage function
$f_s([\![u]\!], \bar{\alpha}) = \sqrt{E[\![u]\!]^2} - \bar{\alpha}$ (58)
Loading-unloading conditions
$\dot{\bar{\alpha}} \geq 0 \quad ; \quad f_s \leq 0 \quad ; \quad \dot{\bar{\alpha}} f_s = 0$ (59)
Softening law
$\dot{q}_s = \bar{H} \dot{\bar{\alpha}} \; ; \quad \bar{H} = \dfrac{1}{2} \dfrac{f_t^2}{G_f E} = \text{constant} < 0$ (60)
initial conditions
$\bar{\alpha}\vert_{t=t_b} = 0 \quad ; \quad q_s\vert_{t=t_b} = q(t_b)$ (61)

Remark: The response of the law described in Table 3 is plotted in Figure 6-c. Initially, it shows a rigid response, with: $\lim_{[\![u]\!] \to 0} t_s = q_s \sqrt{E} = f_t$ (observe that $\lim_{[\![u]\!] \to 0} \bar{\alpha} = 0$). After the activation of the displacement jump, the stiffness becomes $\bar{H} E$.

Solution of the Bar Problem Provided by the CSDA We solve the 1D bar problem of the previous section by means of the CSDA. In the softening regime (post critical regime), the solution obtained using this methodology assumes a loading condition in S and elastic unloading in $\Omega \backslash S$. Then, by denoting with subindex s and $\Omega \backslash S$ the bar domain where variables are evaluated, the following equations must be verified:

i) displacement compatibility condition:

$$\Delta u_T = \Delta u_{\Omega \setminus S} + [\![u]\!] \qquad (62)$$

where Δu_T is the total displacement of the load application point, $[\![u]\!]$ is the displacement jump in S, and $\Delta u_{\Omega \setminus S}$ the difference between both displacements;

ii) equilibrium condition:

$$t_s = \sigma_{\Omega \setminus S} = \frac{P}{A} \qquad (63)$$

iii) constitutive relations:
The discrete damage model in S, equations in Table 3, provides a traction t_s versus jump displacement $[\![u]\!]$ given by (see Figure 6-c):

$$t_s = \bar{H} E [\![u]\!] + f_t \qquad (64)$$

while in $\Omega \setminus S$, the elastic stress is: $\sigma_{\Omega \setminus S} = E \Delta u_{\Omega \setminus S} / \ell$.

Solution of these equations provides the load P as a unique function of the displacement Δu_T, see Figure 6-d:

$$P = \frac{\bar{H} E A}{1 + \bar{H} \ell} \left(\Delta u_T + \frac{f_t}{\bar{H} E} \right) \quad ; \quad \text{for } \Delta u_T \geq \frac{f_t \ell}{E} \qquad (65)$$

This solution gives a post critical structural response which depends on a characteristic length determined by the fracture energy G_f through the parameter $\bar{H}$.

2.3 The Continuum-Strong Discontinuity Approach in 3D Problems

Let us consider now a BVP corresponding to a body $\Omega \in \mathbb{R}^3$, displaying a displacement solution with a jump across the surface S. This surface S, with its normal vector $\boldsymbol{n}$, divides the body in two disjoint parts Ω^+ and Ω^-, as shown in Figure 7-a. The displacement jump vector, i.e., the difference of displacements on both sides of the surface S, is denoted by $[\![\boldsymbol{u}]\!] = \boldsymbol{u}^+ - \boldsymbol{u}^-$. Thus, the displacement field can be written as follows:

$$\boldsymbol{u}(\boldsymbol{x}) = \bar{\boldsymbol{u}}(\boldsymbol{x}) + \mathcal{H}_s(\boldsymbol{x}) [\![\boldsymbol{u}]\!] \qquad (66)$$

where $\bar{\boldsymbol{u}}(\boldsymbol{x})$ is a smooth field and $\mathcal{H}_s(\boldsymbol{x})$ the Heaviside's step function on S ($\mathcal{H}_s(\boldsymbol{x}) = 0$ in $\boldsymbol{x} \in \Omega^-$ and $\mathcal{H}_s(\boldsymbol{x}) = 1$ in $\boldsymbol{x} \in \Omega^+$). Considering the generalized gradient of the function $\mathcal{H}_s$: $\quad \partial_x \mathcal{H}_s = \delta_s(\boldsymbol{x}) \boldsymbol{n}$, where $\delta_s(\boldsymbol{x})$ is the generalized Dirac's delta function on S, the strain field is given by:

$$\boldsymbol{\varepsilon}(\boldsymbol{x}) = \bar{\boldsymbol{\varepsilon}}(\boldsymbol{x}) + \delta_s(\boldsymbol{x}) (\boldsymbol{n} \otimes [\![\boldsymbol{u}]\!])^{sym} \qquad (67)$$

where the first term is the smooth part $\bar{\varepsilon}$ and is given by: $\bar{\varepsilon}(x) = \nabla^{sym}\bar{u}$, while the second term is singular.

A regularized version of the strain field (67), is convenient for its subsequent mathematical and numerical treatment. Let a band Ω_s^k of finite width k including the surface S be introduced, and such that $\lim_{k \to 0} \Omega_s^k = S$, as shown in Figure 7-b. Similar to the 1-D case, it is possible to define the k-regularized sequence of functions

$$\delta_s^k = \frac{1}{k}\mu(x), \qquad (68)$$

where: $\mu(x) = 1$ in $x \in \Omega_s^k$ and 0 otherwise, and such that: $\lim_{k \to 0} \delta_s^k = \delta_s$.

Replacing (68) in (67), we obtain the h-sequence of regularized strains:

$$\varepsilon(x) = \bar{\varepsilon}(x) + \frac{1}{h}\mu(x)(n \otimes [\![u]\!])^{sym} \qquad (69)$$

Thus, this kinematics setting captures either a weak ($h \neq 0$) or a strong discontinuities ($h = k \to 0$), respectively.

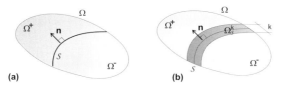

Figure 7. (a) Strong discontinuity kinematics, 3D case; (b) regularization of the strong discontinuity (weak discontinuity) in Ω^k.

Traction-Separation Law in 3D Problems Utilizing an identical procedure to that of the 1D case, we can determine the traction-separation law projected by the continuum damage model defined in Table 1.

Variables evaluated in the domains Ω_s^k ($\lim_{k \to 0} \Omega_s^k = S$) or $\Omega \backslash \Omega_s^k$ are denoted with subindex s or $\Omega \backslash \Omega_s^k$ respectively.

The procedure is as follows:

i) the equilibrium condition requires the traction vector continuity across the surface bounding Ω_s^k and $\Omega \backslash \Omega_s^k$:

$$\boldsymbol{\sigma}_s \cdot \boldsymbol{n} = \boldsymbol{\sigma}_{\Omega \backslash \Omega_s^k} \cdot \boldsymbol{n} \qquad (70)$$

From this equation, and given the bounded character of $\boldsymbol{\sigma}_{\Omega \backslash \Omega_s^k}$, the stresses $\boldsymbol{\sigma}_{\Omega_s^k}$ should be bounded, even when $k \to 0$.

ii) replacing the strain $\varepsilon_{\Omega_s^k}$, from equation (69), into the damage constitutive relation of the continuum model (equation (17) in Table 1, and taking the limit for k going to zero:

$$\begin{aligned}\boldsymbol{\sigma}_s \cdot \boldsymbol{n} &= \lim_{k\to 0} \frac{q_s}{r_s}\left[\mathbb{E}:\left(\bar{\varepsilon}(\boldsymbol{x}) + \frac{1}{k}(\boldsymbol{n}\otimes[\![\boldsymbol{u}]\!])^{sym}\right)\right]\cdot\boldsymbol{n} = \\ &= \lim_{k\to 0}\frac{q_s}{kr_s}(\boldsymbol{n}\cdot\mathbb{E}\cdot\boldsymbol{n})[\![\boldsymbol{u}]\!] = \frac{q_s}{\bar{\alpha}}\boldsymbol{Q}^e[\![\boldsymbol{u}]\!] \end{aligned} \qquad (71)$$

In the last identity $\boldsymbol{Q}^e$ is the elastic acoustic tensor (equation (25)). Using a similar argument as in the 1D case, and considering that q_s, $\boldsymbol{Q}^e$ and $[\![\boldsymbol{u}]\!]$ are bounded, as also the traction vector, then, $\bar{\alpha} = \lim_{k\to 0} kr_s$ must be a bounded term.

iii) by redefining the softening modulus, $\bar{H} = \frac{H}{k}$, the evolution of the internal variable $\dot{q}_s$ is bounded and given by:

$$\dot{q}_s = \lim_{k\to 0} H\dot{r}_s = \bar{H}\dot{\bar{\alpha}} \qquad (72)$$

iv) introducing the strain $\boldsymbol{\varepsilon}$ given by equation (69), into the damage function from equation (18), and taking the limit of expression:

$$\lim_{k\to 0} kf(\boldsymbol{\varepsilon},r) = 0; \qquad (73)$$

a discrete damage criterion can be determined by:

$$f_\alpha([\![\boldsymbol{u}]\!]) - \bar{\alpha} = 0 \quad ; \quad f_\alpha([\![\boldsymbol{u}]\!]) = \sqrt{[\![\boldsymbol{u}]\!]\cdot(\boldsymbol{Q}^e)\cdot[\![\boldsymbol{u}]\!]} \qquad (74)$$

Remark: the traction-separation law is automatically induced once the strong discontinuity kinematics and the softening regularization are introduced in the model. Thus, the cohesive traction is determined by a degeneration of the continuum constitutive relation.

3 Finite Elements with Embedded Discontinuities

Finding good numerical solutions of concrete fracture problems using the CSDA depends on the discretization technique and its capacity for adequately capturing one of the most salient aspects of this approach: the strong discontinuity kinematics. A well adapted technique to reach this objective is finite elements with embedded discontinuities.

In recent years, these finite elements have been the object of increasing study and development. Its rising popularity comes from the fact that, a displacement discontinuity model can be introduced in the bulk of the element (having an arbitrary direction regardless of the orientation of the element mesh), in combination

Table 4. Discrete 3D traction-separation law as a continuum damage model projection (t_b is the activation time of the strong discontinuity mode).

Traction-separation law
$$\boldsymbol{t}_s = \boldsymbol{\sigma}_s \cdot \boldsymbol{n} = \frac{q_s}{\bar{\alpha}} \boldsymbol{Q}^e \cdot [\![\boldsymbol{u}]\!] \qquad (75)$$
Discrete Damage function
$$f_\alpha([\![\boldsymbol{u}]\!], \bar{\alpha}) = \sqrt{[\![\boldsymbol{u}]\!] \cdot \boldsymbol{Q}^e \cdot [\![\boldsymbol{u}]\!]} - \bar{\alpha} \qquad (76)$$
Loading-unloading conditions
$$\dot{\bar{\alpha}} \geq 0 \quad ; \quad f_\alpha \leq 0 \quad ; \quad \dot{\bar{\alpha}} f_\alpha = 0 \qquad (77)$$
Softening law
$$\dot{q}_s = \bar{H}\dot{\bar{\alpha}} \; ; \quad \bar{H} = \frac{1}{2}\frac{f_t^2}{G_f E} = \text{constant} < 0 \qquad (78)$$
initial conditions
$$\bar{\alpha}|_{t=t_b} = 0 \quad ; \quad q_s|_{t=t_b} = q(t_b) \qquad (79)$$

with appropriate propagation mechanisms. Moreover, it has been shown that, in strain localization scenarios, some of those elements jointly with the CSDA or discrete cohesive laws, completely overcome the well-known problem of the spurious mesh size and mesh orientation dependence of the results.

A quite large family of finite elements with strong discontinuities has been presented in the literature. We are addressing only two particular classes of them, as also, their numerical performances related with accuracy and convergence of solutions.

3.1 Strong Discontinuities: The Local Form of the BVP Governing Equations

Let us consider the body Ω, of Figure 8-a, that is subjected to displacements $\boldsymbol{u}^\star$ on the boundary, Γ_u, and imposed traction loads $\boldsymbol{t}^\star$ on Γ_σ. Without loss of generality, it is considered that there are no volume forces. Let us also consider that the body is undergoing a displacement discontinuity $[\![\boldsymbol{u}]\!](\boldsymbol{x};t)$ across the surface S, and that this surface splits the body into Ω^+ (pointed by the unit normal $\boldsymbol{n}$ to S) and Ω^-.

The corresponding quasi-static boundary value problem (BVP) can be described, in rate form, through the following set of equilibrium equations:

$$\begin{aligned}
\nabla \cdot \dot{\boldsymbol{\sigma}} &= \boldsymbol{0} & &\text{in } \Omega \backslash S & &\text{internal equilibrium} \\
\dot{\boldsymbol{\sigma}} \cdot \boldsymbol{\nu} &= \dot{\boldsymbol{t}}^\star & &\text{on } \Gamma_\sigma & &\text{external equilibrium} \\
\dot{\boldsymbol{\sigma}}_{\Omega^+} \cdot \boldsymbol{n} - \dot{\boldsymbol{\sigma}}_{\Omega^-} \cdot \boldsymbol{n} &= \boldsymbol{0} & &\text{on } S & &\text{outer traction continuity} \\
\dot{\boldsymbol{\sigma}}_{\Omega^+} \cdot \boldsymbol{n} - \dot{\boldsymbol{\sigma}}_S \cdot \boldsymbol{n} &= \boldsymbol{0} & &\text{on } S & &\text{inner traction continuity}
\end{aligned} \quad (80)$$

Equations (80) are the classical equilibrium equations. The equations (80-c:d) are the equilibrium conditions on the surface S. $\boldsymbol{\sigma}_{\Omega^+}$ and $\boldsymbol{\sigma}_{\Omega^-}$ are the stresses evaluated on Ω^+ and Ω^-, respectively, while $\boldsymbol{\sigma}_s$ stands for the stresses arising in the strain localization zone, S. It was previously shown that the evaluation of $\boldsymbol{\sigma}_s$ using the continuum constitutive relation is a basic assumption of the CSDA. Additional equations of the BVP are the following:

$$\begin{aligned}
\dot{\boldsymbol{\varepsilon}} - \nabla^{sym} \dot{\boldsymbol{u}} &= \boldsymbol{0} & &\text{in } \Omega & &\text{kinematical compatibility} \\
\dot{\boldsymbol{u}} &= \dot{\boldsymbol{u}}^\star & &\text{on } \Gamma_u & &\text{kinematical boundary conditions} \\
\dot{\boldsymbol{\sigma}} &= \dot{\Sigma}(\boldsymbol{\varepsilon}, \boldsymbol{\alpha}) & &\text{in } \Omega & &\text{constitutive relation}
\end{aligned} \quad (81)$$

where the constitutive relation (81-c) depends on the strain ε, as also, on a set of internal variables $\boldsymbol{\alpha}$.

3.2 A Variational Consistent Formulation of the BVP with Strong Discontinuities

For the variational formulation of a problem displaying strong discontinuities, it is convenient to describe the velocity field as follows:

$$\dot{\boldsymbol{u}}(\boldsymbol{x}) = \dot{\bar{\boldsymbol{u}}}(\boldsymbol{x}) + \mathcal{M}_s(\boldsymbol{x}) [\![\dot{\boldsymbol{u}}]\!] \quad (82)$$

where $\dot{\bar{\boldsymbol{u}}}$ is a smooth velocity field and $\mathcal{M}_s(\boldsymbol{x})$ is a conveniently defined function that presents a jump in S ($[\![\mathcal{M}_s]\!] = 1$) and its support is Ω^k, being Ω^k a domain like a band covering the discontinuity line S, see Figure 8-b and c. The function $\mathcal{M}_s(\boldsymbol{x})$ is built as follows:

$$\mathcal{M}_s(\boldsymbol{x}) = \mathcal{H}_s(\boldsymbol{x}) - \varphi(\boldsymbol{x}) \quad (83)$$

Crack Models with Embedded Discontinuities

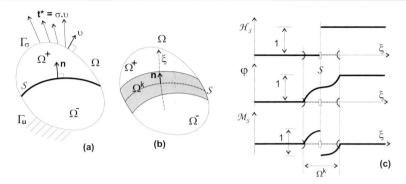

Figure 8. A mechanical problem with strong discontinuity.

where $\mathcal{H}_s$ is the previously defined Heaviside's step function on S, and $\varphi(\boldsymbol{x})$ is a smooth function such that $\varphi = 0$ in $\Omega^- \backslash \Omega^k$ and $\varphi = 1$ in $\Omega^+ \backslash \Omega^k$.

The reason of defining a new function $\mathcal{M}_s$ for describing the velocity jump, lies on the reduced support of this function. In fact, by assuming that Ω^k does not intersect a boundary Γ_u ($\mathcal{M}_s \equiv 0$ outside Ω^k), the velocity boundary conditions on Γ_u are only imposed on the smooth function $\bar{\boldsymbol{u}}$, as it is done in classical approaches.

The generalized gradient of $\mathcal{M}_s$ is:

$$\nabla \mathcal{M}_s = \delta_s \boldsymbol{n} - \nabla \varphi \tag{84}$$

The BVP Weak Form. Let us consider the following variational formulation:

$$\int_\Omega \dot{\boldsymbol{\sigma}} : \nabla \boldsymbol{\eta} \, d\Omega = \int_{\Gamma_\sigma} \dot{\boldsymbol{t}}^* \cdot \boldsymbol{\eta} \, d\Gamma_\sigma \quad ; \quad \forall \boldsymbol{\eta} \in \mathcal{V}_o = \{\boldsymbol{\eta} = \bar{\boldsymbol{\eta}} + \mathcal{M}_s \boldsymbol{\beta}\} \tag{85}$$

where $\mathcal{V}_o$ is the space of the test functions, $\bar{\boldsymbol{\eta}}$ is defined as a smooth function in Ω verifying the homogeneous essential boundary conditions ($\bar{\boldsymbol{\eta}} = \boldsymbol{0}$ in Γ_u), and $\boldsymbol{\beta}$ is defined in S. The variational equation (85) is equivalent to the local form of the equilibrium equations (80). This statement is briefly demonstrated in the following, additional details can be seen in Simo and Oliver (1994).

i) Choosing smooth test functions $\boldsymbol{\eta}$ (with $\boldsymbol{\beta} = \boldsymbol{0}$), the variational formulation (85) yields, via the Green's theorem in the regular parts Ω^+ and Ω^-, plus standard arguments, the equations (80)-a, b and c.

ii) Choosing test functions $\boldsymbol{\eta}$, such that $\bar{\boldsymbol{\eta}} = \boldsymbol{0}$, then: $\nabla \boldsymbol{\eta} = [\boldsymbol{\beta} \otimes (\delta_s \boldsymbol{n} - \nabla \varphi)] + \mathcal{M}_s \nabla \boldsymbol{\beta}$ and $\boldsymbol{\eta} = \boldsymbol{0}$ in Γ_σ (due to the assumption that the $\mathcal{M}_s$-function has a reduced support) and substituting η in equation (85) yields:

$$\int_\Omega \dot{\boldsymbol{\sigma}} : \nabla \eta \, d\Omega = \int_\Omega \delta_s(\dot{\boldsymbol{\sigma}} \cdot \boldsymbol{n}) \cdot \boldsymbol{\beta} \, d\Omega + \int_\Omega (\dot{\boldsymbol{\sigma}} \cdot \nabla \varphi) \cdot \boldsymbol{\beta} \, d\Omega + $$
$$+ \int_\Omega \mathcal{M}_s (\dot{\boldsymbol{\sigma}} : \nabla \boldsymbol{\beta}) \, d\Omega = \boldsymbol{0} \qquad (86)$$

which results in the following equation:

$$\int_S (\dot{\boldsymbol{\sigma}} \cdot \boldsymbol{n}) \cdot \boldsymbol{\beta} \, dS - \int_S (\dot{\boldsymbol{\sigma}}_{\Omega^+} \cdot \boldsymbol{n}) \cdot \boldsymbol{\beta} \, dS = \boldsymbol{0} \qquad (87)$$

The first term in (87) results from the first term in the left hand side in equation (86), while the last term is derived from the second and third terms, considering: *a)* the Green's theorem on the domain Ω^k, *b)* the strong form of the equilibrium equation in the regular slice $\Omega^k \backslash \Omega^+$ and the jump condition across S obtained in the item *(i)*, *c)* $\varphi = 0$ in S^- ; $\varphi = 1$ in S^+ and that *d)* the edges of the slice Ω^k intersecting the body boundary are free of tractions.

Equation (87) finally proves the equivalence between the weak, equation (85), and the strong form, equation (80), of the equilibrium of bodies with strong discontinuities.

3.3 Finite Elements with Embedded Discontinuities

Let us consider the body Ω discretized with a linear (CST) triangular finite element mesh. Also, let us assume that an available discontinuity path detection algorithm (the so called tracking algorithm, to be explained in the next section) determines the subset $\mathcal{J}$ of the elements that are crossed by S at the considered time, as shown in Figure9-a. For every element of $\mathcal{J}$, the tracking algorithm also provides the position of the element discontinuity interface S_e (see Figure9-b) of length l_e which defines the domains Ω_e^- and Ω_e^+ and the nodes j^1, j^2 and j^{sol}, with j^{sol} being the node that lies in Ω_e^+.

Kinematics. The following interpolation of the velocity field, $\dot{u}$, inside a given element e is considered:

$$\dot{u}(x) = \sum_{i=1}^{3} N_i(x)\dot{d}_i + \mathcal{M}_s^e(x)\dot{\boldsymbol{\beta}} \qquad (88)$$
$$\mathcal{M}_s^e(x) = [\mathcal{H}^e(x) - N^{sol}(x)] \qquad \forall e \in \mathcal{J}$$

where $N_i(x)$ are the linear standard shape functions of the triangular finite element and $\dot{d}_i$ the nodal velocities. The second term in (88-a) can be considered an enrichment velocity mode that captures the discontinuous part of the velocity field.

The unit jump shape function $\mathcal{M}_s$ has a support of one finite element, with $\mathcal{H}^e$ being the Heaviside's function in the element e, and N^{sol} the linear shape function that corresponds to the node j^{sol} ($N^{sol} = N_{j^{sol}}$).

Figure9-c displays the function $\mathcal{M}_s^e$. The parameter $\dot{\boldsymbol{\beta}} \in \mathbb{R}^{ndim}$, with $ndim$ the space dimension, is a vectorial elemental parameter, which represents the velocity jump of one finite element. This is an elemental parameter, not shared with other finite elements.

From equation (88), the strain rate reads:

$$\dot{\varepsilon} = \sum_{i=1}^{3}(\nabla N_i(\boldsymbol{x}) \otimes \dot{\boldsymbol{d}}_i)^{sym} + \frac{1}{k}\mu(\boldsymbol{n} \otimes \dot{\boldsymbol{\beta}})^{sym} - (\nabla N^{sol} \otimes \dot{\boldsymbol{\beta}})^{sym} \qquad (89)$$

where, in order to compute the singular term arising from the generalized derivative of the discontinuous velocity, we have replaced the Dirac's delta function by the k-regularized sequence: $\delta_s^e = (1/k)\mu$, where: $\mu = 1$ in Ω^k and 0 otherwise, see Figure9-b and d.

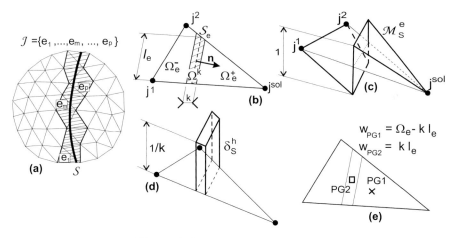

Figure 9. Finite element approach.

In order to integrate the terms defined in Ω^k, we add a second sampling point, as shown in Figure9-e, that represents the domain Ω^k, and whose weight is the area of the domain Ω^k : area $\Omega^k = kl_e$, and the regularization parameter k is an arbitrarily small factor.

a) Symmetric FE: Variationally Consistent Formulation. The variationally consistent FE formulation utilizes a similar velocity field interpolation to that proposed in equation (88); while the test function space is chosen as follows:

$$\mathcal{V}_o = \{\boldsymbol{\eta} \mid \boldsymbol{\eta} = N_i(\boldsymbol{x})\bar{\boldsymbol{\eta}}_i + \mathcal{M}_s^e(\boldsymbol{x})\delta\boldsymbol{\beta}^e\} \qquad \boldsymbol{\eta} = \boldsymbol{0} \text{ on } \Gamma_u$$

where $\bar{\eta}_i$ and $\delta\beta^e$ are nodal and element parameters of the interpolation functions, respectively. Introducing this discrete functional space in the variational formulation (85), and after performing the respective variations of parameters, it yields the following system of equations:

$$\int_{\Omega} \dot{\boldsymbol{\sigma}} \cdot \nabla N_i \, d\Omega - \int_{\Gamma_\sigma} \boldsymbol{t}^* N_i \, d\Gamma_\sigma = \boldsymbol{0} \qquad (90)$$

$$\int_{\Omega^e} \frac{1}{k}\mu \left(\dot{\boldsymbol{\sigma}} \cdot \boldsymbol{n}\right) d\Omega^e - \int_{\Omega^e} \dot{\boldsymbol{\sigma}} \cdot \nabla N^{sol} \, d\Omega^e = \boldsymbol{0} \qquad (91)$$

Equations (90) are the standard variational finite element equation arising in classical problems without strong discontinuities. Thus, there are $ndim$ equations of this type for every node "i" of the mesh. Equations (91) imposes a weak equilibrium across the discontinuity line S. There are $ndim$ equations of this type for every element "e" in $\mathcal{J}$.

2D implementation. For the two-dimensional case, in a Cartesian coordinate system (x,y), using the Voigt's notation for the strains $\{\varepsilon\} = [\varepsilon_{xx}, \varepsilon_{yy}, 2\varepsilon_{xy}]^T$ and the stresses $\{\sigma\} = [\sigma_{xx}, \sigma_{yy}, \sigma_{xy}]^T$ (where $(\cdot)^T$ stands for the transpose of $(\cdot)$), considering the linear triangle as underlying element and using the standard finite element $\boldsymbol{B}$-*matrix* format (Zienkiewicz and Taylor (2000)), equations (90)-(91) can be written as follows:

$$\bigcup_{e=1}^{n_{elem}} \left[\int_{\Omega_e} \boldsymbol{B}^{(e)T} \cdot \dot{\boldsymbol{\sigma}}(\{\varepsilon\}^{(e)}) \, d\Omega - \dot{\boldsymbol{F}}^{ext(e)}\right] = \boldsymbol{0} \qquad (92)$$

where $\dot{\boldsymbol{F}}^{ext(e)}$ stands for the classical element external forces vector and:

$$\begin{aligned} \{\dot{\varepsilon}\}^{(e)} &= \boldsymbol{B}^{(e)} \cdot \dot{\boldsymbol{d}}^{(e)} \\ \dot{\boldsymbol{d}}^{(e)} &= \begin{bmatrix} \dot{d}_1, & \dot{d}_2, & \dot{d}_3, & \dot{d}_4, & \dot{\beta}_e \end{bmatrix}^T \end{aligned} \qquad (93)$$

$$\boxed{\begin{aligned} \boldsymbol{B}^{(e)} &= \begin{bmatrix} \boldsymbol{B}_1^{(e)}, & \boldsymbol{B}_2^{(e)}, & \boldsymbol{B}_3^{(e)}, & \boldsymbol{G}^{(e)} \end{bmatrix} \\ \boldsymbol{B}_i^{(e)} &= \begin{bmatrix} \partial_x N_i^{(e)} & 0 \\ 0 & \partial_y N_i^{(e)} \\ \partial_y N_i^{(e)} & \partial_x N_i^{(e)} \end{bmatrix} \\ \boldsymbol{G}^{(e)} &= \frac{1}{k}\mu \begin{bmatrix} n_x & 0 \\ 0 & n_y \\ n_y & n_x \end{bmatrix} - \begin{bmatrix} \partial_x N^{sol(e)} & 0 \\ 0 & \partial_y N^{sol(e)} \\ \partial_y N^{sol(e)} & \partial_x N^{sol(e)} \end{bmatrix} \end{aligned}} \qquad (94)$$

where $\bigcup$ stands for the classical assembling operator and $\boldsymbol{n} = [n_x, n_y]^T$.

Remark: The structure of equations (90-91) suggests the introduction of an internal additional fourth node for each element e, that is activated only for the elements crossed by the discontinuity interface ($e \in \mathcal{J}$) and whose corresponding degrees of freedom and associated shape function are, respectively, the displacement jumps $\boldsymbol{\beta}_e$ and $\mathcal{M}_S^{(e)}$ in equation (89). Since the support of $\mathcal{M}_S^{(e)}$ is only Ω_e, those internal degrees of freedom can be eventually condensed at the element level and removed from the global system of equations.

Remark: The integration of equation (91) is evaluated using an additional integration point that accounts for those terms in Ω^k. The weight of the additional integration point is the area assigned to Ω^k: kl_e. Furthermore, at the additional integration point $\mu = 1$, and zero otherwise, see Figure 9-e.

b) Non-symmetric FE. An alternative formulation to the symmetric FE approach of the previous subsection, can be deduced by replacing the weak equilibrium condition on S, equation (91), by one determined through the average terms of the identity: $\dot{\boldsymbol{\sigma}}_{\Omega^+} \cdot \boldsymbol{n} = \dot{\boldsymbol{\sigma}}_s \cdot \boldsymbol{n}$, as follows:

$$\underbrace{\frac{1}{\Omega^e} \int_{\Omega^e} (\dot{\boldsymbol{\sigma}} \cdot \boldsymbol{n}) \, d\Omega^e}_{mean\ value\ on\ \Omega \backslash S} = \underbrace{\frac{1}{l_e} \int_S (\dot{\boldsymbol{\sigma}}_s \cdot \boldsymbol{n}) \, dS}_{mean\ value\ on\ S} \qquad (95)$$

Convergence Test for the Symmetric FE Formulation

The equivalence of the BVP weak form (92) with the strong form equations (80-81) can be rigorously proven, see for instance, Simo and Oliver (1994)). Therefore, using the symmetric formulation, it is expected that mesh refinement will determine a correct trend (convergence) to the fulfillment of those equations.

Taking the linear triangle as the underlying element, the test in Figure 10 constitutes a simple corroboration of this fact and provides an assessment of the order of convergence of the symmetric finite element with embedded discontinuities.

The test consists of a homogeneous rectangular strip pulled from the right end with a force P, imposing a displacement Δ, up to the formation of an inclined failure line and, then, continued to the total failure and release of the stresses. Due to the induced constant stress field, the considered bilinear stress-strain law for the constitutive model should translate into a bilinear force-displacement $(P - \Delta)$ curve. To check the convergence to the right slope of the descending branch and, therefore, to the right energy dissipation G_f, a homogeneous mesh refinement, parameterized in terms of the element size h, is performed. Figure 10-a shows the results, for increasingly fine meshes, converging to the exact solution (the curve limiting the gray zone). The errors in the dissipated energy (fracture energy) for the different levels of discretization are displayed in Figure 10-b, where the linear convergence of the element is observed.

It is worth mentioning that the symmetric variational formulation, although convergent, exhibits accuracy smaller than the non-symmetric formulation of subsection 3.3-b, which, for this type of homogeneous/constant stress problem, provides the exact solution with only one element.

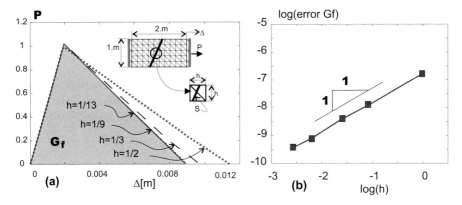

Figure 10. Symmetric finite element formulation: convergence analysis.

3.4 An Embedded Strong Discontinuity FE not Needing the Crack Path Continuity Enforcement

For the correct implementation of both finite element technologies described previously, the crack path across the finite element mesh must be known, and furthermore, it should be continuous. The crack path continuity condition is necessary to discriminate the relative position of the finite element nodes with respect to the discontinuity line S, and to know whether they stay on Ω^+ or Ω^-. A wrong selection of their relative position produces severe numerical locking.

Furthermore, in the non-symmetric formulation, for capturing a correct dissipation of energy it is required an exact evaluation of the length l_e (as shown in Figure 9-b), which is only possible if the intersection points of the discontinuity line with the element sides, are known. The symmetric formulation does not need this value.

In this section we present a finite element technique with embedded strong discontinuity not needing the crack path evaluation. The formulation was proposed by Sancho et al. (2005) and its implementation is very simple, providing good results in concrete fracture problem simulations.

Crack Models with Embedded Discontinuities

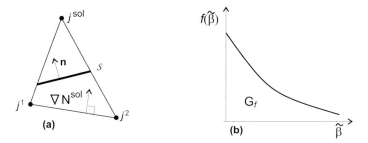

Figure 11. Finite element approach without enforcing crack path continuity.

Kinematics. It is based on the strong discontinuity kinematics with a velocity field interpolation similar to that described in equation (88) using CST (linear triangles) parent elements. Thus, using an identical notation, the velocity field is given by:

$$\dot{\boldsymbol{u}}(\boldsymbol{x}) = \sum_{i=1}^{3} N_i(\boldsymbol{x})\dot{\boldsymbol{d}}_i + [\mathcal{H}^e(\boldsymbol{x}) - N^{sol}(\boldsymbol{x})]\dot{\boldsymbol{\beta}} \quad ; \quad \forall e \in \mathcal{J} \quad (96)$$

An important issue of this technique is referred to the way that cohesive forces, at the discontinuity interface, are defined and handled. These cohesive forces are provided by a discrete law, independent of the continuum constitutive model. In this case, the stresses and strains in S are meaningless and the strains in $\Omega \backslash S$ are computed as follows:

$$\dot{\boldsymbol{\varepsilon}} = \underbrace{\dot{\tilde{\boldsymbol{\varepsilon}}}}_{(\nabla N_i \otimes \dot{\boldsymbol{d}}_i)^{sym}} - (\nabla N^{sol} \otimes \dot{\boldsymbol{\beta}})^{sym}; \quad (97)$$

thus, only regular terms define (97). Similar to the previous FE techniques, N^{sol} is the standard shape function of the CST finite element related with the node lying in Ω^+, see Figure 11-a. However, we will see in the next paragraph that the choice of the node lying on Ω^+ does not require an algorithm to trace the crack path.

Cohesive Model. The cohesive model is defined using a discrete law of central forces. Thus, considering two bodies Ω^+ and Ω^-, the interaction forces $\boldsymbol{t}$, are proportional to the displacement jumps $\boldsymbol{\beta}$, and its magnitude depends on the (historical maximum) separation between them:

$$\boldsymbol{t} = f(\tilde{\beta})\frac{\boldsymbol{\beta}}{\tilde{\beta}} \quad ; \quad \tilde{\beta} = \max_{t>0} \|\boldsymbol{\beta}(t)\| \quad (98)$$

where $f(|\tilde{\beta}|)$ is a monotonic decreasing function of its argument, and must be defined such that the area enclosed by the curve: $f(\beta)$ vs. β, is the fracture energy G_f, see Figure 11-b.

Kinematically Consistent Variational Formulation. The equilibrium of the body is imposed through the following discrete variational approach:

$$\int_\Omega \sigma : \delta\varepsilon \, d\Omega + \int_S t \cdot \delta\beta \, dS = \int_{\Gamma_\sigma} t^\star \cdot \delta u \, d\Gamma_\sigma \qquad (99)$$

$$\forall \, \delta u \in \mathcal{V}_o; \quad \delta\varepsilon = \nabla^{sym}\delta u$$

$$\mathcal{V}_o = \{\delta u \mid \delta u = N_i(x)\delta d_i + (\mathcal{H}_s - N^{sol})\delta\beta\} \; ; i = 1, 2, 3$$

by performing the variations of both parameters, $(\delta d_i, \delta\beta)$, defining the interpolation of virtual displacements field, the following system of equations is obtained:

$$\int_\Omega (\sigma \cdot \nabla N_i) \, d\Omega - \int_{\Gamma_\sigma} N_i t^\star \, d\Gamma_\sigma = 0 \qquad (100)$$

$$-\int_\Omega (\sigma \cdot \nabla N^{sol}) \, d\Omega + \int_S t \, dS = 0 \qquad (101)$$

Equation (101) is implemented through the local form:

$$t = n \cdot \sigma \qquad (102)$$

Replacing the material bulk constitutive relation, which is assumed to be linear elastic, the stress can be written:

$$\sigma = \mathbb{E} : \varepsilon = \mathbb{E} : (\bar{\varepsilon} - (\nabla N^{sol} \otimes \beta)^{sym}) \qquad (103)$$

and, finally, introducing (103) and the traction-displacement jump relation (98) in (102), the relation between the strain ε and the displacement jump β results:

$$\left[\frac{f(|\tilde{\beta}|)}{|\tilde{\beta}|}\mathbb{1} + (n \cdot \mathbb{E} \cdot \nabla N^{sol})\right]\beta = n \cdot \mathbb{E} : (\bar{\varepsilon} - (\nabla N^{sol} \otimes \beta)^{sym})) \qquad (104)$$

which must be solved in conjunction with the equilibrium equation (100) and the constitutive relation (103).

Remark: This approach is derived from a non-symmetric variational formulation (trial and test functions are different). Therefore, the discrete stiffness matrix is non-symmetric.

Selection of N^{sol}. In order to choose the shape function N^{sol}, in the context of the CST finite element, Sancho et al. (2005) have proposed a very simple methodology. It consists of taking the gradient vector ∇N_i, $(i = 1, 2, 3)$ which is as parallel as the normal vector n to the crack:

$$\nabla N^{sol} = \max_{i=1,2,3} \frac{|\nabla N_i \cdot \boldsymbol{n}|}{|\nabla N_i|} \qquad (105)$$

An important ingredient of the methodology is that, after the onset of the crack activation, the normal vector to the crack, $\boldsymbol{n}$, is allowed to rotate for a small angle before the final crack direction becomes fixed. This artifact is called crack adaptability by Sancho et al. (2005).

4 Algorithmic Aspects of the CSDA

Some finite elements with embedded strong discontinuities, explained in the previous Section, require the use of procedures for computing the position of the discontinuity surface (or crack path in fracture problems) by assuming its geometrical continuity as it crosses the finite element mesh.

In general, the selection of the deformation modes capturing the displacement jumps crucially depends on the way that the element is crossed by the discontinuity.

The strategies devoted to predict and capture the geometrical position of the discontinuity surface, are termed tracking strategies. There are two basic tracking strategies:

i) those based on a local (or propagating) procedure, where the crack path is tracked element by element, see for example Oliver (1996b), Oliver et al. (1999), Oliver et al. (2002). Handling this algorithm, when several cracks should be computed, is complex;

ii) alternatively, global tracking algorithms are based on the information provided by the propagation field of the complete finite element mesh.

All crack tracking algorithms are based on two types of data that must be determined, in every point of the body,using a convenient material failure criterion. They are: *1)* a condition for detecting the onset of crack propagation, and *2)* a direction for the crack propagation. In the following subsection, we introduce a summary of these criteria. After that, a global tracking strategy, taken from Oliver et al. (2004), is presented.

Selection of a Local Material Failure Criterion. In the simplest cases, the crack propagation onset condition is marked by the end of the elastic regime, through the fulfillment of some yield or damage criterion. Also, it could provide a crack propagation direction by assuming, for example, that the crack is orthogonal to the maximum tensile stress. A criterion like this one, works reasonably well in some quasi-brittle fracture problems, such as concrete fracture.

However, rigorous procedures to determine the onset of local material failure should be based on the material stability concept, as it was explained in Section 1, associated to the singularity of the so called localization tensor $Q(n, H) = n \cdot \mathbb{D}_T \cdot n$. The critical value of the softening modulus, H^{crit}, signaling the onset of failure, and the corresponding propagation directions, n^{crit}, are then determined from (the problem) $\det Q^{tg} = 0$.

4.1 A Global Tracking Algorithm

Consider the vector field $n(x; t)$ and the orthogonal vector field $T(x; t)$, which indicates the direction of propagation of a discontinuity line at the point x. In the context of a finite element analysis we assume that there is one available vector $T(e)$ for every sampling point in the element e, see figure 12.

We determine a scalar field $\theta(x; t)$, such that, its iso-level contour lines are the envelopes of the vector field $T(x; t)$. Therefore, these lines represent the set of possible crack paths across the body.

Evaluation of the Vector Field Envelopes. Heat conduction-like problem. Let us now focus on a procedure to compute the envelopes of the vector field $T(x; t)$ in a two-dimensional domain Ω, Its extension to the general 3-D situation is straightforward. We shall assume that T is a unit vector field ($\|T\| = 1$), whose sense is not relevant. A scalar field $\theta(x)$ whose iso-level contour lines are the envelopes of the vector field T, must verify:

$$T \cdot \nabla \theta = 0 \quad ; \quad \text{in } \Omega \tag{106}$$

Multiplying equation (106) by T, it results:

$$(T \cdot \nabla \theta) T = K \cdot \nabla \theta = 0 \quad ; \quad K = (T \otimes T)$$

where we introduce the rank-one tensor K. Defining the vector q by:

$$q = -K \cdot \nabla \theta \quad \text{in } \Omega, \tag{107}$$

then, its divergence is zero everywhere. Equation (107) can be written in terms of q, as a BVP in Ω, see Figure 12:

$$\begin{aligned} \nabla \cdot q &= 0 & \text{in } \Omega \\ q \cdot \nu &= 0 & \text{in } \Gamma_q \\ \theta &= \theta^* & \text{in } \Gamma_\theta \end{aligned} \tag{108}$$

where ν is the outward normal to the boundary Γ_Ω of the body Ω. Newman boundary conditions are prescribed in the boundary Γ_q and Dirichlet boundary conditions in Γ_θ.

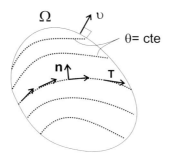

Figure 12. The pseudo steady-state heat conduction problem.

This BVP defined in Ω is similar to a steady-state heat conduction problem with no internal heat sources and null heat flux input in the boundary Γ_q (adiabatic boundary). In this case, θ plays the role of a pseudo-temperature field, q is the heat flux-like vector and K is a point dependent rank-one anisotropic thermal conductivity tensor.

Finite element formulation of the heat conduction-like BVP. The problem (108) can be numerically solved by means of well-known, very simple and computationally cheap, finite element procedures (Oliver et al., 2004). These features and the fact that all possible crack paths could be determined once the field θ is known, are the most important advantages of the present tracking algorithm method.

Remark: at every point, the global tracking algorithm requieres an adequte material failure criterion providing the local crack propagation direction. This criterion must predict the propagation direction even in those points where an elastic process is taking place. Once this condition is furnished, the algorithm can be used to simulate different kind of fracture problems, such as ductile or brittle fracture; no matter the material model type.

Representative Numerical Simulations. Figure 13 displays the application of this algorithm to a concrete fracture problem: the four-point beam test. Also, in the same Figure, an application to a slope stability analysis is shown. In this case, the material response is simulated using a J2 elastoplastic constitutive relation. The material failure mode corresponds to a shear band formation, which is typical for this type of constitutive response.

In both examples, it can be observed the distribution of the vector field T through the complete loading process, in every body point, even in regions which are far away from the fracture process or shear band zones.

This methodology is easily extensible to 3-D problems, as shown in Oliver et al. (2004).

4.2 Factors Influencing the Stability and Accuracy of the Numerical Method

Very often, due to a deficient numerical strategy, the numerical procedures for solving fracture problems provide non-physical responses. Spurious solutions are intimately related with algorithms and discrete formulations which are unstable. Furthermore, this type of algorithms is an important source of numerical troubles for the evaluation of solutions.

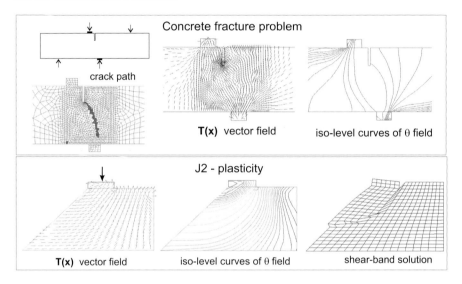

Figure 13. The pseudo steady-state heat conduction problem.

There are a number of algorithmic issues that have a strong influence on the numerical stability. Two of them are: *i)* the FE technology and *ii)* the numerical time-integration scheme of the constitutive material model:

i) the comparison presented in Section 3 shows, through a simple example, that the non-symmetric FE formulation is more accurate than the symmetric one. It is mainly due to the fact that the non-symmetric approach, in linear triangular or tetrahedral FE, imposes exactly (in strong form) the traction continuity condition on the discontinuity line, while, in the symmetric approach, it is weakly imposed. However, if considering the numerical stability property, the second approach is more stable than the non-symmetric approach. The reason of this behavior has been explained in the papers of Jirasek (2000) and Oliver et al. (2005).

ii) the strain softening of the constitutive relation is the responsible for the appearance of negative eigenvalues in the material constitutive modulus $\mathbb{D}_T$.

In these cases, however, the algorithmic constitutive modulus $\mathbb{D}_{alg}$ does not necessarily will display negative eigenvalues (observe the difference between the tensors $\mathbb{D}_T$ and $\mathbb{D}_{alg}$; $\mathbb{D}_T$ is the tensor that associates $\dot{\sigma}$ with $\dot{\varepsilon}$; while, $\mathbb{D}_{alg}$ is defined as: $\mathbb{D}_{alg} = \partial_{\varepsilon_{n+1}} \sigma_{n+1}$, where σ_{n+1} is the stress value at time $n+1$ obtained by an time-integration scheme as a function of the strains ε_{n+1}. The stability of the numerical approach depends on $\mathbb{D}_{alg}$.

For example, the use of classical implicit time-integration algorithms (Backward-Euler scheme) for evaluating constitutive models equipped with strain softening, leads to accurate results, even for large time step increments. However, it also determines algorithmic constitutive tangent tensors, $\mathbb{D}_{alg}$, with negative eigenvalues provoking ill-conditioned stiffness matrices in the associated discrete problem.

Alternatively, if the positive definite character of the algorithmic constitutive modulus $\mathbb{D}_{alg}$ is ensured by the time-integration scheme at any point of the body, this source of numerical instability would be removed. This observation motivates the implicit-explicit integration procedure presented in the following.

The IMPL-EX (Implicit–Explicit) Time-Integration Scheme. Le us recall the damage model depicted in Table 1. In the present section, we show a time integration scheme such that, given the strain ε_{n+1} in the time step $n+1$, determines the stresses σ_{n+1} and the internal variables r_{n+1} and q_{n+1}. This scheme, called IMPL-EX integration algorithm (Oliver et al., 2008), determines a positive definite algorithmic tangent tensor, and therefore, displays a more robust behavior for solving problems where the material constitutive relation is equipped with strain softening.

The IMPL-EX integration algorithm is based on the following two stages performed in the time step $(n+1)$:

i) In a first stage an explicit evaluation of the stresses, $\tilde{\sigma}_{n+1}$, and the stress-like internal variable, $\tilde{q}_{n+1}$, is performed in terms of the implicit values at the previous time step n and extrapolated values of the strain-like internal variable, $\tilde{r}_{n+1}$. Details of the operations performed in this stage are given in next subsection.

i) In a second stage, a standard implicit Backward-Euler integration of the constitutive model is performed and the implicitly integrated stresses, σ_{n+1}, are obtained. Details of the evaluations performed in this stage for the damage model are given in Table 5.

Remark: in addition, fulfillment of the momentum balance equation, equations (90–91), at time step $n+1$ is imposed in terms of the IMPL-EX stresses $\tilde{\sigma}_{n+1}$.

Table 5. Implicit backward-Euler integration scheme for the isotropic continuum damage model.

DATA: given $\varepsilon_{n+1}, r_n, q_n$

Compute effective stresses and trial values:

$\bar{\sigma}_{n+1} = \mathbb{E} : \varepsilon_{n+1}$; $\tau_{n+1} = \sqrt{\bar{\sigma}_{n+1} : (\mathbb{E})^{-1} : \bar{\sigma}_{n+1}}$

$r^{trial} = r_n$; $q^{trial} = q_n$

Compute loading-unloading condition:

IF $\tau_{n+1} < r^{trial}$ **THEN**

    Elastic unloading:

    $r_{n+1} = r^{trial}$; $q_{n+1} = q^{trial}$

    $\sigma_{n+1} = \frac{q_{n+1}}{r_{n+1}} \bar{\sigma}_{n+1}$

ELSE

    Damage evolution:

    $r_{n+1} = \tau_{n+1}$; $q_{n+1} = q_n + H(r_{n+1} - r_n)$

    $\sigma_{n+1} = \frac{q_{n+1}}{r_{n+1}} \bar{\sigma}_{n+1}$

ENDIF

Compute algorithmic modulus (used for determining the material bifurcation condition):

    Unloading condition: $\mathbb{D}_{alg} = \frac{q_{n+1}}{r_{n+1}} \mathbb{E}$

    Loading condition: $\mathbb{D}_{alg} = \frac{q_{n+1}}{r_{n+1}} \mathbb{E} - \frac{q_{n+1} - H r_{n+1}}{(r_{n+1})^3} (\bar{\sigma}_{n+1} \otimes \bar{\sigma}_{n+1})$

The explicit stage. For the isotropic damage model, the explicit stage is summarized in Table 6. Let us consider the strain-like internal variable flow, $\dot{r}$, defining the evolution of the plastic damage, as shown in Table 1. Also, consider that at the beginning of the time step computations, $n+1$, the implicit integration results of r, at the previous time steps, $(n, n-1, ...)$ are available. Thus, the Taylor series expansion of r in the time reads:

$$r_{n+1} = r_n + \frac{\Delta t_{n+1}}{\Delta t_n}(r_n - r_{n-1}) + \mathcal{O}(\Delta t_{n+1}^2) \tag{109}$$

where we have approximated $\dot{r}|_{t_n}$ by the expression $\dot{r}(t_n) \approx \frac{(r_n - r_{n-1})}{\Delta t_n}$, being $\Delta t_n = t_n - t_{n-1}$; $\Delta t_{n+1} = t_{n+1} - t_n$ the time step increment at the steps n and $n+1$, respectively.

Considering the truncation of expansion (109) to the first order term, it provides the second formula (step 1) in Table 6.

Steps 3 and 4, in Table 6, are obtained in terms of the extrapolated value of r: $\tilde{r}_{n+1}$, yielding the IMPL-EX integrated values of the remaining variables $\tilde{q}_{n+1}$ and $\tilde{\sigma}_{n+1}$.

Crack Models with Embedded Discontinuities

Table 6. Explicit extrapolation (first stage of the Implex algorithm), isotropic continuum damage model.

DATA: *given ε_{n+1}, and the implicit values of the previous time step:* $(r_n, q_n, r_{n-1}, q_{n-1})$

Compute explicit extrapolation:

$$\tilde{r}_{n+1} = r_n + \frac{\Delta t_{n+1}}{\Delta t_n}(r_n - r_{n-1}) \qquad \text{step 1}$$

Update internal and damage variables:

$$\tilde{q}_{n+1} = q_n + H \Delta \tilde{r}_n \qquad \text{step 2}$$

$$\tilde{\sigma}_{n+1} = \frac{\tilde{q}_{n+1}}{\tilde{r}_{n+1}}(\mathbb{E} : \varepsilon_{n+1}) \qquad \text{step 3}$$

Compute the Impl-Ex algorithmic tangent operator:

$$\tilde{\mathbb{D}}_{alg} = \frac{\tilde{q}_{n+1}}{\tilde{r}_{n+1}}\mathbb{E} \quad ; \quad \frac{\tilde{q}_{n+1}}{\tilde{r}_{n+1}} > 0 \qquad \text{step 4}$$

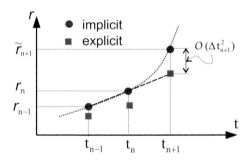

Figure 14. IMPL-EX integration algorithm.

The values obtained with the IMPL-EX scheme $(\tilde{r}_{n+1}, \tilde{q}_{n+1}, \tilde{\sigma}_{n+1})$ are then substituted in equations (90–91) to fulfill the momentum balance equation.

Remark: as the equilibrium equations, at step $n+1$, are evaluated with the IMPL-EX stresses $\tilde{\sigma}_{n+1}$, the structural stiffness matrix, which is necessary to compute the Newton-Raphson iterations for the convergence of the residual forces, must be evaluated with the IMPL-EX constitutive tangent tensor: $\partial_{\varepsilon_{n+1}} \tilde{\sigma}_{n+1} = \tilde{\mathbb{D}}_{alg}$.

Remark: unlike in the standard implicit integration, the values $\tilde{r}_{n+1}$ and $\tilde{q}_{n+1}$ are independent of the current value of the strains, ε_{n+1}, and therefore, they are known at the beginning of time step $n+1$. It means that $\tilde{r}_{n+1}$ and $\tilde{q}_{n+1}$ remains constant during the current time step. This furnishes relevant properties to the constitutive tangent tensor $\tilde{\mathbb{D}}_{alg}$:

i) In all, it is symmetric and positive definite. The arguments are trivial for the isotropic damage model. As shown in step 4 of Table 6, $\mathbb{E}$ is symmetric and positive definite, while the factor $\frac{\tilde{q}_{n+1}}{\tilde{r}_{n+1}}$ is positive.

ii) For the present damage model, the algorithmic tangent operator is constant (independent of the current strain, ε_{n+1}).

Property *(i)* could be easily extended to other constitutive models equipped with strain softening, such as plasticity, see for example Oliver et al. (2008). Also, property *(ii)* can be verified in more general cases. However, it is noted that, while the first property is an important requirement to reach numerical stability, the second one is not. Stable integration schemes could be developed without preserving the property *(ii)*.

Time-Integration Accuracy Assessment of the IMPL-EX Method. The Double Cantilever Beam (DCB) test of Figure 15-a, with material parameters shown in the same Figure, is solved in order to evaluate the influence of the IMPL-EX incremental time step length Δs on the accuracy of the obtained numerical solution.

We analyze the solution convergence in Figure 15-b through different equilibrium solutions when $\Delta s \to 0$ ($G_f = 50 N/m$). Also, Figure 15-c shows the convergence rate of computed fracture energy, G_f, with Δs.

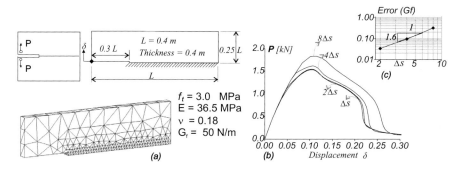

Figure 15. Accuracy assessment of the IMPL-EX method: DCB test. (a) Geometry and finite element mesh; (b) Load P vs. vertical displacement δ curves using different integration-time steps that are proportional to Δs; (c) Error of the computed G_f as a function of the integration-time step length.

5 Applications of the CSDA Methodology to Concrete Fracture Problems

In this Section, three applications of the CSDA methodology for solving typical concrete fracture problems are shown. They are: *a)* the fracture process zone analysis of two-well known concrete structure tests (Subsection 5.1); *b)* the capture of the size effect observed in fracture of concrete structures (Subsection 5.2); and *c)* the simulation of a dynamic fracture problem (Subsection 5.3);

Emphasis is given to show the capability of the computational tool for analyzing such problems, and not to the physical aspects involved in every case. Thus, detailed explanations of the size effect phenomena or the phenomenology involved in dynamics fracture mechanics are not provided here, The interested reader is addressed to the specific bibliography, in the referenced works, for additional details.

5.1 Modeling the Fracture Process Zone

A salient feature of the CSDA for modeling crack problems is that it can simulate the complete fracturing process phenomena, from the continuum to the discrete ones, using a unified methodology. In this sense, inelastic deformations with no macro crack formation in the material bulk, a typical phenomenon well-described by continuum models, can be represented using the same methodological framework as that utilized in the simulation of macro-crack propagation processes. This feature is remarked in this section: an accurate description of the FPZ in two standard problems is shown.

As it was explained in Section 1, during the structural loading process in fracture problems, a material point subjected to cracking displays successive stages of material behavior that can be idealized as follows:

  i) initially, the material shows an elastic response;
  ii) at the time t_d, the material starts displaying an inelastic response, that, in the particular case of concrete fracture problems, could be described by a damage model;
  iii) in a subsequent time, t_{loc}, the singularity of the localization tensor, $\mathbb{Q}^{tg}$, defined in terms of the tangent constitutive operator, $\mathbb{D}_T$ as:

$$\mathbb{Q}^{tg} = \boldsymbol{n} \cdot \mathbb{D}_T \cdot \boldsymbol{n} \quad ; \quad \mathbb{D}_T = \frac{\partial \boldsymbol{\sigma}}{\partial \boldsymbol{\varepsilon}} \qquad (110)$$

signals the onset of material instability, strain localization, and eventually, the development of discontinuities in the displacement field that characterizes the macro-crack formation.

For a given material model, the delay between t_d and t_{loc} depends on the material properties and the stress state. For a given stress state, the interval $[t_d, t_{loc}]$ becomes larger with larger (less negative) softening modulus, H. Therefore, from a phenomenological approach to crack modeling, one can take advantage of this fact in order to consider the amount of volumetric energy dissipation in front of the crack tip, as also, the size and influence of the fracture process zone. The involved phenomenology that can be simulated with the CSDA is sketched in Figure 16: the initial value of the continuum softening modulus, H (see Figure 16-a), rules the stable damage production at the considered point. Material points in this stage, in front of the crack tip, are displayed in the gray zone in Figure 16-b, between points D (standing for onset of damage) and L (standing for onset of localization). The length ℓ_d of this stable damaged zone is governed by the value H as said before.

As, at time t_{loc}, material instability is detected, the continuum softening modulus is no longer determined as a material property but regularized from the intrinsic softening modulus $\bar{H}$ (in turn depending on the fracture energy) according to equation (78) in Table 4; see Figure 16-a. At the same time, the strong discontinuity kinematics, equations (66–67), is activated, this ensuring fulfillment of the degenerated traction/separation law and the subsequent correct dissipation of the fracture energy G_f up to the complete stress release. Points in this stage are signaled, in Figure 16-b, as the cohesive zone between points L (standing for onset of localization) and R (standing for stress-free zone limit).

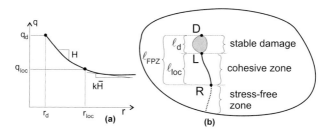

Figure 16. The fracture process zone.

Both, the stable damage and the cohesive zone, constitute the generalization, in a CSDA setting, of the fracture process zone (FPZ), classically considered in non-linear fracture mechanics (Bazant and Planas, 1998), as the locus of material points where the fracture is processed by dissipative mechanisms. As it will be shown in the following, the total size of this fracture process zone, FPZ, basically depends on the amount of fracture energy G_f, and the length of the stable damage zone, ℓ_d (and the corresponding dissipation) is ruled by the continuum softening modulus H. Therefore, the size of the cohesive zone, ℓ_{loc}, can be expressed as:

$$\ell_{loc}(G_f, H) = \ell_{FPZ}(G_f) - \ell_d(H) \tag{111}$$

with the conditions:
$$\frac{d\ell_{FPZ}}{dG_f} \geq 0; \qquad \frac{d\ell_d}{d|H|} \leq 0$$

5.1.1 The Three-Point Bending Test. The crack propagation problem in the notched concrete beam supported on three points, shown in Figure 17-a, is used to assess several aspects of the fracture process zone analyzed with the CSDA methodology, additional details of this analysis have been reported in in Huespe et al. (2006a).

The continuum constitutive relation adopted for this problem, is the damage model presented in Table 1. From equation (18) in this Table, note that damage evolution is inhibited in compressive stress states. The damage model is specified for the plane stress case.

a) Cohesive Force Induced by the Continuum Damage Model. Figures 17-c and d show the evolution, in terms of the pseudo-time u_y (the monotonically increasing vertical displacement of the load application point), of the normal and tangential components of the traction vector ($t_S = \sigma_S \cdot n$) at four different points of the crack path (points A, B, C and D at a distance ζ from the notch tip, Figure 17-b). There, it can be observed that both components of t_S are continuous along the time, irrespective of the transition between different fracture process stages.

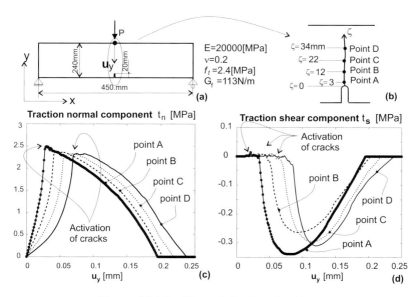

Figure 17. The Three-Point Bending Test.

b) Study of the Fracture Process Zone Characteristic Lengths. Let us focus on the role played by the initial continuum softening modulus, H, the softening rule in Figure 16-a, and its influence on the fracture process zone structure.

Curves in Figure 18-a show the dependence of the fracture process zone length, ℓ_{FPZ}, on the fracture energy, G_f, as a function of the material characteristic length parameter ($\ell_{ch} = \frac{EG_f}{f_t^2}$) characterizing the material brittleness (see Elices et al. (2002)). In the figure, it can be observed that ℓ_{FPZ} depends clearly on G_f but it barely changes with substantial changes of H. Note that the solution called "with elastic bulk" means that H is a very large negative parameter, such that, strain localization conditions are immediately detected at onset of the inelastic deformation. Therefore, no stable damage is induced in the FPZ.

Thus, the experiment assesses the dependence of the total FPZ length with the fracture energy, G_f. In Figure 18-b, it is shown the dependence of the stable damage zone size, ℓ_d, with the H and G_f values. Results are parameterized in terms of the normalized length, χ, ($\chi = -H\ell_{ch}$). Plots of curves $\ell_d(\chi)$, for different values of $\ell_{ch}(G_f)$, are then presented. In agreement to the assertion in Bazant (1996), about the changing length of the micro cracking zone in front of the crack tip, the numerical simulations show a similar behavior. Therefore, all lengths in Figure 18-b have been measured at the time that stresses are completely released at the notch tip (complete stress release).

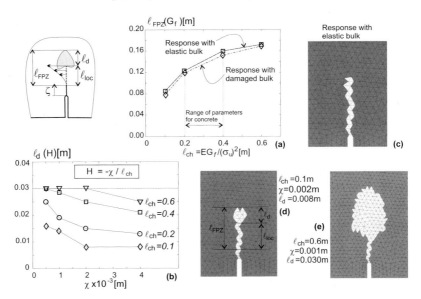

Figure 18. The Three-Point Bending Test, fracture process zone analysis.

It can be observed that, for a given value of G_f, increasing values of $|H|$, diminish the stable damage zone size ℓ_d (the lower bound is, approximately, the size of one finite element 0.004–0.007 m). Typical developments of the FPZ, for different material parameter values, are displayed in Figure 18-c:e (finite elements undergoing dissipation are plotted in light gray color). There, the stable damage zone (round shaped in front of the crack tip) can be clearly differentiated from the remaining domains of the FPZ.

5.1.2 Double Cantilever Beam (DCB Test). In this subsection we show additional details of the fracture process zone through the classical double cantilever beam test analysis (DCB Test) of Figure 19-a.

a)**Predicted Peak Load Values.** This test emphasizes the effects that the stable dissipation zone, as a function of the continuum softening modulus H, induces on the predicted peak loads.

The problem is solved by using two different procedures. In Procedure A, we impose an elastic response of the bulk material, like in standard cohesive models. It is defined a very large (negative) value of the initial softening modulus, H (the stable dissipation zone does not exist). In Procedure B, bulk dissipation is admitted by specifying a softening modulus $H \approx 0$ (large stable dissipation zone). In both cases the same value for the fracture energy G_f is considered. Figure 19-b displays the corresponding force-displacement (CMOD) curves. It can be checked that allowing dissipation on the bulk has a non-negligible effect on the predicted peak load.

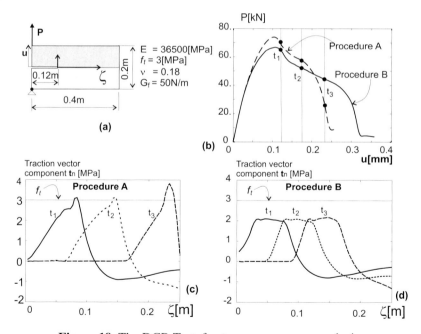

Figure 19. The DCB Test, fracture process zone analysis.

Figure 19-c plots the distribution of the crack normal stress component, t_n,

along the crack path (stress profile), at three different times, t_1, t_2 and t_3, specified in the equilibrium curves of Figure 19-b. Those curves computed with the Procedure A (standard cohesive model) show peak values greater than the material tensile strength ($f_t = 3MPa$). This inconsistency has also been detected by other authors using a similar procedure, where it is claimed that this unwelcome effect disappears, or diminishes, with mesh refining. Results obtained with Procedure B (bulk dissipation is allowed) show, on the contrary, consistent stress profiles. They are not spike-shaped, as in the previous case, and the maximum stress level remains always below f_t, even for the relatively coarse mesh used in the analysis.
b) Study of the Fracture Process Zone Characteristic Lengths. Using a similar technique of that presented in subsection 5.1, we evaluate the sensitivity of the FPZ length with the fracture energy G_f parameter. Thus, we determine the size ℓ_{FPZ} for three different values of G_f: ($= 25, 50$ and $75 N/m$). The results are compared with the material characteristic length $\ell_{ch} = EG_f/f_t^2$ in Figure 20 for both tests: the three-point bending test of section 5.1 and the present DCB test.

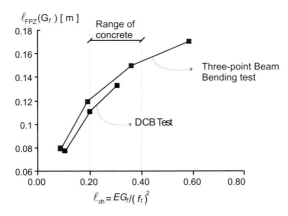

Figure 20. Comparison of the fracture process zone lengths vs. the material characteristic length for both tests: Three-point bending and DCB test.

5.2 Size Effect Analysis

The influence of the structure size on the nominal strength, when proportional (or geometrically similar) structures are compared, is called the size effect, see Bazant and Planas (1998). The nominal strength is the ratio between the ultimate structural load and a structure representative area.

The size effect is only significant in those specimens with a characteristic size not being so small, if compared with FPZ size. There are several laws describing this effect in concrete structures subjected to fracture problems. Some of them are based on Non-Linear Fracture Mechanics principles while others are empirical laws. However, they are not general laws in the sense that they cannot be applied to a wide range of geometries or loading conditions.

Computational Failure Mechanics, and particularly the CSDA approach, is an alternative tool for evaluating the size effect phenomenon in structural analysis. We show in the following examples an application to this problem extracted from Blanco (2006).

5.2.1 The Brazilian Test. The Brazilian test has been included in several building standards and design codes as a valid procedure to determine the tensile strength of quasi-brittle materials.

One version of this test corresponds to a cylindrical specimen that is compressed by two diametrically opposed loads, as it is shown in Figure 21-a. The tensile stresses produced by these loads induce the formation of a vertical planar crack along the plane defined by the opposite loads, which, eventually, produces the splitting of the specimen.

The main specimen dimensions are specified in Figure 21-a. The test provides the splitting tensile strength value, f_{st}, given by:

$$f_{st} = \frac{2P_u}{\pi BD} \qquad (112)$$

where B and D are specimen dimensions and P_u the ultimate load. The value f_{st} decreases with larger specimen sizes and with smaller bearing strip sizes b. However, f_{st} tends, asymptotically, to the material tensile strength.

We have reproduced the f_{st} values obtained with different specimen sizes. The experimental values of these test were reported by Rocco et al. (1999), who have analyzed granite cylindrical specimens with the following dimensions: $B = 30mm$ and diameters: $D = 30, 60, 120$ and $240mm$. Figure 21-b presents the material parameters adopted for the numerical model.

Figure 21-c displays several experimental results, taken from Rocco et al. (1999), for the case $b/D = 0.16$. The plots correspond to the structural response: load P vs. diametric opening w_d. Three characteristic points are displayed in each curve: Point 1 is the maximum load coinciding with the vertical diametric crack onset. The crack formation causes a marked structural load carrying capacity loss. The Point 2 is the minimum value attained after 1. After the point 2, several cracks show up below the load-bearing strips and evolve until reaching a second maximum in the Point 3.

The performed nine numerical tests were run until reaching the point 2, Thus, the formation of the second pattern of cracks, below the load-bearing strips, were not captured. Figure 21-d displays the finite element mesh for cases 1, 4, and 7

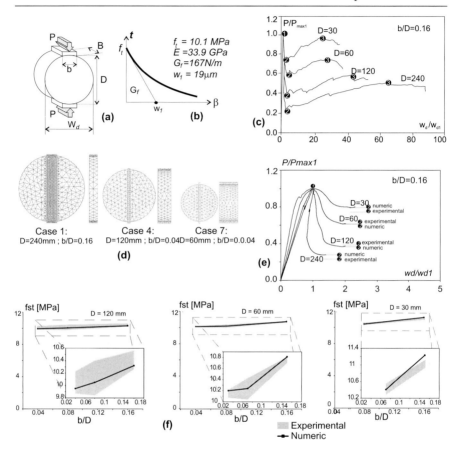

Figure 21. Size effect analysis: the Brazilian test.

that we have modeled, and Figure 21-e,f the numerical solutions obtained with the CSDA methodology that are compared, in all cases, with the experimental results.

We remark the capacity of the numerical model for capturing the effect observed in the experimental test: the larger is the specimen the greater is the brittleness.

5.3 Dynamic Fracture Simulation

An application of the CSDA approach to simulate dynamics fracture problems, reported in Huespe et al. (2006b), is here presented. It captures the most intriguing aspects of the problem.

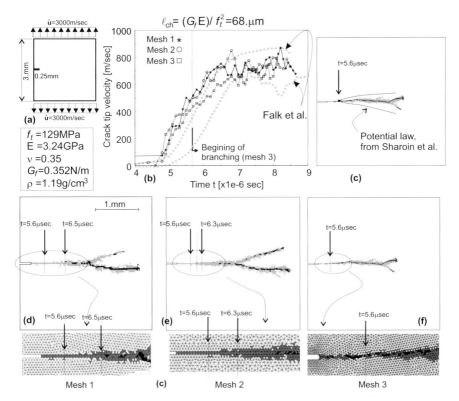

Figure 22. Prediction ofn dynamic fracture: (a) specimen geometry; (b) crack propagation velocity; (d) crack pattern morphology; (d-e-f)Crack paths for Mesh 1, for Mesh 2 and for Mesh 3.

Numerical Methodology. The FE technology with embedded strong discontinuities which does not need enforcing the crack path continuity, reported by Sancho et al. (2005) and explained in Section 3, is adopted. As it is mentioned in that Section, an important advantage of this procedure is that it does not need a crack path tracking algorithm. Therefore, this technique results more convenient in order to capture possible crack bifurcations, which are common in dynamic fracture problems.

Additional details of the implementation aspects, that were specially addressed to capture crack branching, could be found in Huespe et al. (2006b).

a) Prediction of Dynamic Fracture in PMMA. It is analyzed the crack propagation problem in a square specimen with $3.mm$ sides and a notch $0.25mm$ long (see Figure 22-a). In the analysis, it is imposed a vertical constant velocity of

$3000.m/sec$ on the top and bottom edges. The material is characterized with parameters depicted in the same Figure. With these values, the dilatational wave speed is: $c_d = 2090m/sec$, the shear wave speed: $c_s = 1004.m/sec$, and the Rayleigh wave speed: $c_R = 938, m/sec$. A numerical simulation, using cohesive interface finite elements, was presented by Falk et al. (2001).

a.1) Crack Tip Velocity. In Figure 22-b, it is plotted the crack tip velocity, which has been evaluated as the numerical derivative of the crack tip position as a function of time. Also, in this figure it is presented the results reported by Falk et al. where two different implementations of a discrete cohesive model have been used.

a.2) Crack Pattern Morphology. Also, with this methodology, it is possible to capture the crack branching effect, as it is shown in Figure 22-d-f. We depict the distribution of cracks at the end of the simulation process for the three meshes (1,2 and 3). The black zone corresponds to active elements, or cracks in opening mode, while the gray zone represents elements that previously have been activated but that at the end of the simulation process are arrested.

Other crack distribution pattern morphological features were reported in the literature, and it was concluded that the branches of the cracks are distributed following a potential law like that plotted in Figure 22-c. It can be observed that CSDA simulation agrees reasonably well with this experimental fact.

6   A Model for Reinforced Concrete Fracture via CSDA and Mixture Theory

Reinforced concrete, constituted by concrete with long fibers (reinforcements) oriented in different directions embedded in it, could be analyzed following two different conceptual models implying different length scales: *i)* a mesoscopic scale model describing the response of every composite material component (matrix and reinforcement) as an independent subsystem that is mechanically interacting with the neighbor components. The numerical simulation of mesoscopic models requires high computational costs; *ii)* a macroscopic scale model describing the response of the composite material via a homogenized constitutive model. In this case, the success of the model relies on the homogenization procedure, which becomes the key issue in this type of conceptual approach.

Following the second approach, it is possible to develop a rather simple homogenization procedure based on combining CSDA with mixture theory, which provides a computational model that captures the most salient phenomena governing the failure of reinforced concrete structures. An important feature of this model is that it requires a reduced computational effort. The resulting model was presented in Linero (2006).

In the present Section, it is explained the FE model which follows this proposal.

6.1 A Mixture Theory for Reinforced Concrete

Reinforced concrete is assumed to be a composite material made of a matrix (concrete) and long fibers (steel bars) arranged in different directions, as shown in Figure 23.

According to the basic hypothesis of mixture theory, the composite is a continuum in which each infinitesimal volume is occupied by all the constituents with volumetric fractions given by the factor $k^i \leq 1$ (for the i-th constituent). Assuming a parallel layout, all constituents are subjected to the composite deformation ε. The composite stresses, σ, are obtained by adding the stresses of each constituent, weighted according to their corresponding volumetric participation. Thus, the matrix strains, ε^m, coincide with the composite strains, ε:

$$\varepsilon^m = \varepsilon \tag{113}$$

The extensional strain of a fiber f in direction $\boldsymbol{r}$, ε^f, see Figure 23, is equal to the component ε_{rr} of the composite strain field in that direction, that is:

$$\varepsilon^f = \boldsymbol{r} \cdot \boldsymbol{\varepsilon} \cdot \boldsymbol{r} \tag{114}$$

In order to take the dowel action into consideration, the fiber shear strains, γ, are obtained as the shear components of the composite strain field. In a local 2-D orthogonal reference system $(\boldsymbol{r}, \boldsymbol{s})$, the fiber shear component ε^f_{rs} is given by:

$$\varepsilon^f_{rs} = \frac{\gamma^f_{rs}}{2} = \boldsymbol{r} \cdot \boldsymbol{\varepsilon} \cdot \boldsymbol{s} \tag{115}$$

The stresses of a composite with nf fibers (or fiber bundles) oriented in different directions $\boldsymbol{r}^f$ ($f = 1, 2, .., nf$) can be obtained using the following weighted sum of each contribution:

$$\boldsymbol{\sigma} = k^m \boldsymbol{\sigma}^m(\boldsymbol{\varepsilon}^m) + \sum_{f=1}^{nf} k^f [\sigma^f(\varepsilon^f)(\boldsymbol{r}^f \otimes \boldsymbol{r}^f) + 2\tau^f_{rs}(\gamma^f_{rs})(\boldsymbol{r}^f \otimes \boldsymbol{s}^f)^{sym}] \tag{116}$$

where k^m and k^f are the matrix and the fiber f volumetric fraction, respectively, $\boldsymbol{\sigma}^m$ is the matrix stress tensor, σ^f is the fiber normal (axial) stress, and τ^f_{rs} is the fiber shear stress component.

For the sake of simplicity, in equation (116), it is assumed that the normal and tangential stress components of the fibers are related to the corresponding strains by means of specific constitutive equations in a completely decoupled behavior of the matrix response.

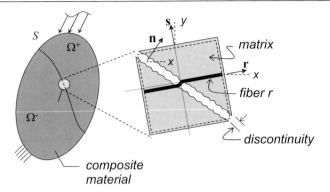

Figure 23. Reinforced concrete model.

The incremental form of the composite constitutive equation can be written as:

$$\dot{\boldsymbol{\sigma}} = \mathbb{D}_T : \dot{\boldsymbol{\varepsilon}}; \tag{117}$$

$$\mathbb{D}_T = k^m \mathbb{D}_T^m + \sum_{f=1}^{nf} k^f [E_T^f (\boldsymbol{r}^f \otimes \boldsymbol{r}^f) \otimes (\boldsymbol{r}^f \otimes \boldsymbol{r}^f) +$$

$$+ \ 4 G_{rs}^f (\boldsymbol{r}^f \otimes \boldsymbol{s}^f)^{sym} \otimes (\boldsymbol{r}^f \otimes \boldsymbol{s}^f)^{sym}]$$

where $\mathbb{D}_T^m = \partial \boldsymbol{\sigma}^m / \partial \boldsymbol{\varepsilon}^m$; $E_T^f = \partial \sigma^f / \partial \varepsilon^f$ and $G_{rs}^f = \partial \tau_{rs}^f / \partial (\gamma^f / 2)$ are the tangent operators for the involved constitutive relations of concrete and fiber, respectively.

6.2 Constitutive Model for the Composite: Regularization Procedure based on the CSDA

Crack onset and growth in the composite can be modeled in the context of the CSDA. Therefore, a strong discontinuity kinematics, like that presented in Section 2, equations (66–67), is assumed.

a) Constitutive Model for the Concrete. The concrete shows very different responses either in tensile or compressive stress regimes. In each case, the most striking difference is observed in the failure mode. Under tensile stresses, the concrete displays a much lower strength than in compressive states. Also, it shows a higher brittleness in tensile stress conditions. In fact, formation of cracks are expected only if tensile states are observed while in compressive states, the concrete behaves like a plastic material, sometimes displaying a failure mechanisms like shear bands.

Additionally, in reinforced concrete structures, due to the reinforcement, the concrete is generally subjected to high confinement stress regimes, which plays a very important role in the structural strength, suggesting that this effect must be considered in the concrete model.

Therefore, it is advisable to use a concrete constitutive relation having the ability to capture the phenomenology observed under both tensile and compressive stress conditions. This motivates the concrete model presented in Sánchez et al. (2011), which is based on the damage constitutive relation, depicted in Table 1 and regularized via CSDA, for positive mean stress states; and a plastic response for negative mean stress states. Additional details of this model can be found in the mentioned reference.

b) Constitutive Model for the Steel Fibers. Steel fibers are regarded as one-dimensional elements embedded in the matrix. They can contribute to the composite mechanical behavior introducing axial or shear strength and stiffness.

The axial contribution of each fiber bundle depends on its mechanical properties and the matrix fiber bond/slip behavior. The combination of both mechanisms is modeled by the slipping-fiber model described below. In this framework, the dowel action can be provided by the fiber shear stiffness contribution in the crack zone.

Slipping-fiber Model. The fiber axial contribution can be modeled through one-dimensional constitutive relations, relating extensional strains with normal stresses. The assumed compatibility between matrix and fiber strains allows capturing the slip effect due to the bond degradation by means of a specific strain component associated with the slip. Thus, the fiber extensional strain, ε^f, can be assumed as a composition of two parts: one due to fiber mechanical deformation, ε^d, and the other related to the equivalent relaxation due to the bond-slip in the matrix-fiber interface, ε^i :

$$\varepsilon^f = \varepsilon^d + \varepsilon^i \tag{118}$$

Assuming a serial composition between fiber and interface, as illustrated in Figure 24, the normal stress of the slipping-fiber model, σ^f, is equal to each component stress:

$$\sigma^f = \sigma^d = \sigma^i \tag{119}$$

The stress associated to the fiber elongation, as well as the one associated with the matrix-fiber slip effect, can be related to the corresponding strain component by means of a uniaxial linearly elastic/perfectly plastic constitutive model

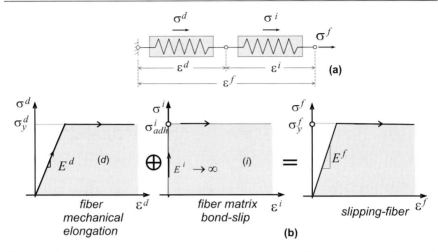

Figure 24. Slipping-fiber model: the composition ($\oplus$) of both elements (d) and (i) must be understood as a serial mechanical system, in the sense that, deformations are additive ($\varepsilon^f = \varepsilon^d + \varepsilon^i$) and stresses are not ($\sigma^f = \sigma^d = \sigma^i$).

6.3 Representative numerical simulations

The capability of the model to predict reinforced concrete structure failure modes is shown through two typical problems which solutions are contrasted with experimental results. The first one is the simulation of a reinforced concrete corbel, a typical 3D test, and the second one is a reinforced concrete beam assumed a plane strain case.

a) Reinforced Concrete Corbel. The concrete corbel of Figure 25 is simulated and results are contrasted with the experimental test published by Mehmel and Freitag (1967) and the numerical solution of Hartl (2002). The test description, geometry and reinforcing bar distribution, are supplied in Figure 25-a,b, see also Manzoli et al. (2008) where additional details of the numerical simulation are presented.

Symmetry conditions are assumed; therefore, one fourth of the structure is discretized. The finite element mesh is shown in Figure 25-d. Each rebar and the surrounding concrete were modeled as a composite material with equivalent mechanical properties. The corbel was then divided into sub-regions involving the steel bars. In each sub-region, the fiber direction and volumetric fraction corresponds to the embedded steel bars in the real problem. The assumed mechanical properties for the comonent materials are depicted in the Figure.

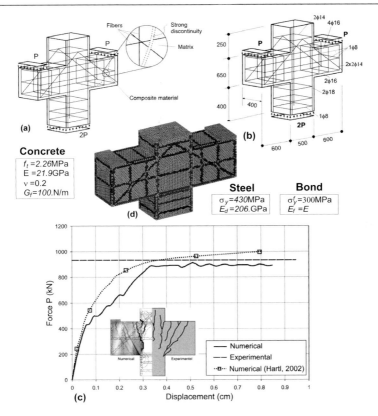

Figure 25. Reinforced concrete corbel: (a-b) model description; (c) load vs. vertical displacement of the load application point curves, insert fracture crack pattern, simulated (left) and experimental (right); (d) finite element mesh.

Figure 25-c compares the structural action-response curves (load P vs. vertical displacement of the load application point) obtained numerically, with the proposed methodology, and the ones obtained using a model based on the smeared crack model with embedded representation of the rebar, presented by Hartl (2002). Both methodologies provide a reasonable prediction of the experimental ultimate load capacity. In the insert of Figure 25-c (left), it is shown the simulated crack pattern that are represented by the iso-displacement contours at the end of the analysis. The predicted crack pattern is in good correspondence with that observed in the experimental test, as shown in the right half symmetric part of the same picture.

b) Application to Corroded Reinforcement Concrete Beams. The corrosion phenomenon observed in Reinforced Concrete structures (RC structures) is a very important failure mode that limits the service life of these structures. In the present example, taken from Sánchez et al. (2010), it is shown an application of the above presented numerical approach to simulate the mechanical consequences, typically the structural load carrying capacity loss, due to the reinforcement corrosion. The mechanism by which the steel corrosion causes a loss of the structural strength is due to an expansion of the corroded rebar, which induces cracking in the concrete cover, loss of steel-concrete bond, as also net area reduction of the steel fiber. The effects of the above mentioned mechanisms on the structural load carrying capacity can be analyzed as a function of the reinforcement corrosion degree. Therefore, the model makes possible to determine the influence and sensitivity of this key variable, the reinforcement corrosion level, in the structural deterioration problem.

The proposed numerical strategy can be applied to beams, columns, slabs, etc., through two successive and coupled 2-D mesoscopic mechanical analyzes, as follows:

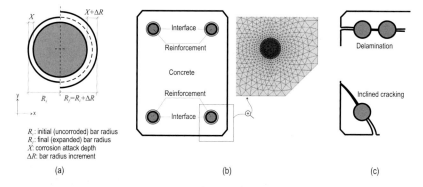

Figure 26. Analysis of the rebar corrosion: cross section analysis of the structural member.

(i) At the structural member cross section level, we simulate the reinforcement expansion due to the volume increase of the steel bars as a consequence of corrosion product accumulation. Damage and crack pattern distribution, in the concrete bulk and cover, is evaluated, which (indirectly) defines the concrete net section loss in the structural member.

(ii) A second macroscopic (homogenized) model at the structural level, considering the results of the previous analysis, evaluates the mechanical response of the structural member subjected to an external loading system. This evaluation determines the global response and the macroscopic mechanisms of failure.

b.1) Cross Section Analysis of the Structural Member (expansion mode). Let us consider the cross section of an arbitrary RC structural member, as displayed in Figure 26-b, whose reinforcement is experiencing a corrosion process. The products derived from the steel bar corrosion, such as ferric oxide rust, reduce the net steel area and accumulates, causing volumetric expansion of the bars (see Figure 26-a), what induces a high hoop tensile stress state in the surrounding concrete. As a consequence the cover concrete undergoes a damage and a degradation process displaying two typical fracture patterns: (i) inclined cracks and (ii) delamination cracks, as observed in Figure 26-c. Obviously, these induced cracks can increase the rate of corrosion process in the structural member.

The two-dimensional plane strain mesoscopic model, as idealized in Figure 26-b, considers three different domains of analysis: (i) the concrete matrix, (ii) the steel reinforcement bars and (iii) the steel-concrete interface. Each of them are characterized by a different constitutive response, and FE technology, that takes into account the main mechanisms involved in the corrosion process.

A steel-concrete interface model is considered in order to capture the possible slipping movement between both components once the concrete fractures.

In Figure 27, we display two different numerical solutions related to the expansion mechanism of steel bars for a predefined corrosion attack depth, and the degradation induced in concrete cover at the cross section level. From a qualitative point of view, it can be observed that the proposed mesoscopic plane strain numerical model captures physically admissible failure mechanisms. The introduction of friction-contact (interface) finite elements in the simulations is a key aspect in order to obtain consistent crack patterns, which match very well with the semi-analytical predictions published in the literature.

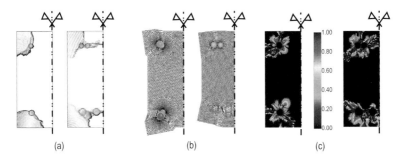

Figure 27. Analysis of the rebar corrosion problem for two different reinforcement bar distributions. Plane strain expansion analysis: (a) iso-displacement contour lines (crack pattern); (b) scaled deformed configuration; (c) damage contour fill.

b.2) Macroscopic Model to Simulate the Structural Load Carrying Capac-

ity (**Flexure Mode**). The model presented in this subsection provides qualitative information related to the concrete degradation mechanisms due to the steel expansion. However, it does not give additional information about the mechanical behavior of a deteriorated RC structure subjected to external loads.

Therefore, a 2-D macroscopic homogenized composite model, as it was shown in the previous sections, could be proposed in order to evaluate the residual load carrying capacity of the corroded RC member. The idealized scheme of the discrete model is shown in Figure 28. It is modeled the plain concrete by means of the CSDA approach with the constitutive relation described in Section 6.2.

b.3) Coupling Strategy Between the Cross Section and the Structural Member Analysis. Finally, a coupling strategy must be provided in order that the results of the cross-section model (Subsection 6.3) could be transferred, as a data set, to the macroscopic analysis of the structural member. It consists of projecting, from one domain of analysis to the other, the average value of the damage variable d across horizontal slices of the cross section model. This projection is consistent because both analyses use the same continuum isotropic damage model for simulating the concrete domain. Thus, the final degradation state of concrete, induced by the steel bar volumetric deformation process, is considered to be the initial damage condition for the subsequent structural analysis. This means that we are assuming that the two models are coupled in only one direction.

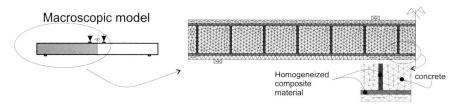

Figure 28. Macroscopic FE model to simulate the structural load carrying capacity of a structural member (flexure mode).

b.4) Numerical results of the corroded structural member. The structural analysis of the RC beam with the model of Subsection 6.3, where an initial damage distribution was adopted following the results of the analysis in Subsection 6.3, provides structural responses which change with the rebar corrosion attack depth. For a typical beam subjected to different levels of corrosion attack, these responses are plotted in Figure 29-a, depicting several loads vs. mid-span vertical displacement plots. It can be observed the structural strength sensitivity with the rebar corrosion level.

The same structural analysis provides the damage distribution that is displayed in Figure 29-b for one case of corrosion level. Also, Figure 29-c shows the iso-displacement contour lines which exhibit the final failure mechanism and the active cracks at the end of analysis.

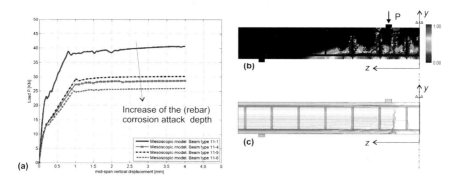

Figure 29. Numerical results of the macroscopic FE model to simulate the structural load carrying capacity: (a) load vs. mid-span vertical displacement plots, curves correspond to the same RC beam subjected to different corrosion attack depths. (b) final contour fill of damage. (c) iso-displacement contour lines in z-direction.

Bibliography

Z.P. Bazant. Mechanics of distributed cracking. *Appl. Mech. Rev.*, 39:675–701, 1986.

Z.P. Bazant. Analysis of work of fracture method for measuring fracture energy of concrete. *J. Eng. Mech. Div. ASCE*, pages 138–144, 1996.

Z.P. Bazant and J. Planas. *Fracture and size effect in concrete and other quasibrittle materials*. CRC Press, Boca Raton, FL, 1998.

S. Blanco. *Contribuciones a la simulación numérica del fallo material en medios tridimensionales mediante la metodología de discontinuidades fuertes de continuo*. PhD thesis, E.T.S. Enginyers de Camins, Canals i Ports, Technical University of Catalonia (UPC), Barcelona, 2006. In spanish.

M. Elices, G.V. Guinea, J. Gómez, and J. Planas. The cohesive zone model: advantages, limitations and challenges. *Engineering Fracture Mechanics*, 69: 137–163, 2002.

M.L. Falk, A. Needlemann, and J.R. Rice. A critical evaluation of dynamic fracture simulation using cohesive surfaces. *J. de Physique IV*, pages Pr–5–43 to Pr–5–50, 2001.

H. Hartl. *Development of a continuum-mechanics based toll for 3D finite element analysis of reinforced concrete structures and application to problems of soil-structure interaction*. PhD thesis, Graz University of Technology, Graz, Austria, 2002.

K. Hellan. *Introduction to Fracture Mechanics*. McGraw-Hill, New York, 1984.

R. Hill. General theory of uniqueness and stability in elasto-plastic solids. *J. Mech. Phys. Solids*, 6:236–249, 1958.

A.E. Huespe, J. Oliver, M.D.G Pulido, S. Blanco, and D.L. Linero. On the fracture models determined by the continuum-strong discontinuity approach. *Int. J. of Fracture*, 137:211–229, 2006a.

A.E. Huespe, J. Oliver, P.J. Sánchez, S. Blanco, and V. Sonzogni. Strong discontinuity approach in dynamic fracture simulations. In *Mecánica Computacional Vol XXV*, pages 1997–2018, 2006b. http://www.cimec.org.ar/ojs/index.php/mc/article/viewFile/594/565.

M. Jirasek. Conditions of uniqueness for finite elements with embedded cracks. In *ECCOMAS 2000*, Barcelona, Spain, 2000.

J. Lemaitre and R. Desmorat. *Engineering Damage Mechanics: Ductile, Creep, Fatigue and Brittle Failures*. Springer-Verlag, Berlin, 2005.

D.L. Linero. *A model of material failure for reinforced concrete via Continuum Strong Discontinuity Approach and mixing theory*. PhD thesis, E.T.S. Enginyers de Camins, Canals i Ports, Technical University of Catalonia (UPC), Barcelona, 2006. Cimne Monograph M106.

O. L. Manzoli, J. Oliver, G. Diaz, and A. E. Huespe. Three-dimensional analysis of reinforced concrete members via embedded discontinuity finite elements. *Revista IBRACON de Estrut. e Mater.*, 1:58–83, 2008. http://www.ibracon.org.br/publicacoes/revistas_ibracon/riem/volume1.asp.

A. Mehmel and W. Freitag. Tragfahigkeitsversuche an stahlbetonkonsolen. *Bauingenieur*, 42:362–369, 1967.

J. Oliver. Modelling strong discontinuities in solid mechanics via strain softening constitutive equations. Part 1: Fundamentals. *Int. J. Num. Meth. Eng.*, 39(21): 3575–3600, 1996a.

J. Oliver. Modelling strong discontinuities in solid mechanics via strain softening constitutive equations. Part 2: Numerical simulation. *Int. J. Num. Meth. Eng.*, 39(21):3601–3623, 1996b.

J. Oliver. On the discrete constitutive models induced by strong discontinuity kinematics and continuum constitutive equations. *Int. J. Solids Struct.*, 37: 7207–7229, 2000.

J. Oliver. *Topics on Failure Mechanics*. Cimne Monograph M68, 2002.

J. Oliver and A.E. Huespe. Theoretical and computational issues in modelling material failure in strong discontinuity scenarios. *Comput. Meth. App. Mech. Eng.*, 193:2987–3014, 2004.

J. Oliver, M. Cervera, and O. Manzoli. Strong discontinuities and continuum plasticity models: the strong discontinuity approach. *Int. J. Plasticity*, 15(3):319–351, 1999.

J. Oliver, A. E. Huespe, M. D. G. Pulido, and E. Chaves. From continuum mechanics to fracture mechanics: the strong discontinuity approach. *Engineering Fracture Mechanics*, 69:113–136, 2002.

J. Oliver, A. E. Huespe, E. Samaniego, and E. W. V. Chaves. Continuum approach to the numerical simulation of material failure in concrete. *Int. J. Num. Anal. Meth. Geomech.*, 28:609–632, 2004.

J. Oliver, A.E. Huespe, S. Blanco, and D.L. Linero. Stability and robustness issues in numerical modeling of material failure with the strong discontinuity approach. *Comput. Meth. App. Mech. Eng.*, 195:7093–7114, 2005.

J. Oliver, A. E. Huespe, and J.C. Cante. An implicit/explicit integration scheme to increase computability of non-linear material and contact/friction problems. *Comput. Meth. App. Mech. Eng.*, 197:1865–1889, 2008.

J. Rice. The localization of plastic deformation. In W. Koiter, editor, *Theoretical and Applied Mechanics, 14th IUTAM Congress*, pages 207–220, Amsterdam, North-Holland, 1976.

C. Rocco, V. Guinea, J. Planas, and M. Elices. Size effect and boundary conditions in the brazilian test: theoretical analysis. *Materials and Structures*, 32:437–444, 1999.

P.J. Sánchez, J. Oliver A.E. Huespe, and S. Toro. Mesoscopic model to simulate the mechanical behavior of reinforced concrete members affected by corrosion. *Int. J. of Solids and Struct.*, 47:559–570, 2010.

P.J. Sánchez, J. Oliver A.E. Huespe, G. Díaz, and V. Sonzogni. A macroscopic damage-plastic constitutive law for modeling quasi-brittle fracture and ductile behavior of concrete. *Int. J. Numer. Anal. Meth. Geomech.*, 2011. on-line, DOI: 10.1002/nag.1013.

J.M. Sancho, J. Planas, A. Fathy, J. Galvez, and D. Cendon. Three-dimensional simulation of concrete fracture using embedded crack elements without enforcing crack path continuity. *Int. J. Numer. Anal. Meth. Geomech*, 5:1–9, 2005.

T. Siegmund and W. Brocks. A numerical study on the correlation between the work of separation and the dissipation rate in ductile fracture. *Eng. Frac. Mech.*, 67:139–154, 2000.

J. Simo and T. Hughes. *Computational inelasticity*. Springer-Verlag, 1998.

J. Simo and J. Oliver. A new approach to the analysis and simulation of strong discontitnuities. In Z.B. Bazant, Z. Bittnar, M. Jirásek, and J. Mazars, editors, *Fracture and Damage in Quasi-brittle Structures.*, pages 25–39. E & FN Spon, 1994.

O.C. Zienkiewicz and R.L. Taylor. *The Finite Element Method*. Butterworth-Heinemann, Oxford, UK, 2000.

Plasticity based crack models and applications

G. Hofstetter[*], C. Feist[†], H. Lehar[*], Y. Theiner[*], B. Valentini[*], and B. Winkler[‡]

[*] University of Innsbruck, Technikerstraße 13, A-6020 Innsbruck, Austria
[†] CENUMERICS - Consulting Engineers, A-6020 Innsbruck, Austria
[‡] HILTI Corporation, FL-9494 Schaan, Principality of Liechtenstein

1 Introduction

Models for the numerical simulation of concrete cracking are traditionally based either on the smeared crack approach or the discrete crack approach.

In the former, the cracked material is still viewed as a continuum with a continuous displacement field and the crack is taken into account by modifying the strength and stiffness of concrete for the cracked parts of a structure. Typically, the smallest cracked region corresponds to the region associated with a sampling point for numerical integration over the volume of a finite element.

The latter is characterized by accounting for the displacement discontinuity across the crack by introducing discontinuity interfaces within the solid. Its behavior is governed by a discrete traction-separation law. In a finite element mesh, discontinuity interfaces are placed at element boundaries. Hence, adaptive remeshing is required to represent an arbitrary crack path. In the last decade the strong discontinuity approach (SDA) has gained wide popularity in numerical simulations of concrete cracking. It starts from the formulation of the strong discontinuity kinematics, which idealize a macroscopic crack as a discontinuity in the displacement field. In contrast to the discrete crack models, displacement discontinuities are not constrained to element boundaries anymore but are rather introduced within the element domains. Hence, the macroscopic crack path may cross a given spatial discretization in a more or less arbitrary way without the need for adaptive procedures. In this context two different FE-methods dealing with the strong discontinuities for representing cracks can be distinguished: The extended finite element method (XFEM) and the concept of elements with embedded discontinuities.

Which type of crack models is better suited in a numerical simulation depends on the given problem. If the structural behaviour is governed by a few dominant cracks with large crack widths, the discrete approach or the strong discontinuity approach will be the better model. The smeared approach, however, is simpler and often allows a good approximation of the mechanical behaviour of properly designed reinforced concrete structures, since the behaviour of such structures is characterized by many cracks with small crack widths due to a suitable arrangement of the reinforcement.

One option for formulating crack models is the theory of plasticity. The basic concepts of plasticity theory are described, e.g., in [Lubliner(1990), Hofstetter(1995)]. Constitutive models formulated within the framework of plasticity theory, consist of (i) a yield function, (ii) a flow rule and (iii) a hardening and/or softening law. In this chapter two crack models formulated within the theory of plasticity will be presented. The first one is a concrete model for two-dimensional stress states with smeared cracks and the second one a crack model based on the strong discontinuity approach and formulated within the framework of elements with embedded discontinuities. First, the basic features of the respective models are presented, followed by examples for the validation of the models and, finally, by applications to problems encountered in engineering practice.

2 Plasticity Based 2D Concrete Model with Smeared Cracks

The concrete model for biaxial stress states, described subsequently, is based on the model proposed by Feenstra and de Borst [Feenstra(1996)]. It is characterized by a composite yield function and suitable softening laws to describe tensile and compressive failure. Cracking is represented in a smeared manner. This model was enhanced by Winkler [Winkler(2004)] by coupling damage due to tensile stresses with damage due to compressive stresses for mixed tension-compression loading, by introducing an isotropic scalar damage model for unloading and reloading and by extending it to reinforced concrete. The model was implemented into the FE-program ABAQUS [ABAQUS] and was applied to the numerical simulation of structures, made of plain and reinforced concrete.

2.1 Formulation for plain concrete

Yield function Starting from the initial yield curve at the limit of linear elastic material behaviour (dashed curve in Figure 1) the yield curve expands until the yield curve at ultimate load (solid curve in Figure 1) is attained, followed by shrinking of the yield curve due to softening. The expressions for the yield curve are chosen such that good correspondence with the experimental results for the ultimate strength envelope by Kupfer [Kupfer(1969)] is achieved for biaxial tension and biaxial compression. The yield curve is composed of the Rankine principal stress criterion

$$f_1(\boldsymbol{\sigma}, q_1) = F_1(\boldsymbol{\sigma}) - q_1(\alpha_1) \tag{1}$$

to limit the tensile stress and of a Drucker-Prager yield function

$$f_2(\boldsymbol{\sigma}, q_2) = F_2(\boldsymbol{\sigma}) - c_2 \, q_2(\alpha_2) \tag{2}$$

to describe the compression regime. In (1) and (2)

$$F_1(\boldsymbol{\sigma}) = \sigma_1 \, , \qquad F_2(\boldsymbol{\sigma}) = \sqrt{\frac{3}{2} s_{ij} s_{ij}} + c_0 \, I_1^\sigma \tag{3}$$

with σ_1, I_1^σ and s_{ij} denoting the maximum principal stress, the first invariant of the stress tensor $\boldsymbol{\sigma}^T = \{\sigma_{11}, \sigma_{22}, \sigma_{12}\}$ for biaxial stress states and the deviatoric stress tensor, respectively. $q_1(\alpha_1)$ and $q_2(\alpha_2)$ are equivalent stresses, which represent the uniaxial tensile strength and the uniaxial compressive strength in terms of the internal variables α_1 and α_2.

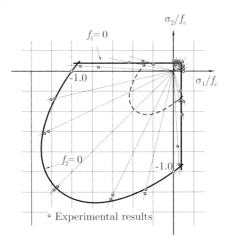

Figure 1. Comparison of the composite yield function at the limit of linear elastic behaviour (dashed curve) and at ultimate strength (solid curve) with the experimental results by Kupfer [Kupfer(1969)].

The parameters

$$c_0 = \frac{\beta_c - 1}{2\beta_c - 1} \, , \qquad c_2 = \frac{\beta_c}{2\beta_c - 1} \tag{4}$$

are given in terms of the ratio β_c of the biaxial compressive strength and the uniaxial compressive strength. According to [Kupfer(1969)] $\beta_c \approx 1,16$ holds.

Constitutive relations For small strains, the strain rate $\dot{\varepsilon}$ can be subdivided into an elastic part $\dot{\varepsilon}^e$ and a plastic part $\dot{\varepsilon}^p$ by

$$\dot{\varepsilon} = \dot{\varepsilon}^e + \dot{\varepsilon}^p . \tag{5}$$

From the elastic strain rate the stress rate follows from Hooke's law as

$$\dot{\boldsymbol{\sigma}} = \mathbb{E} : \dot{\varepsilon}^e = \mathbb{E} : (\dot{\varepsilon} - \dot{\varepsilon}^p) , \tag{6}$$

where $\mathbb{E}$ denotes the elasticity tensor.

Flow rule The evolution of the plastic strain rate is determined by means of Koiter's generalized flow rule for multi-surface plasticity

$$\dot{\varepsilon}^p = \sum_{i=1}^{2} \dot{\lambda}_i \frac{\partial g_i}{\partial \boldsymbol{\sigma}} \tag{7}$$

with the plastic multipliers

$$\begin{aligned} \dot{\lambda}_i > 0 \quad &\text{if} \quad \dot{\varepsilon}^p \neq \mathbf{0} , \\ \dot{\lambda}_i = 0 \quad &\text{if} \quad \dot{\varepsilon}^p = \mathbf{0} , \end{aligned} \tag{8}$$

serving as scaling parameters for the magnitude of the plastic strain rate, and the plastic potential functions g_i. Application of an associated flow rule, i.e. choosing $g_1 = f_1$ and $g_2 = f_2$, commonly results in an overestimation of the volumetric part of the plastic strains. This shortcoming may be accepted for models restricted to biaxial stress states.

The plastic multipliers and the yield functions have to fulfill the loading/unloading conditions [Lubliner(1990), Hofstetter(1995)]

$$f_i \leq 0 , \quad \dot{\lambda}_i \geq 0 , \quad \dot{\lambda}_i f_i = 0 \tag{9}$$

and the consistency condition

$$\dot{\lambda}_i \dot{f}_i = 0 . \tag{10}$$

Evolution of the internal variables Instead of describing the evolution of the internal variables α_1 and α_2 in terms of an equivalent plastic strain

$\dot{\varepsilon}_{eq} = \sqrt{\dot{\varepsilon}^p : \dot{\varepsilon}^p}$, in the present model the evolution of α_1 and α_2 is described according to [Feenstra(1996)] by means of the work-hardening hypothesis

$$\dot{W}_i^p = \boldsymbol{\sigma} : \dot{\boldsymbol{\varepsilon}}^p = q_i(\alpha_i)\,\dot{\alpha}_i \qquad \text{(no summation over } i\text{)} . \qquad (11)$$

In (11) it is assumed that the rate of plastic work of the biaxial stress state is equivalent to the respective rate of an equivalent uniaxial stress state. Extending the rate of the plastic work for the biaxial stress state in (11) by the scalar quantities b_{ij} and the ratio of the equivalent stresses q_i/q_j in order to account for possible coupling effects between α_1 and α_2 and making use of (7) yields

$$\dot{W}_i^p = \sum_{j=1}^{2} b_{ij} \frac{q_i}{q_j} \dot{\lambda}_j\, \boldsymbol{\sigma} : \frac{\partial g_j}{\partial \boldsymbol{\sigma}} = q_i(\alpha_i)\,\dot{\alpha}_i \qquad \text{(no summation over } i\text{)} . \qquad (12)$$

Using an associated flow rule, i.e. for $g_i = f_i$ in (12), from the property of F_i in (3) representing homogeneous functions of degree 1 and from Euler's theorem for homogeneous functions follows

$$\boldsymbol{\sigma} : \frac{\partial f_j}{\partial \boldsymbol{\sigma}} = \boldsymbol{\sigma} : \frac{\partial F_j}{\partial \boldsymbol{\sigma}} = F_j . \qquad (13)$$

From (1) and (2) follows $F_j = c_j q_j$, where $c_1 = 1$ and c_2 is given according to (4). Substituting this result into (13) and the so-obtained expression into (12) yields

$$\dot{\alpha}_i = \sum_{j=1}^{2} b_{ij} c_j \dot{\lambda}_j \qquad (14)$$

i.e.,

$$\dot{\alpha}_1 = b_{11}\, c_1\, \dot{\lambda}_1 + b_{12}\, c_2\, \dot{\lambda}_2 ,$$
$$\dot{\alpha}_2 = b_{21}\, c_1\, \dot{\lambda}_1 + b_{22}\, c_2\, \dot{\lambda}_2 . \qquad (15)$$

In contrast to (12), in (14) the rate of the internal variables does not depend on the stresses. Hence, the evolution of the internal variables is determined by the plastic strain rate. They can be considered as a measure of the progressive damage of the material.

According to the experimental data of [Kupfer(1969)] damage of the material due to compressive loading in one direction results in a reduction of the maximum attainable tensile stress in lateral direction, such that the material cannot sustain any tensile stresses in lateral direction when the uniaxial compressive strength is reached. This effect can be modelled by coupling of the internal variables α_1 and α_2 by non-vanishing values of the scalar quantities b_{12} and b_{21}. Figure 2 shows the comparison of the computed ultimate strength curve, considering coupling of the internal variables, with the experimental results of [Kupfer(1969)].

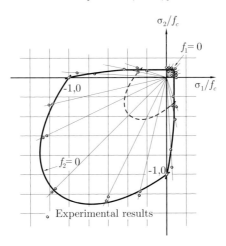

Figure 2. Comparison of the yield function at ultimate strength (solid curve) considering coupling of the internal variables with the experimental results by Kupfer [Kupfer(1969)].

Softening law for tensile loading In uniaxial tension, linear elastic material behavior is assumed up to the tensile strength f_t. Initiation of cracking is predicted by means of the yield function $f_1 = 0$. After attaining the tensile strength in a displacement controlled uniaxial tension test, a decreasing tensile stress q_1 is observed at increasing crack width ζ_1 (Figure 3(a)). As shown in [van Mier(1984)] (Figure 3(a)) this relationship is independent of the specimen's length. It can be approximated by

$$q_1(\zeta_1) = f_t \exp(-\zeta_1/\zeta_{1,0}) \tag{16}$$

with $\zeta_{1,0}$ determining the tangent to the $q_1\zeta_1$-relationship at $\zeta_1 = 0$ (Figure 3(b)).

Within the concept of smeared cracks the crack width ζ_1 is distributed (or smeared) over the respective finite element. Hence, it can be interpreted as a fictitious crack strain ε_1^f multiplied by an equivalent element length h_e, i.e.

$$\zeta_1 = \varepsilon_1^f h_e \quad \text{and} \quad \zeta_{1,0} = \varepsilon_{1,0}^f h_e \ . \tag{17}$$

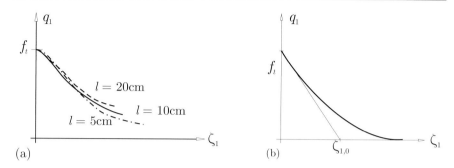

Figure 3. Relationship between the tensile stress q_1 and the crack width ζ_1 in the tension softening domain: (a) experiments with different specimen length, (b) analytical approximation.

h_e depends on the crack direction, the element type, the size and shape of the finite element and on the order of the numerical integration [Oliver(1989)]. However, simple approximations of h_e are also used. For 2D problems a popular approximation is $h_e = \sqrt{A_e}$ with A_e denoting the area of the respective finite element. Since the crack direction is ignored in this approximation, especially long and narrow finite elements should be avoided.

Substitution of (17) into (16) yields the stress-strain relationship for tensile softening

$$q_1(\varepsilon_1^f) = f_t \exp(-\varepsilon_1^f/\varepsilon_{1,0}^f) . \tag{18}$$

In order to guarantee objective results of a FE-analysis - at least in an approximate manner - the specific fracture energy

$$G_f = \int_0^\infty q_1(\zeta_1)\, d\zeta_1 \tag{19}$$

is introduced as an additional material parameter. G_f represents the energy dissipated during formation of a crack of unit area. G_f can be determined by direct tension tests or by three-point bending tests according to the RILEM recommendations [RILEM(1985)]. An extensive experimental program for determining the dependence of G_f on the size of the employed specimens is documented in [Trunk(2000)]. In this study wedge-splitting tests with dimensions ranging from $100 \times 100 \times 200$ mm to $3200 \times 3200 \times 400$ mm and maximum aggreagte sizes from 32 mm to 125 mm were used showing that with increasing specimen dimensions G_f approaches a constant value, which can be viewed as a material parameter.

Alternatively, an approximation of G_f can be determined according to [CEB-FIP(1990)] as
$$G_f = G_{f0}(f_c/10)^{0.7} \qquad (20)$$
with $G_{f0} = 25\,\text{Nm}/\text{m}^2$ for the maximum aggregate size $d_{max} = 8\,\text{mm}$, $G_{f0} = 30\,\text{Nm}/\text{m}^2$ for $d_{max} = 16\,\text{mm}$ and $G_{f0} = 38\,\text{Nm}/\text{m}^2$ for $d_{max} = 32\,\text{mm}$ and f_c as the cylindrical compressive strength in N/mm^2. Typical values of G_f for normal concrete are within the range of $50\,\text{Nm}/\text{m}^2$ and $150\,\text{Nm}/\text{m}^2$.

Substitution of (16) into (19) results in $G_f = f_t\,\zeta_{1,0}$. Inserting $\zeta_{1,0} = G_f/f_t$ into the relation for the fictitious crack strain $\varepsilon^f_{1,0}$ in (17) yields
$$\varepsilon^f_{1,0} = \zeta_{1,0}/h_e = G_f/(h_e f_t) , \qquad (21)$$
i.e. the fictitious crack strain is adjusted to the size of the respective finite element. Using (21) in (18) assures that during formation of a crack of unit area the same amount of energy is dissipated irrespective of the dimensions of the respective finite element.

For the present concrete model $\varepsilon^f_1 \equiv \alpha_1$ and $\varepsilon^f_{1,0} \equiv \alpha_{1,0}$, which yields the tension softening law shown in Figure 4a.

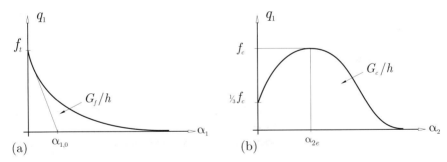

Figure 4. Hardening and softening laws for plain concrete sujected to (a) uniaxial tension, (b) uniaxial compression.

Hardening and softening law for compressive loading For uniaxial compressive loading linear elastic material behavior is assumed for stresses up to one third of the uniaxial compressive strength (Figure 4(b)). For stresses exceeding $f_c/3$ up to the uniaxial cylindrical compressive strength f_c, the equivalent stress is described by means of the isotropic hardening law
$$q_2(\alpha_2) = \frac{f_c}{3}\left[1 + 4\frac{\alpha_2}{\alpha_{2e}} - 2\frac{\alpha_2^2}{\alpha_{2e}^2}\right] \quad \text{for} \quad 0 \leq \alpha_2 \leq \alpha_{2e} . \qquad (22)$$

In the post-peak region the equivalent stress is given by the isotropic softening law

$$q_2(\alpha_2) = f_c \exp\left(-\frac{(\alpha_2 - \alpha_{2e})^2}{\alpha_{2u}^2}\right) \quad \text{for} \quad \alpha_{2e} \leq \alpha_2 . \tag{23}$$

α_{2e} and α_{2u} are given in terms of the modulus of elasticity E, the specific fracture energy G_c for compressive failure and the equivalent element length h_e as

$$\alpha_{2e} = 0.0022 - \frac{f_c}{E} , \quad \alpha_{2u} = \frac{2G_c}{\sqrt{\pi}h_e f_c} , \tag{24}$$

where the total strain at the uniaxial compressive strength is chosen as 0.0022 according to [CEB-FIP(1990)]. α_{2u} follows from

$$\frac{G_c}{h_e} = \int_{\alpha_{2e}}^{\infty} q_2(\alpha_2)\, d\alpha_2 = \frac{\sqrt{\pi}}{2} f_c \alpha_{2u} . \tag{25}$$

By analogy to tension softening G_c assures an objective description of compressive softening. $G_c \approx 100\, G_f$ can serve as a crude approximation for the specific fracture energy for compressive failure.

Unloading and reloading A weak point of elastic-plastic models is given by the fact that unloading and reloading paths are characterized by the initial elastic stiffness. Thus, for unloading in the tension softening region even for a small decrease of the crack width compressive stresses across the crack face will be predicted.

In order to avoid such unrealistic predictions the degradation of the modulus of elasticity is described within the framework of damage mechanics in terms of the scalar damage parameter D. It relates the degraded modulus of elasticity E_d to the modulus of elasticity E for the undamaged material by

$$E_d = (1 - D)E \tag{26}$$

with $D = 0$ for the undamaged material and $D = 1$ for the totally damaged material. Within the framework of the smeared crack approach the current state of damage can be estimated according to [Crisfield(1989)] by the relation between the equivalent stress and the equivalent strain. Assuming complete crack closure, the damage parameter D can be determined from (Figure 5(a)) as

$$D = 1 - \frac{E_d}{E} = 1 - \frac{\sigma_{1n}}{\sigma_{1n} + E\alpha_1} . \tag{27}$$

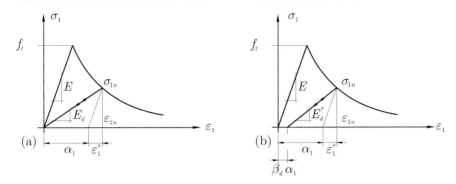

Figure 5. Modelling of unloading and reloading by means of a scalar damage model with (a) no and (b) partly irreversible plastic strains.

However, the assumption of complete crack closure is not realistic. Rather, a permanent crack width remains even at complete unloading. Introducing a permanent crack width, which is defined in terms of α_1 and a scaling factor $0 \leq \beta_d \leq 1$ yields the damage parameter D according to Figure 5(b) as

$$D = 1 - \frac{E_d^*}{E} = 1 - \frac{\sigma_{1n}}{\sigma_{1n} + E\,\alpha_1(1-\beta_d)} \;. \qquad (28)$$

For the special case $\beta_d = 1$ unloading and reloading follows the initial elastic stiffness, whereas for $\beta_d = 0$ complete unloading results in complete crack closure.

2.2 Extension to reinforced concrete

After crack initiation the behavior of reinforced concrete differs considerably from the material behavior of plain concrete. If cracking occurs then tensile stresses are transferred between reinforcement and concrete by bond, resulting in the tension stiffening effect. A uniaxial tension stiffening relationship for an embedded reinforcing steel bar is specified in [CEB-FIP(1990)] (Figure 6(a)). For a smeared crack model the tension stiffening effect can be considered by modifying either the stress-strain relationship of steel or the one of concrete. For the present model, the latter option is chosen.

Tension stiffening can be described as the stress difference $\sigma_{s,TSE}$ between the steel stress σ_s of the reinforced concrete member and the stress $\sigma_{s,bare}$ of a bare steel bar at a given strain. According to [Meiswinkel(1999)] the stress increase $\sigma_{s,TSE}$ can be replaced by an equivalent concrete stress

$\sigma_{c,TSE}$. Assuming that in a cracked reinforced concrete specimen tensile stresses will be transmitted between the cracks from the steel to the surrounding concrete, the equivalent concrete stress $\sigma_{c,TSE}$ (Figure 6(b)) can be determined as

$$\sigma_{c,TSE} = \rho_{eff}\, \sigma_{s,TSE} \qquad (29)$$

with $\rho_{eff} = A_s/A_{c,eff}$ as the effective reinforcement ratio, defined by the sectional area A_s of the steel and the area $A_{c,eff}$ of the effective zone of the concrete. The latter is computed as proposed in [CEB-FIP(1990)]. If the tension stiffening effect is related to the constitutive relation for the reinforcing steel, then the modified steel stress σ_{sr} at crack initiation (Figure 6(a)) is computed from $\sigma_{sr}\, A_s = f_t\, A_{c,eff} + f_t\, n\, A_s$ with $n = E_s/E$ as

$$\sigma_{sr} = f_t \frac{1 + n\, \rho_{eff}}{\rho_{eff}}, \qquad (30)$$

where E_s represents the modulus of elasticity of the reinforcing steel. The values of the strain, characterizing the stress-strain relations in Figure 6 (a) and (b) follow as

$$\varepsilon_{sr} = \frac{f_t}{E}, \qquad \varepsilon_{srn} = \frac{\sigma_{srn}}{E_s} - \beta_t\, \Delta\varepsilon_{sr},$$

$$\varepsilon_{sry} = \frac{f_{sy}}{E_s} - \beta_t\, \Delta\varepsilon_{sr}, \qquad \varepsilon_{sy} = \frac{f_{sy}}{E_s} \qquad (31)$$

with f_{sy} as the steel yield stress and $\beta_t\, \Delta\varepsilon_{sr}$ as the mean value of the difference between the steel strain of an embedded steel bar and a bare steel bar.

The strain difference $\Delta\varepsilon_{sr}$ between a bare steel bar and an embedded steel bar at crack initiation is obtained from Figure 6(a) as

$$\Delta\varepsilon_{sr} = \frac{\sigma_{sr}}{E_s} - \varepsilon_{sr} = \frac{f_t}{E_s\, \rho_{eff}}, \qquad (32)$$

where use of (30) has been made. The stress increase $\sigma_{s,TSE}$ is determined as

$$\sigma_{s,TSE} = \sigma_{srn} - E_s\, \varepsilon_{srn} = E_s\, \beta_t\, \Delta\varepsilon_{sr}, \qquad (33)$$

where use of $\varepsilon_{srn} = (\sigma_{srn}/E_s - \beta_t\, \Delta\varepsilon_{sr})$ has been made according to Figure 6(a). Following [CEB-FIP(1990)] $\sigma_{srn} = 1.30\, \sigma_{sr}$ and $\beta_t = 0.4$ can be assumed.

Substituting (33) into (29) yields the respective concrete stress $\sigma_{c,TSE} = \beta_t\, f_t$ for a tension stiffening model related to the concrete (Figure 6(b)).

The stress-strain relationship of Figure 6(b) has to be adapted to a relationship within the theory of plasticity (Figure 6(c)) by subtracting the

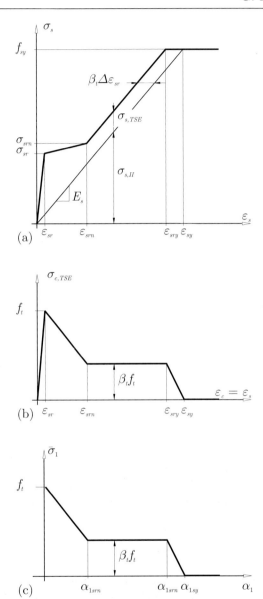

Figure 6. Modelling of the tension stiffening effect by modifying the constitutive relation for (a) the reinforcing steel bar, (b) the concrete, (c) equivalent concrete stress-strain relationship adapted to the elastic-plastic concrete model.

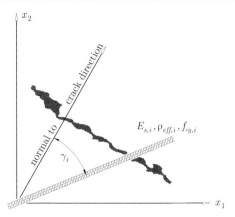

Figure 7. Reinforcing bar enclosing the angle γ_i with the crack normal.

elastic concrete strain $\beta_t f_t/E$ from the total strain values ε_{srn} and ε_{sry}. Finally, the equivalent concrete stress-strain relationship for the elastic-plastic concrete model is obtained as

$$q_1(\alpha_1) = f_t \left[1 - (1-\beta_t) \frac{\alpha_1}{\alpha_{1srn}} \right] \quad \text{for} \quad \alpha_1 < \alpha_{1srn},$$
$$q_1(\alpha_1) = \beta_t f_t \quad \text{for} \quad \alpha_{1srn} \leq \alpha_1 \leq \alpha_{1sry}, \quad (34)$$
$$q_1(\alpha_1) = \beta_t f_t \frac{\alpha_{1sy} - \alpha_1}{\alpha_{1sy} - \alpha_{1sry}} \quad \text{for} \quad \alpha_{1sry} < \alpha_1 \leq \alpha_{1sy}.$$

This uniaxial tension stiffening model can be extended to reinforced concrete subjected to biaxial stress states by assuming that the effective extensional stiffness per unit area of the reinforcement in the direction normal to the crack can be computed by the transformation of a fourth order tensor

$$E_s \rho_{eff} = \sum_{i=1}^{n} E_{s,i} \cos^4 \gamma_i \, \rho_{eff,i}. \quad (35)$$

γ_i denotes the angle between the direction normal to the crack and the direction of the reinforcing bar i (Figure 7) and n is the number of reinforcing bars with different directions. Since the tension stiffening effect will be neglected when the yield strength of the reinforcing steel has been reached, α_{1sy} has to be modified by $\alpha_{1sy} = (f_{sy}/E_s)/\cos^2\gamma$. In case of several reinforcement bars α_{1sy} is determined from the reinforcement bar, which plays the dominant role for the tension stiffening effect. Finally, the transition between reinforced concrete and plain concrete is modelled by using the minimum reinforcement ratio according to [CEB-FIP(1990)]. Whereas the starting point for deriving the tension softening law of plain concrete was the stress-displacement relation (16), the tension stiffening relation (34) is directly formulated in terms of a stress-strain law. This is a consequence of the fact that in a plain concrete specimen, irrespective of the specimen length, only one crack will be formed, whereas in a reinforced concrete specimen the number of cracks increases with increasing specimen length according to the bond properties between the reinforcing bar and the concrete. Hence, only in the latter case crack widths can be directly distributed or smeared along the specimen length, yielding a fictitious crack strain. For reinforced concrete with less than the minimum reinforcement ratio according to [CEB-FIP(1990)] the steel is not able to carry the tensile forces after the tensile strength of concrete has been attained. Hence, failure of such slightly reinforced concrete members is governed by the material behavior of plain concrete. Including the minimum reinforcement ratio in the computation of the effective extensional stiffness per unit area of the reinforcement as a lower limit for tension stiffening allows to model the transition from plain to reinforced concrete, taking into account the direction of the crack with respect to the direction of the reinforcement.

2.3 Validation

Biaxial concrete tests In the tests on plate type concrete specimens with dimensions of $200 \times 200 \times 50$ mm, documented in [Kupfer(1969)], monotonically increasing biaxial loads with constant stress ratios σ_1/σ_2 were applied. The material parameters are taken as $f_c = 31.10$ MPa, $f_t = 2.84$ MPa, $E = 33000$ MPa and $\nu = 0.195$. The fracture energies for tensile and compressive failure are assumed as 0.1 N/mm and 10 N/mm, respectively. The results of the numerical simulation of some tests with combined tension-compression loading are shown in Figure 8.

The comparison of the experimentally and numerically determined compressive stress-strain curves demonstrates good correspondence between numerical and experimental results. Beginning with the stress ratio $\sigma_1/\sigma_2 = 0.0/-1.0$ and ending with the ratio $\sigma_1/\sigma_2 = 0.204/-1.0$ the ultimate compressive stress decreases as the applied tensile stress in lateral direction is increased (Figure 8(a)). The influence of damage coupling can be seen in Figure 8(b) in terms of the maximum attainable tensile stress, which decreases with increasing compressive stress. If the coupling of compressive and tensile damage was neglected, then irrespective of the stress ratio σ_1/σ_2 for mixed tension-compression loading the uniaxial tensile strength would

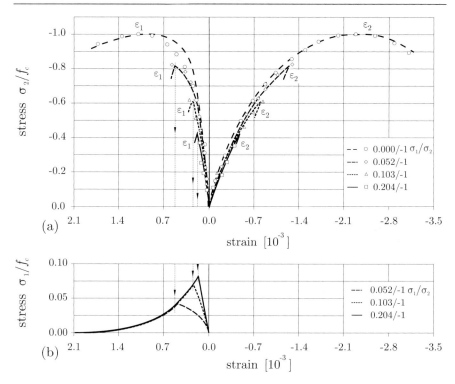

Figure 8. Comparison of computed stress-strain curves for combined tension-compression loading with experimental results by [Kupfer(1969)]: (a) compressive stresses, (b) tensile stresses.

be reached, which would be in contradiction with the observed material behaviour.

L-shaped panel tests Laboratory tests on L-shaped panels, made of plain and reinforced concete, were performed at the University of Innsbruck. They are documented in [Winkler(2001a)] and [Winkler(2001b)]. In Figure 9 the geometric properties of the L-shaped panels and the boundary conditions are shown. The long and the short edges are given as 500 mm and 250 mm, the thickness is 100 mm. The lower horizontal edge of the vertical leg is fixed. A vertical line load, acting opposite to the direction of gravity, was applied on the lower horizontal surface of the horizontal leg at a distance of 30 mm from the vertical end face. Actually, the experiments were conducted by controlling the vertical displacement of the load point

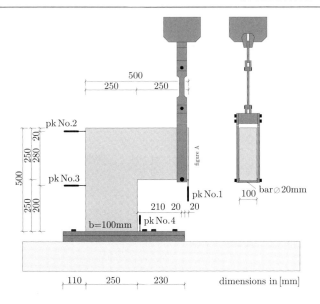

Figure 9. Test setup for the laboratory tests on L-shaped panels made of plain and reinforced concrete.

and the corresponding force was measured. The constant displacement rate was chosen as 0.02 mm/min. As shown in Figure 9 the displacements were measured at four different points by inductive pick-ups during the execution of the tests.

Four different series of experiments with three tests on identical panels for each series were performed. In series A the behaviour of plain concrete was investigated whereas in the other three series L-shaped panels made of reinforced concrete with different layouts of the reinforcement were tested. In test series B, the reinforcement consisted of four steel bars, in test series C a welded orthogonal reinforcing grid was used with the bars aligned to the edges of the panel and in test series D the bars of the reinforcing grid enclosed an angle of 45° with the edges of the panel. The nominal diameter of the applied reinforcing steel bars was 6 mm. The wire meshes, made of heat-treated, ductile ribbed steel, had a 50 mm grid spacing. The reinforcement was placed in the middle surface of the L-shaped panels with a concrete cover of 25 mm with respect to the edges. The test setup and the experimental results for the four test series are given in [Winkler(2001b)].

In the following, results of the numerical simulations for test series A, C and D are reported. The material parameters for those tests, determined from 3 samples for each test series, are summarized in Table 1. In the latter f_{st} respresents the tensile strength of the reinforcing steel.

Table 1. Uniaxial material parameters for concrete and steel.

	test series A	test series C	test series D
concrete			
f_c	31.0 MPa	38.9 MPa	29.45 MPa
f_t	2.7 MPa	2.7 MPa	2.65 MPa
E	25850 MPa	29300 MPa	26075 MPa
G_f	0.065 N/mm	0.074 N/mm	0.063 N/mm
steel			
f_{sy}		533.2 MPa	526.3 MPa
f_{st}		597.2 MPa	584.5 MPa
E_s		201580 MPa	179073 MPa

For the numerical simulations three different regular meshes are generated, the second and the third mesh representing consistent refinements of the first and the second mesh, respectively. For the L-shaped panel made of plain concrete the first mesh consists of 300 quadrilateral four-node elements with dimensions of 25×25 mm and, consequently, the second and third mesh contain 1200 and 4800 elements, respectively. In addition, meshes with triangular three-node elements are employed. The latter meshes are chosen such that two triangular elements are equivalent with one quadrilateral element.

For the reinforced L-shaped panels coarser regular meshes are employed, consisting of 75, 300 and 1200 four-node quadrilateral elements. Hence, the elements of the coarse mesh are characterized by dimensions of 50×50 mm. The reinforcement is modelled by means of rebar-elements, i.e. one-dimensional elements, taking into account elastic-plastic material behaviour.

Concerning test series A with the L-shaped panels made of plain concrete, the ultimate load is reached when the first crack is formed (Figure 10). Further increase of the vertical displacement of the point of load application led to a decrease of the load. The ultimate load, computed on the basis of the three meshes, is insensitive to the element size (Figure 11). In addition, the computed post-peak behaviour shows both good correspondence with the experimental results and good correspondence between the computed results. Hence, the numerical results can be viewed as objective with respect to the chosen discretization.

The predicted smeared cracking can be displayed by the internal variable

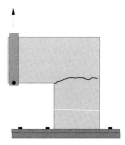

Figure 10. Observed crack pattern for the tests on L-shaped panels made of plain concrete (test series A).

α_1, which represents an indicator for tensile damage. For test series A α_1 is shown in Figure 12. The material behaviour of those elements, which are shown in black colour, is characterized by tension softening (i.e. $\alpha_1 > 0$), thus indicating cracking. It follows from this figure that there is a clear tendency to predict crack paths following a particular element row (this shortcoming of smeared crack models is denoted as mesh bias) and that very fine meshes are required to achieve a satisfactory representation of an individual crack.

For the reinforced panels experimental and numerical results are compared in Figures 13 and 14, respectively. In contrast to the L-shaped panels made of plain concrete the steel bars of the reinforced panels of test series C and D carry the tensile forces after the first crack in the concrete had developed. Subsequently, further cracks form until the ultimate load is attained by yielding of the reinforcement. The computed crack propagation agrees well with the experimental one. The L-shaped panels of test series D failed by rupture of a reinforcing steel bar, the obtained ultimate load being lower than for the L-shaped panels reinforced with an orthogonal grid aligned to the edges of the panel. As soon as the first crack was formed, the structural behaviour of the panels of test series D was softer than for the panels of test series C. This effect, which is due to the acute angle between the reinforcing bars and the cracks in test series D, is well represented by the employed material model for reinforced concrete.

2.4 Applications

Precast segmental tunnel lining The constitutive model for concrete is applied to the analysis of the installation process of a permanent tunnel lining, consisting of hexagonal precast segments. The tunnel lining is com-

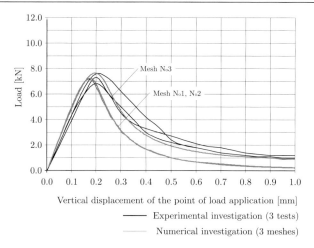

Figure 11. Load-displacement curves for the tests on L-shaped panels, made of plain concrete (test series A).

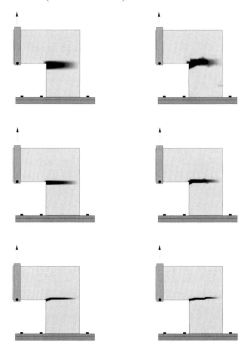

Figure 12. Computed regions of tensile damage for the L-shaped panels made of plain concrete (i.e. $\alpha_1 > 0$) on the basis of quadrilateral elements (left: meshes 1, 2 and 3) and triangular elements (right: meshes 1, 2 and 3).

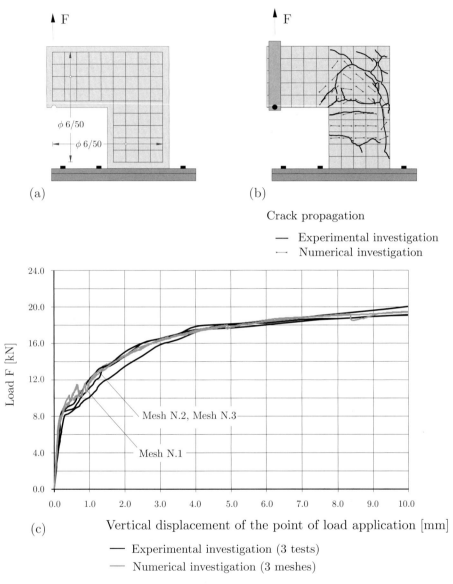

Figure 13. Results of tests on reinforced L-shaped panels with bars aligned to the specimen edges (test series C): (a) specimen geometry and reinforcement layout, (b) computed versus observed crack pattern, (c) load-displacement curves.

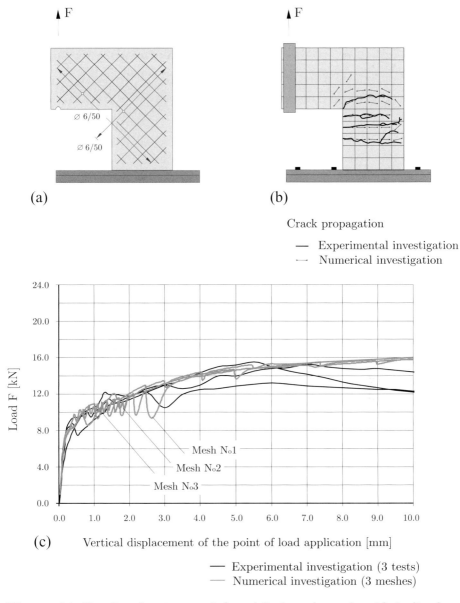

Figure 14. Results of tests on reinforced L-shaped panels with inclined bars (test series D): (a) specimen geometry and reinforcement layout, (b) computed versus observed crack pattern, (c) load-displacement curves.

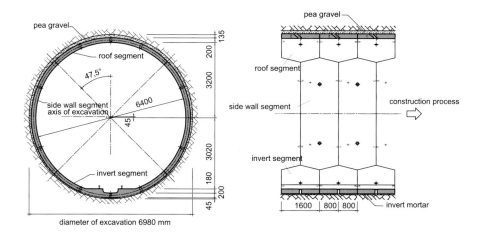

Figure 15. Cross section (left) and longitudinal section (right) of the tunnel lining (Reference: Archives of the Vorarlberger Illwerke AG).

posed of four segments per ring of uniform hexagonal shape with a thickness of 200 mm [Vigl(2000)]. The invert segment contains an invert culvert, invert pads and an invert grouting hole. The side wall segments and the roof segment are nearly identical, differing only in the arrangements of the required grouting holes (Figure 15).

The segments are installed under the protection of the tail shield of a double shield tunnel boring machine (TBM) and are kept in position by the help of the contact pressure of the TBM. The invert section of the lining is backfilled with mortar in order to stabilize the circular lining ring. After installation the gap between the remaining part of the lining and the rock mass is backfilled with pea gravel. The backfilled gravel provides a certain instantaneous bedding stiffness between the installed segments and the rock mass. Such a segmental lining system is designed to act as an arch load bearing system, which predominantly carries axial compression forces. Hence, only a minimum quantity of reinforcement is needed for the loads, carried during production, handling, transport and installation.

During the construction work hairline cracks were detected on the inner surface of some precast segments, running parallel to the longitudinal axis of the tunnel lining. They were supposed to be mainly caused by the installation process of the lining. In order to gain more information about the origin of these cracks, a nonlinear numerical analysis of the installation process of the lining is performed.

Taking advantage of symmetry only half of the permanent tunnel lining is modelled using continuum elements for the precast segments and the backfilled gravel (Figure 16a). Concerning the width of the precast segments of 1600 mm and the gaps between the single segments in the longitudinal direction of the tunnel lining a plane stress state is assumed. The diameter of the excavation amounts to 6.98 m, the inner diameter of the tunnel lining to 6.40 m. The axis of the excavation is located 45 mm above the axis of the tunnel lining.

The generated mesh consists of 1158 finite elements, using 822 elements for the discretization of the precast segments and 336 finite elements for the discretization of the backfilled gravel (Figure 16a). The employed elements are eight-node quadrilateral elements with reduced integration. The contour of the rock mass is modelled as a rigid surface. The mechanical connection between the precast segments and the backfilled gravel and between the backfilled gravel and the rock mass, respectively, is modelled by contact surfaces, allowing relative displacements between the precast segments and the gravel and between the gravel and the rock mass, respectively, disregarding possible effects due to friction. Assuming a considerable stiffness and stabilization effect due to the gearing of the segments with respect to the direction parallel to the longitudinal axis of the tunnel lining, the joints between the roof segments and the side wall segments and between the side wall segments and the invert segments are not modelled.

The reinforcement of the segments is represented by means of rebar-elements, taking into account elastic-plastic material behaviour. However, since only a minimum quantity of reinforcement is provided, required for the loads carried during production, handling, transport and installation, failure is governed by the material behaviour of plain concrete.

For the precast segments concrete qualities C30/37 and C35/45 are used. The uniaxial tensile strength is given as 3.00 MPa. According to [CEB-FIP(1990)], the fracture energy for tensile failure is assumed as 0.095 N/mm.

Concerning the installation process there are two essential loadings, namely the dead weight of the segments, given as 25 kN/m^3, and the dead weight of the backfilled gravel, given as 19 kN/m^3.

Considering the installation procedure of the segments, in a first step the segments are kept in position by the help of the contact pressure of the TBM. This is modelled by fixing the continuum elements, representing the

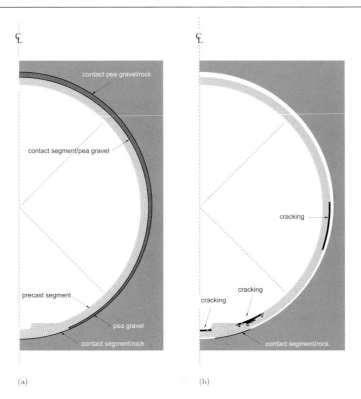

Figure 16. Tunnel lining: (a) discretization, (b) computed tensile damage.

precast segments. During this fixation step the dead weight of the segments and the dead weight of the backfilled gravel are activated. In the following step the fixations of the segments are removed. Thus, the dead weight of the segments and the dead weight of the backfilled gravel are stressing the already stabilized tunnel lining. Neglecting geometric imperfections of the lining this load case results only in small bending moments acting in the invert segment and in the roof segment, respectively. Assuming a completely backfilled and compacted gravel layer and the regularly bedding of the segments the predicted tensile stresses are smaller than the uniaxial tensile strength. Hence, no cracks should appear.

However, the observed hairline cracks on the inner surface of some precast segments, running parallel to the longitudinal axis of the tunnel lining, indicate additional loads acting during the installation process, which cause the mentioned cracks.

Thus, a further numerical analysis is performed dealing with the situation before the gap between the lining and the rock mass will be backfilled with gravel. Concerning the installation of the segments they are kept in position by the help of the contact pressure of the TBM. Due to this pressure and the gearing of the segments with respect to the direction parallel to the longitudinal axis of the tunnel lining a considerable part of the dead weight of the segments will be transferred to the neighbouring ring of segments, which already has been stabilized by the backfilled gravel. The remaining part of the dead weight acts on the invert segment, which is not embedded at that stage. Considering this load case, additional bending moments will arise in the invert segment.

For the numerical simulation of this load case the already described finite element model for the tunnel lining is used, just disregarding the backfilled gravel and the mortar bedding of the invert segment. Neglecting the load transfer to the neighbouring ring of segments due to the gearing effects with respect to the longitudinal direction, the structure is loaded by the total dead weight of the segments, applying the latter in incremental steps.

In Figure 16b the material damage after application of the total dead weight is displayed in terms of the internal variable α_1, which represents the indicator for tensile damage. The material behaviour of those elements, which are shown in black colour, is characterized by tension softening, i.e., $\alpha_1 > 0$.

In the numerical simulation the first cracks are predicted at the inner surface of the invert segment. These cracks are observed at 52% of the total dead weight of the segments and are located at the cantilevering part of the invert segment. At 55% of the total dead weight additional cracks are predicted. These cracks appear at the inner surface of the invert segment in the region of the invert culvert. Since they are caused by stress concentrations at the corner of the invert culvert, they will be prevented in practice by rounding off the corner region. Finally, at 95% of the total dead weight of the segments, cracks are predicted at the outer surface of the side wall segment.

Concerning the results of the FE-analysis the numerically determined crack propagation corresponds well with the cracks observed at the construction site. The computed cracking loads are in agreement with the crude assumption, that approximately half of the dead weight of the segments will be transferred to the already stabilized neighbouring ring of segments, otherwise cracks will occur. This explains, why hairline cracks are detected only for some precast segments.

Reactor safety containment structure The Bhabha Atomic Research Centre (BARC) in Trombay, India, conducted a Round Robin analysis program for predicting the structural response of a model of a 540 MWe pressurized heavy water reactor primary containment. This primary (inner) containment structure consists of a pre-stressed concrete shell, which is composed of a cylindrical wall and a spherical dome. In the case of malfunctions in which pressurized gases or steam may be discharged, it prevents the release of radioactive materials into the atmosphere. A design specification requires that for prescribed design accidents and earthquakes no tensile stresses in the concrete are admitted at the interior surface of the containment structure. However, if the internal pressure increases beyond the design value, then the concrete will be subjected to tensile stresses, followed by the initiation of cracks in the concrete, which will propagate from the inner to the outer surface. Upon further increase of the internal pressure, plastic strains will develop in the reinforcement and the pre-stressing tendons. Finally, the containment structure will collapse due to rupture of pre-stressing tendons.

The containment model on a scale of 1:4 has an inner diameter of 12.376 m and a height of 15.750 m excluding the foundation (Figure 17). The concrete is characterized by a compressive strength of 45 MPa (determined on cubic specimens), a direct tensile strength of 2.78 MPa and a modulus of elasticity of 33540 MPa. The maximum aggregate size is 12.5 mm. The thicknesses of the cylindrical wall and of the spherical dome are 0.188 m and 0.164 m, respectively. At the base of the cylindrical wall up to 1.2 m from the floor its thickness is increased to 0.350 m. At the intersection of the cylindrical wall and the spherical dome the structure is stiffened by a ring beam with a rectangular cross section of 0.350×1.231 m.

The containment structure is prestressed with mono-strand tendons. The tendons have a cross-sectional area of 142.8 mm^2 and the modulus of elasticity, the yield strength at 0.1% permanent strain and the tensile strength are given as 189600 N/mm^2, 1649 MPa and 1848 MPa, respectively. For each of the tendons the prestressing force amounts to 210 kN. In the dome the tendons are placed in orthogonal directions at distances of 100 mm. They are anchored in the ring beam. In the cylindrical wall vertical tendons are placed at distances of 224 mm and tendons in circumferential direction are characterized by distances of 112 mm. The vertical tendons in the cylindrical wall are anchored in the ring beam and in the foundation. The tendons placed in hoop direction of the cylindrical wall are anchored in four vertical buttresses (Figure 17).

The material parameters of the reinforcement are given as $f_{sy} = 415$ MPa, $f_{st} = 590$ MPa and $E_s = 200000$ MPa, respectively. In the cylindri-

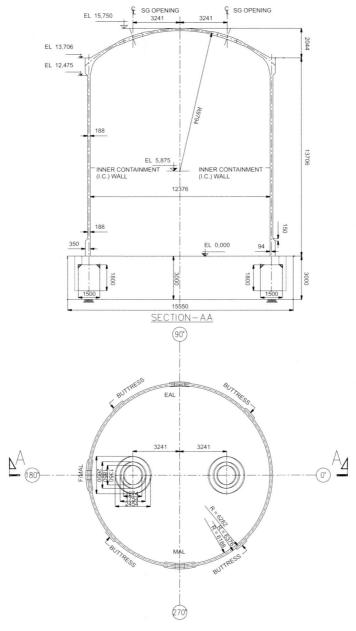

Figure 17. BARC Containment model (taken from the model documents by BARC): Vertical section (top) and plan view (bottom).

cal wall the basic reinforcement consists of an inner and outer layer with orthogonal bars of 12 mm diameter at distances of 200 mm. In the dome the basic reinforcement consists of an inner and outer layer with orthogonal bars of 12 mm diameter at distances of 100 mm.

Several openings are provided in the shell structure. The cylindrical wall contains the main air lock (MAL in Figure 17), the fuelling machine air lock (F/MAL in Figure 17) and the emergency air lock (EAL in Figure 17). In the spherical dome there are two steam generator openings. In the vicinity of the openings the shell thickness is increased and additional reinforcement is provided.

The containment model is tested by BARC for a design pressure of $p_d = 140$ kN/m^2 by applying compressed air. The tests are planned by BARC as follows: (i) calibration of sensors during a low pressure test up to $0.5\,p_d$, (ii) proof test up to $1.1\,p_d$, (iii) integrated leakage rate test up to $1.0\,p_d$, (iv) over pressure test up to functional failure, (v) ultimate load test.

The complexity of the containment model results from different section thicknesses, different section layouts, different offsets of stiffeners from the shell midsurface, different directions of the reinforcement and tendons, different diameters of the reinforcing bars and different distances between reinforcing bars and tendons. To manage all these different geometric and material parameters within one model, the structure is subdivided into 82 regions, assuming uniform properties for each region, i.e. the section thickness, the material parameters, the amount and spacing of the reinforcement bars and tendons do not change within a particular region. In Figure 18 the 82 regions are illustrated with different grey shades. Each region is discretized by finite elements. Layered 8-node shell elements with 9 integration points through the thickness are used for the cylindrical wall and the spherical dome. The reinforcement and the tendons in the cylindrical wall and the spherical dome are modelled with rebar layers embedded into the shell elements. The prestressing forces are taken into account by applying initial stresses in the tendons prescribed as 1478 MPa, amounting to 80% of the tensile strength. These rebar layers are parallel to the reference surface of the shell elements. Hence, stirrups are not considered in the model. The ring beam is modelled by 3-node beam elements with 5×5 integration points through the cross-section and 3-node truss elements representing the tendons placed in the ring beam. Figure 18 shows the finite element mesh of the whole containment model. It consists of about 37000 elements with 875000 degrees of freedom in total.

At the bottom of the containment, between the stiffened base of the cylindrical wall and the foundation, the displacements and rotations are assumed to be fully constrained. This assumption is based on the layout of

Figure 18. Containment model: 82 regions (left) and FE-mesh (right).

the reinforcement extending from the foundation into the cylindrical wall.

In the first step of the numerical simulation, the dead load and pre-stressing forces are applied. In the second step, the containment is loaded by a uniformly distributed pressure at the inner surface, which is increased step by step. Cracks are predicted in the numerical simulation already for the containment structure subjected to dead-load and pre-stressing forces. Cracked regions are located at the outer surface of the cylindrical wall in the vicinity of the base and the ring beam (Figure 19 top, left). At the design pressure small additional cracks are predicted in the vicinity of the openings (Figure 19 top, right). If the pressure is increased to twice the design pressure, then the cracked regions around the openings propagate and new cracks are predicted at the inner surface of the dome next to the ring beam (Figure 19 middle, left). At the pressure of 2.5 p_d, the cracked regions are propagating from the main air lock towards the vertical buttresses. Cracks are also present at the inner surface of the ring beam and the adjacent regions of the inner surface of the dome (Figure 19 middle, right). Further increase of the internal pressure results in cracking of the vertical buttresses on each side of the main air lock. At 2.75 p_d, the cylindrical wall is almost completely cracked through the thickness of the wall (Figure 19 bottom, left). Upon further small increase of the internal pressure the deformations are increasing significantly and the predicted crack strains are propagating rapidly (Figure 19 bottom, right) until the failure load is reached at about 2.91 p_d.

Figure 19. Cracked regions at six different pressure states: 0 p_d and 1.0 p_d (top), 2.0 p_d and 2.5 p_d (middle), 2.75 p_d and 2.91 p_d (bottom); the displacements are magnified by a factor of 100.

3 Plasticity Based Crack Model with Embedded Discontinuities

Two different FE-methods dealing with strong discontinuities for representing cracks can be distinguished: The extended finite element method (XFEM) and the elements with embedded discontinuities. The former is characterized by a nodal enrichment of the displacement field in order to represent the discontinuous displacements at a crack, whereas the latter is characterized by an elemental enrichment, typically with elimination of the additional degrees of freedom at element level. A systematic comparison of both methods is contained in [Jirasek(2002)] and [Oliver(2006)]. In the following a crack model is proposed, which is based on the strong discontinuity approach and formulated within the framework of elements with embedded discontinuities. A simple linear triangular element is used as the underlying finite element. This embedded crack model is then combined with a smeared rotating crack model. In the resulting model, the early stages of crack development, characterized by smeared microcracking, are described by the smeared crack model and a discontinuity is embedded in the displacement field of the respective finite element only, if the crack opening, predicted by the smeared crack model, attains a prescribed threshold value [Jirasek(2001)]. This so-called delayed embedded discontinuity model allows considering crack opening in the direction normal to the crack and relative tangential displacements of the crack faces with transfer of shear forces across the crack faces.

3.1 Strong Discontinuity Kinematics

Consider a body $\mathcal{B}$ occupying the closed domain Ω with the boundary denoted by $\partial\Omega$ (Figure 20a). The boundary can be decomposed into $\partial_u\Omega$ with prescribed displacements $\boldsymbol{u}^*$ and $\partial_t\Omega$ with prescribed tractions $\boldsymbol{t}^*$. The domain Ω is subdivided into the subdomains Ω^+ and Ω^- by a smooth (i.e., C^1-continuous) internal discontinuity surface Γ, defined by its unit normal vector $\boldsymbol{n}$ pointing to Ω^+.

The displacement field $\boldsymbol{u}$ of the body $\mathcal{B}$ can be subdivided into two parts

$$\boldsymbol{u}(\boldsymbol{x}) = \hat{\boldsymbol{u}}(\boldsymbol{x}) + H_\Gamma(\boldsymbol{x}) [\![\boldsymbol{u}]\!](\boldsymbol{x}) . \tag{36}$$

$\hat{\boldsymbol{u}}(\boldsymbol{x})$ represents the regular part of the displacement field, i.e. the displacement field without the presence of a discontinuity, and $H_\Gamma(\boldsymbol{x})[\![\boldsymbol{u}]\!](\boldsymbol{x})$ represents the displacement jump due to the discontinuity Γ. Both functions $\hat{\boldsymbol{u}}(\boldsymbol{x})$ and $[\![\boldsymbol{u}]\!](\boldsymbol{x})$ are smooth, continuous functions on Ω [Samaniego(2003)], the discontinuity of the displacement field in (36) results from the use of the Heaviside function $H_\Gamma(\boldsymbol{x})$ centered on the discontinuity surface Γ, i.e.

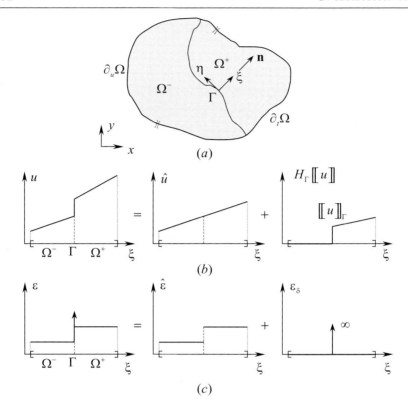

Figure 20. Strong discontinuity kinematics: (a) body occupying the domain $\Omega = \Omega^+ \cup \Omega^-$ with a discontinuity surface Γ; decomposition of (b) the displacements and (c) the strains.

$H_\Gamma(\boldsymbol{x}) = 0 \ \ \forall \ \boldsymbol{x} \in \Omega^-$ and $H_\Gamma(\boldsymbol{x}) = 1 \ \ \forall \ \boldsymbol{x} \in \Omega^+$. The magnitude of the displacement jump across the discontinuity surface $[\![\boldsymbol{u}]\!]_\Gamma$ is given by the value of $[\![\boldsymbol{u}]\!](\boldsymbol{x})$ at the discontinuity as

$$[\![\boldsymbol{u}]\!]_\Gamma = [\![\boldsymbol{u}]\!](\boldsymbol{x}) \quad \forall \ \boldsymbol{x} \in \Gamma \ . \tag{37}$$

For a local coordinate system η, ξ aligned to the discontinuity Γ (Figure 20a) the decomposition of the displacement field into the regular part $\hat{\boldsymbol{u}}(\boldsymbol{x})$ and the jump term $H_\Gamma(\boldsymbol{x}) [\![\boldsymbol{u}]\!](\boldsymbol{x})$ is shown for the section $\eta = 0$ across Γ in Figure 20b.

From (36) the strain field follows from the linearized kinematic relations

$\varepsilon = \nabla^{sym} u = [\nabla u + (\nabla u)^T]/2$ as

$$\varepsilon(x) = \nabla^{sym} \hat{u} + H_\Gamma \nabla^{sym} [\![u]\!] + \delta_\Gamma ([\![u]\!] \otimes n)^{sym} \quad (38)$$

with $\nabla H_\Gamma = \delta_\Gamma n$, where δ_Γ represents the Dirac delta function centered on Γ. The strain field (38) can be subdivided into a regular part $\hat{\varepsilon}$ and a singular part ε_δ as

$$\hat{\varepsilon} = \nabla^{sym} \hat{u} + H_\Gamma \nabla^{sym} [\![u]\!] \,, \quad \varepsilon_\delta = \delta_\Gamma ([\![u]\!] \otimes n)^{sym} \,. \quad (39)$$

The unbounded nature of ε_δ emerges from the gradient of the Heaviside function. The decomposition of the strains is shown in Figure 20c. Note that the regular and singular part of the strain tensor (Figure 20c) do not directly emanate from the respective part of the displacement field (Figure 20b).

Next, a subdomain $\Omega_\varphi \subset \Omega$ is defined, which is subdivided by Γ into parts Ω_φ^- and Ω_φ^+ (Figure 21a). Ω_φ serves as the support of a function $\varphi(x)$ chosen as

$$\varphi(x) = \begin{cases} 0 & \forall x \in \Omega^- \setminus \Omega_\varphi^- \\ 1 & \forall x \in \Omega^+ \setminus \Omega_\varphi^+ \\ C^0\text{-continuous in interval } [0,\ 1] & \forall x \in \Omega_\varphi \end{cases} \quad (40)$$

and a continuous displacement field

$$\bar{u}(x) = \hat{u}(x) + \varphi(x) [\![u]\!](x) \quad (41)$$

is introduced. Inserting $\hat{u}(x)$, following from (41), into (36) yields (Figure 21b) [Oliver(1996)]

$$u(x) = \bar{u}(x) + M_\Gamma(x) [\![u]\!](x) \quad \text{with} \quad M_\Gamma(x) = H_\Gamma(x) - \varphi(x) \,, \quad (42)$$

from which the strain field

$$\begin{aligned}\varepsilon(x) = & \nabla^{sym} \bar{u}(x) + M_\Gamma(x) \nabla^{sym} [\![u]\!](x) - ([\![u]\!](x) \otimes \nabla \varphi(x))^{sym} \\ & + \delta_\Gamma(x) ([\![u]\!](x) \otimes n)^{sym}\end{aligned} \quad (43)$$

is obtained.

If in the context of a finite element formulation the displacement jump is assumed to be constant within a finite element, then in (43) the term $\nabla^{sym} [\![u]\!](x)$ vanishes. In this case the regular part $\hat{\varepsilon}$ and the singular part ε_δ of the strain field (43) read as

$$\hat{\varepsilon} = \nabla^{sym} \bar{u}(x) - ([\![u]\!](x) \otimes \nabla \varphi(x))^{sym} \quad (44)$$
$$\varepsilon_\delta = \delta_\Gamma(x) ([\![u]\!](x) \otimes n)^{sym} \,. \quad (45)$$

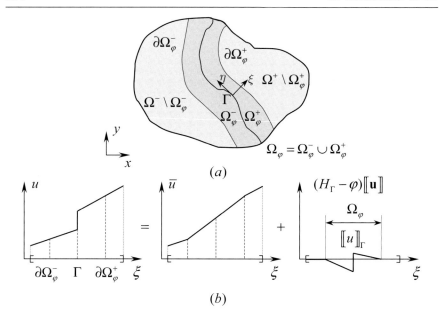

Figure 21. Reformulation of the strong discontinuity kinematics: (a) domain Ω containing a subdomain Ω_φ with a discontinuity surface Γ, (b) decomposition of the displacements.

The second term on the right hand side of the expression for the regular strains (44) can be interpreted as enhanced strains according to the enhanced assumed strain concept.

Figure 22 illustrates the continuous displacement field (41), the total displacement field (42) and the strain field (43) for a simple bar of length ℓ with a displacement discontinuity $[\![\boldsymbol{u}]\!]_\Gamma$ at $\ell/2$, fixed at the left end and subjected to an axial displacement u_r at its right end. In this figure the bar covers the subdomain $\Omega_\varphi^- \cup \Omega_\varphi^+$. Thus, the boundaries $\partial\Omega_\varphi^-$ and $\partial\Omega_\varphi^+$ coincide with the bar ends.

For a 2D problem the displacement jump (37) across the discontinuity surface Γ can be expressed as

$$[\![\boldsymbol{u}]\!]_\Gamma = \zeta_1\,\boldsymbol{n} + \zeta_2\,\boldsymbol{t} \qquad (46)$$

with ζ_1 denoting the crack opening in the direction $\boldsymbol{n}$ and ζ_2 as the relative tangential displacement of the crack faces in the direction of the tangent vector $\boldsymbol{t}$ to the crack, i.e. $\boldsymbol{t}\cdot\boldsymbol{n} = 0$. In the context of the finite element for-

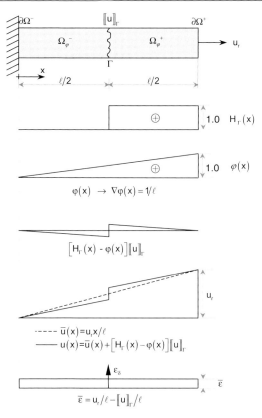

Figure 22. Illustration of the strong discontinuity kinematics for a bar subjected to uniaxial tension.

mulation, presented subsequently, a constant displacement jump is assumed within a finite element.

Substitution of (46) into the regular part of the strains (44) yields

$$\hat{\varepsilon} = \boldsymbol{\nabla}^{sym} \bar{\boldsymbol{u}}(\boldsymbol{x}) - [\boldsymbol{n} \otimes \boldsymbol{\nabla}\varphi(\boldsymbol{x})]^{sym} \zeta_1 - [\boldsymbol{t} \otimes \boldsymbol{\nabla}\varphi(\boldsymbol{x})]^{sym} \zeta_2 . \quad (47)$$

In (47), the first term on the right hand side contains the standard expression for the strains, whereas the second and third term can be interpreted as enhanced strains due to a discontinuity in the displacement field.

3.2 Determination of the displacement jump and the stresses

The Rankine criterion
$$\sigma_1 = f_t \qquad (48)$$
with f_t as the tensile strength and $\sigma_1 > 0$ as the maximum principal tensile stress is used for predicting the initiation and the direction of a crack, the latter given by the normal vector $\boldsymbol{n}_1$ to the crack, which corresponds to the principal direction associated with σ_1. σ_1 and $\boldsymbol{n}_1$ are obtained from the spectral decomposition of the stress tensor

$$\boldsymbol{\sigma} = \sum_i \sigma_i \left(\boldsymbol{n}_i \otimes \boldsymbol{n}_i \right) \qquad (49)$$

with the principal stresses σ_i and the respective principal directions $\boldsymbol{n}_i$ (the latter should not be confused with the normal vector $\boldsymbol{n}$ to the crack surface). Within the framework of the fixed crack concept the crack direction is kept fixed as soon as cracking is initiated.

The displacement jump (46) is determined within the framework of plasticity theory. To this end the interface laws relating the tractions transferred across the discontinuity to the displacement jumps are given in the form of yield functions. The traction separation law relating the traction component t_1 in the direction $\boldsymbol{n}$ normal to the crack surface to the crack opening ζ_1 is given as

$$f_1(\boldsymbol{\sigma}, q_1) = t_1 - q_1(\zeta_1) \qquad (50)$$

with
$$t_1 = (\boldsymbol{n} \otimes \boldsymbol{n}) : \boldsymbol{\sigma} \qquad (51)$$

and q_1 as the residual tensile stress which can be transferred across the crack. It can be described, e.g., by (16).

The traction separation law relating the traction component t_2 in the tangential direction $\boldsymbol{t}$ to the crack surface to the relative tangential displacement ζ_2 of the crack faces is given as

$$f_2(\boldsymbol{\sigma}, q_2) = t_2 - q_2(\zeta_1, \zeta_2) \qquad (52)$$

with
$$t_2 = (\boldsymbol{t} \otimes \boldsymbol{n}) : \boldsymbol{\sigma} \qquad (53)$$

and q_2 as the shear stress, which can be transferred across rough crack faces. The latter is described by the approximate relationships proposed in [Walraven(1981)]. According to [Walraven(1981)], three different crack states are distinguished. The first one is characterized by relative tangential displacements $\zeta_2 \leq \zeta_2^0$ with ζ_2^0 representing a threshold value. In this state, the shear stress transferred across the crack faces is assumed to be equal to zero, i.e.,

$$q_2(\zeta_1, \zeta_2) = 0 . \tag{54}$$

In order to circumvent a vanishing shear stiffness, a linear hardening law with a small fictitious hardening modulus is employed in this case. If the relative tangential displacement ζ_2 exceeds the threshold value ζ_2^0, then the shear stress, given by

$$q_2(\zeta_1, \zeta_2) = -\frac{f_c}{30} + [1.8\zeta_1^{-0.8} + (0.234\zeta_1^{-0.707} - 0.20)f_c]|\zeta_2| \tag{55}$$

is transferred across the crack faces. In (55), f_c denotes the uniaxial compressive strength of concrete (which is defined here as a positive value) and ζ_2^0 is computed by setting $q_2 = 0$ in (55). If the relative tangential displacement exceeds a second threshold value ζ_2^1, then - in addition to shear stresses - compressive stresses will be transferred normal to the crack due to aggregate interlock. The latter are given in terms of the crack opening and the relative tangential displacement of the crack faces as

$$q_1(\zeta_1, \zeta_2) = \frac{f_c}{20} - [1.35\zeta_1^{-0.63} + (0.191\zeta_1^{-0.552} - 0.15)f_c]|\zeta_2| . \tag{56}$$

ζ_2^1 is computed by setting $q_1 = 0$ in (56). In (55) and (56) the required units for the displacements and the stresses are mm and MPa, respectively. The relations (55) and (56) are shown in Figure 23.

The strain tensor consists of the regular part (44) and the singular part (45), the latter referring to the discontinuity surface Γ. However, the stresses remain bounded as a consequence of the yield functions $f_i(\boldsymbol{\sigma}, \boldsymbol{q}) \leq 0$, $i = 1, 2$, where $\boldsymbol{q} = \lfloor q_1 \; q_2 \rfloor^T$ denotes the vector of stress-like internal variables. From the singular distribution of the strains together with the regular distribution of the stresses, the consistency parameters λ_i, appearing in Koiter's flow rule for multi-surface plasticity (7)

$$\dot{\boldsymbol{\varepsilon}}^p = \sum_i \lambda_i \frac{\partial f_i(\boldsymbol{\sigma}, \boldsymbol{q})}{\partial \boldsymbol{\sigma}} \tag{57}$$

are decomposed into regular and singular parts [Simo(1993)]

$$\lambda_i = \bar{\lambda}_i + \delta_\Gamma \lambda_{\delta, i} . \tag{58}$$

The regular parts $\bar{\lambda}_i$ are related to plastic material behaviour in $\Omega \backslash \Gamma$ which may occur irrespective of the presence of discontinuous displacements.

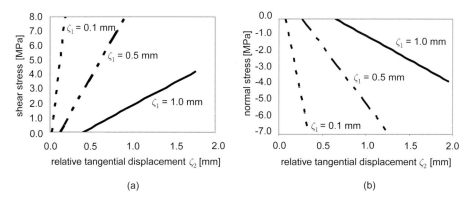

Figure 23. Traction separation laws for the shear stress and the compressive stress tranferred across a crack in terms of ζ_1 and ζ_2 according to [Walraven(1981)].

The singular parts $\delta_\Gamma \lambda_{\delta,i}$ account for singular strains at material points $x \in \Gamma$. The decomposition of the consistency parameters according to (58) is the key for applying the theory of plasticity to the solution of the boundary value problem for a continuum with displacement discontinuities.

In the present context the material in $\Omega \setminus \Gamma$ is assumed to exhibit linear elastic material behavior. Hence, plastic strains are restricted to Γ. Consequently, $\bar{\lambda}_i = 0$ and $\dot{\varepsilon}^p = \mathbf{0} \quad \forall \, x \in \Omega \backslash \Gamma$. Hence, from (57) together with (58), (50) and (52) follows

$$\dot{\varepsilon}^p = \sum_i \delta_\Gamma \lambda_{\delta,i} \frac{\partial f_i(\boldsymbol{\sigma}, \boldsymbol{q})}{\partial \boldsymbol{\sigma}} = \delta_\Gamma \left[\lambda_{\delta,1} \frac{\partial f_1(\boldsymbol{\sigma}, q_1)}{\partial \boldsymbol{\sigma}} + \lambda_{\delta,2} \frac{\partial f_2(\boldsymbol{\sigma}, q_2)}{\partial \boldsymbol{\sigma}} \right] . \quad (59)$$

Comparison of (59) with the rate forms of (43) together with (46) gives

$$\delta_\Gamma \left[\lambda_{\delta,1} \frac{\partial f_1(\boldsymbol{\sigma}, q_1)}{\partial \boldsymbol{\sigma}} + \lambda_{\delta,2} \frac{\partial f_2(\boldsymbol{\sigma}, q_2)}{\partial \boldsymbol{\sigma}} \right] =$$
$$= \delta_\Gamma \left[\dot{\zeta}_1 (\boldsymbol{n} \otimes \boldsymbol{n})^{sym} + \dot{\zeta}_2 (\boldsymbol{t} \otimes \boldsymbol{n})^{sym} \right] . \quad (60)$$

Thus, the stresses for all material points $x \in \Omega \backslash \Gamma$ are computed from the regular part $\hat{\varepsilon}$ (44) of the strain tensor (43) by employing Hooke's law as

$$\dot{\boldsymbol{\sigma}} = \mathbb{E} : \dot{\hat{\varepsilon}} . \quad (61)$$

Equating in (60) the first terms on the left and right hand side and pre-multiplying with $(\partial f_1)/(\partial \boldsymbol{\sigma}) : \mathbb{E}$ and equating in (60) the second terms on the left and right hand side, and pre-multiplying with $(\partial f_2)/(\partial \boldsymbol{\sigma}) : \mathbb{E}$, yields

$$\lambda_{\delta,1} = \dot{\zeta}_1 \frac{(\partial f_1)/(\partial \boldsymbol{\sigma}) : \mathbb{E} : (\boldsymbol{n} \otimes \boldsymbol{n})^{sym}}{(\partial f_1)/(\partial \boldsymbol{\sigma}) : \mathbb{E} : (\partial f_1)/(\partial \boldsymbol{\sigma})} = \dot{\zeta}_1 \, , \quad (62)$$

$$\lambda_{\delta,2} = \dot{\zeta}_2 \frac{(\partial f_2)/(\partial \boldsymbol{\sigma}) : \mathbb{E} : (\boldsymbol{t} \otimes \boldsymbol{n})^{sym}}{(\partial f_2)/(\partial \boldsymbol{\sigma}) : \mathbb{E} : (\partial f_2)/(\partial \boldsymbol{\sigma})} = \dot{\zeta}_2 \, , \quad (63)$$

where - following from (50) to (53) - use of

$$\frac{\partial f_1(\boldsymbol{\sigma}, q_1)}{\partial \boldsymbol{\sigma}} = \boldsymbol{n} \otimes \boldsymbol{n} \, , \qquad \frac{\partial f_1(\boldsymbol{\sigma}, q_1)}{\partial q_1} = -1 \, , \quad (64)$$

$$\frac{\partial f_2(\boldsymbol{\sigma}, q_2)}{\partial \boldsymbol{\sigma}} = \boldsymbol{t} \otimes \boldsymbol{n} \, , \qquad \frac{\partial f_2(\boldsymbol{\sigma}, q_2)}{\partial q_2} = -1 \, , \quad (65)$$

and use of the symmetry properties of $\mathbb{E}$ was made.

Within the framework of plasticity theory the rate of the strain-like internal variables $\boldsymbol{\alpha} = \lfloor \zeta_1 \, , \, \zeta_2 \rfloor^T$ is computed as

$$\dot{\boldsymbol{\alpha}} = \sum_i \lambda_i \frac{\partial f_i(\boldsymbol{\sigma}, \boldsymbol{q})}{\partial \boldsymbol{q}} \, . \quad (66)$$

The stress-like internal variables $\boldsymbol{q}$ are related to $\boldsymbol{\alpha}$ by the plastic moduli tensor $\boldsymbol{H}$ as

$$\dot{\boldsymbol{q}} = -\boldsymbol{H} \cdot \dot{\boldsymbol{\alpha}} \, . \quad (67)$$

Substitution of (66) into (67) and making use of the decomposition of the consistency parameters according to (58), together with $\bar{\lambda}_i = 0$, yields

$$\boldsymbol{H}^{-1} \cdot \dot{\boldsymbol{q}} = -\sum_i \delta_\Gamma \lambda_{\delta,i} \frac{\partial f_i(\boldsymbol{\sigma}, \boldsymbol{q})}{\partial \boldsymbol{q}} \, . \quad (68)$$

The right hand side of (68) is singular due to the Dirac-delta function. Since because of $f_i(\boldsymbol{\sigma}, \boldsymbol{q}) \leq 0$ the stress-like internal variables are bounded, the inverse of the plastic moduli tensor must be singular [Simo(1993)]. Rewriting the latter as $\boldsymbol{H}^{-1} = \delta_\Gamma \hat{\boldsymbol{H}}^{-1}$, where $\hat{\boldsymbol{H}}^{-1}$ is regular, and substituting this expression into (68) yields the evolution law for the stress-like internal variables

$$\dot{\boldsymbol{q}} = -\hat{\boldsymbol{H}} \sum_i \lambda_{\delta,i} \frac{\partial f_i(\boldsymbol{\sigma}, \boldsymbol{q})}{\partial \boldsymbol{q}} \quad \text{with} \quad \hat{\boldsymbol{H}} = \begin{bmatrix} \hat{H}_{11} & \hat{H}_{12} \\ \hat{H}_{21} & \hat{H}_{22} \end{bmatrix} . \quad (69)$$

The components of $\hat{\boldsymbol{H}}$ are computed as

$$\hat{H}_{11} = \frac{\partial q_1}{\partial \zeta_1} \, , \quad \hat{H}_{12} = \frac{\partial q_1}{\partial \zeta_2} \, , \quad \hat{H}_{21} = \frac{\partial q_2}{\partial \zeta_1} \, , \quad \hat{H}_{22} = \frac{\partial q_2}{\partial \zeta_2} \, . \quad (70)$$

The loading/unloading conditions are given as

$$\lambda_i \geq 0, \quad f_i(\boldsymbol{\sigma},\boldsymbol{q}) \leq 0, \quad \lambda_i f_i(\boldsymbol{\sigma},\boldsymbol{q}) = 0 \quad \text{(no summation over } i\text{)} \quad (71)$$

and the consistency condition reads as

$$\lambda_i \dot{f}_i = 0 \quad \text{(no summation over } i\text{)} . \quad (72)$$

For plastic loading the consistency condition for the yield function f_i yields

$$\dot{f}_i = \frac{\partial f_i}{\partial \boldsymbol{\sigma}} : \dot{\boldsymbol{\sigma}} + \frac{\partial f_i}{\partial \boldsymbol{q}} \cdot \dot{\boldsymbol{q}} = 0 . \quad (73)$$

Inserting Hooke's law (61) for the bulk material and the rate equations (69) into (73), yields

$$\frac{\partial f_1}{\partial \boldsymbol{\sigma}} : \mathbb{E} : \dot{\hat{\boldsymbol{\varepsilon}}} - \frac{\partial f_1}{\partial \boldsymbol{q}} \cdot \hat{\boldsymbol{H}} \cdot \left[\lambda_{\delta,1} \frac{\partial f_1}{\partial \boldsymbol{q}} + \lambda_{\delta,2} \frac{\partial f_2}{\partial \boldsymbol{q}} \right] = 0 ,$$

$$\frac{\partial f_2}{\partial \boldsymbol{\sigma}} : \mathbb{E} : \dot{\hat{\boldsymbol{\varepsilon}}} - \frac{\partial f_2}{\partial \boldsymbol{q}} \cdot \hat{\boldsymbol{H}} \cdot \left[\lambda_{\delta,1} \frac{\partial f_1}{\partial \boldsymbol{q}} + \lambda_{\delta,2} \frac{\partial f_2}{\partial \boldsymbol{q}} \right] = 0 , \quad (74)$$

from which

$$\lambda_{\delta,1} = \left(\frac{\hat{H}_{22}}{\det \hat{\boldsymbol{H}}} \frac{\partial f_1}{\partial \boldsymbol{\sigma}} - \frac{\hat{H}_{12}}{\det \hat{\boldsymbol{H}}} \frac{\partial f_2}{\partial \boldsymbol{\sigma}} \right) : \mathbb{E} : \dot{\hat{\boldsymbol{\varepsilon}}} \quad (75)$$

$$\lambda_{\delta,2} = \left(\frac{\hat{H}_{11}}{\det \hat{\boldsymbol{H}}} \frac{\partial f_2}{\partial \boldsymbol{\sigma}} - \frac{\hat{H}_{21}}{\det \hat{\boldsymbol{H}}} \frac{\partial f_1}{\partial \boldsymbol{\sigma}} \right) : \mathbb{E} : \dot{\hat{\boldsymbol{\varepsilon}}} \quad (76)$$

follows.

Making use of $\lambda_{\delta,1} = \dot{\zeta}_1$ and $\lambda_{\delta,2} = \dot{\zeta}_2$ according to (62) and (63), the components of the displacement jump normal and tangential to the discontinuity are obtained by substituting the rate of the regular part of the strains (47) into (75) and (76) and solving the so obtained equations for $\dot{\zeta}_1$ and $\dot{\zeta}_2$.

3.3 Finite element discretization

The body $\mathcal{B}$ in Figure 21 is subdivided into finite elements. Whereas the domain $\Omega \setminus \Omega_\varphi$ is discretized by standard constant strain triangular (CST) elements, the subdomain Ω_φ is dicretized by a band of constant strain triangles enriched with embedded discontinuities. For a particular enriched element e the displacement field (42) is approximated by

$$\boldsymbol{u}^{(e)}(\boldsymbol{x}) \approx \sum_{k=1}^{n_e} N_k^{(e)}(\boldsymbol{x}) \, \boldsymbol{d}_k + [H_\Gamma^{(e)}(\boldsymbol{x}) - \varphi^{(e)}(\boldsymbol{x})] \, [\![\boldsymbol{u}]\!]_\Gamma^{(e)} \quad (77)$$

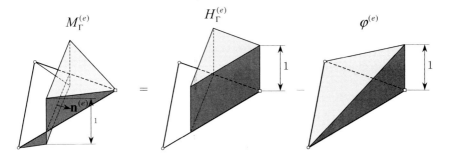

Figure 24. Element with embedded discontinuity: discontinuous interpolation function $M_\Gamma^{(e)}$ for a CST-element.

with $N_k^{(e)}$ as the standard interpolation function for node k of the CST-element and the respective vector of nodal displacements d_k; n_e denotes the number of nodes of a particular finite element, i.e. in the present case $n_e = 3$. The displacement jump $[\![u]\!]_\Gamma^{(e)}$ within a particular finite element is assumed to be constant (hence, $\nabla^s[\![u]\!] = 0$ in (43) and is neglected in the subsequent equations.) Consequently, the vector describing the displacement jump (46) is approximated as

$$[\![u]\!]_\Gamma^{(e)} = \zeta_1^{(e)} n^{(e)} + \zeta_2^{(e)} t^{(e)} \qquad (78)$$

with $\zeta_1^{(e)}$ as the displacement jump in the direction of $n^{(e)}$ and $\zeta_2^{(e)}$ as the displacement jump in the direction of $t^{(e)}$ within element e. In accordance with (40) the function φ is approximated within an element e, located in the subdomain Ω_φ, by

$$\varphi^{(e)} = \sum_{k=1}^{n_e^+} N_k^{(e)} \quad \forall\, e \in \Omega_\varphi \quad \rightarrow \quad \nabla\varphi^{(e)} = \sum_{k=1}^{n_e^+} \frac{\partial N_k^{(e)}}{\partial x}, \qquad (79)$$

where the sum extends over the standard shape functions N_k associated with those nodes of element e located at the "positive" side of $\Gamma^{(e)}$ (Figure 24). The latter is defined by the orientation of $n^{(e)}$. The regular part of the strains follows from (47) for the finite element e as

$$\hat{\varepsilon}^{(e)} \approx \sum_{k=1}^{n_e} \left(\nabla N_k^{(e)}(x) \otimes d_k\right)^{sym} -$$
$$- \left(n^{(e)} \otimes \nabla\varphi^{(e)}(x)\right)^{sym} \zeta_1^{(e)} - \left(t^{(e)} \otimes \nabla\varphi^{(e)}(x)\right)^{sym} \zeta_2^{(e)}. \qquad (80)$$

In (80) the first term on the right hand side contains the standard expression for the strains whereas the second and third term can be interpreted as enhanced strains due to a discontinuity in the displacement field.

A CST-element is decomposed by a discontinuity segment $\Gamma^{(e)}$ such that one single node, i.e. the solitary node, is separated from the remaining two nodes forming the solitary edge. Thus, only one decomposition mode is possible. The gradient (79_2) of function φ is perpendicular to the solitary edge. The uniqueness of the solution poses some restrictions on the angle included between $\boldsymbol{\nabla}\varphi$ and $\boldsymbol{n}$. E.g., for the special case of f_1 as the only active field function, from (74_1) together with (47) follows

$$\dot{\zeta}_1 = \frac{(\boldsymbol{n} \otimes \boldsymbol{n}) : \mathbb{E} : \boldsymbol{\nabla}^{sym}\dot{\boldsymbol{u}}}{(\boldsymbol{n} \otimes \boldsymbol{n}) : \mathbb{E} : (\boldsymbol{n} \otimes \boldsymbol{\nabla}\varphi)^{sym} + \hat{H}_{11}} . \qquad (81)$$

The numerator can be interpreted as the product of the normal to the yield surface (64_1) and the trial stress $\mathbb{E} : \boldsymbol{\nabla}^{sym}\dot{\boldsymbol{u}}$ which in the case of loading is a positive quantity. Thus, $\dot{\zeta}_1 > 0$ requires

$$(\boldsymbol{n} \otimes \boldsymbol{n}) : \mathbb{E} : (\boldsymbol{n} \otimes \boldsymbol{\nabla}\varphi)^{sym} + \hat{H}_{11} > 0 \qquad (82)$$

with $\hat{H}_{11} < 0$ because of tensile softening, following from substituting (16) into (70_1). For the special case of ideal plasticity with $\hat{H}_{11} = 0$ uniqueness will be lost, if $\boldsymbol{\nabla}\varphi$ is perpendicular to $\boldsymbol{n}$. In case of tensile softening, uniqueness will be lost for smaller angles enclosed by $\boldsymbol{\nabla}\varphi$ and $\boldsymbol{n}$. Such decompositions of elements by the discontinuity are denoted as badly decomposed [Borja(2000)] and must be avoided by a suitable mesh layout.

In the three-dimensional case for linear tetrahedron elements two possible decomposition modes are possible: either a single node is separated from the remaining three nodes (mode 1 in Figure 25) or two pairs of nodes are separated from each other (mode 2 in Figure 25).

3.4 Determination of the crack direction

At a material point crack initiation is predicted when – according to the Rankine criterion (48) – the maximum principal tensile stress σ_1 attains the tensile strength for the first time. The crack direction is taken as the principal direction $\boldsymbol{n}_1$ associated with the principal stress σ_1 according to the spectral decomposition of the stresses (49).

However, it has been shown by several authors, e.g. [Jirasek(2001), Mosler(2003)], that the principal stress directions computed at a particular material point are only a crude approximation to the actual ones. Especially for discretizations with finite elements characterized by low order interpolation functions a locally tortuous – although globally correct – crack-path

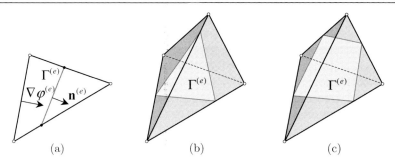

Figure 25. Decomposition modes of elements with embedded discontinuities: (a) mode for linear triangular element, (b) mode 1 and (c) mode 2 for linear tetrahedron elements.

will be obtained, which could lead to severe stress locking and, thus, to incorrect energy dissipation.

The prediction of the crack direction can be improved if the latter is computed from non-local averaged quantities. In this context non-local averaging is not used in the sense of a non-local model providing a localization limiter. It is rather used to improve the numerical representation of the principal strain directions, i.e. to obtain a "smoothing effect" for the strains. Alternatively, this could also be achieved by application of the super convergent patch recovery method [Zienkiewicz(1992)].

In general, the non-local counterpart ($\bar{\bullet}$) of some local material quantity ($\bullet$) is obtained by

$$(\bar{\bullet})(\boldsymbol{x}) = \int_\Omega \alpha(\boldsymbol{x}, \boldsymbol{\xi}) \left[(\bullet)(\boldsymbol{\xi})\right] d\Omega \qquad (83)$$

with $\alpha(\boldsymbol{x}, \boldsymbol{\xi})$ as the non-local weight function. In order to ensure a non-local field corresponding to a constant local field to remain constant even in the vicinity of the boundary $\partial\Omega$ of a finite domain Ω the restriction

$$\int_\Omega \alpha(\boldsymbol{x}, \boldsymbol{\xi}) \, d\Omega = 1 \quad \forall \, \boldsymbol{x} \in \Omega \qquad (84)$$

has to be fulfilled. It is established by rescaling the local weight function $\alpha(\boldsymbol{x}, \boldsymbol{\xi})$ as

$$\alpha(\boldsymbol{x}, \boldsymbol{\xi}) = \frac{\alpha_0(r)}{\Omega_\alpha} \qquad (85)$$

with the weighted volume Ω_α defined as

$$\Omega_\alpha = \int_V \alpha_0(r) \, d\Omega \, . \qquad (86)$$

In the present context the bell-shaped distribution function defined as

$$\alpha_0(r) = \begin{cases} \dfrac{1}{s}\left(1 - \dfrac{r^2}{R^2}\right)^2 & \text{for} \quad 0 \leq r \leq R \\ 0 & \text{for} \quad r > R \end{cases} \tag{87}$$

is employed as the weighting function. In (87) R is a parameter, which defines the volume for the averaging procedure and s is a scaling factor computed from (84) [Rolshoven(2003)]. Within the framework of the finite element method the integral in (83) is replaced by its discretized counterpart

$$(\overline{\bullet})(\boldsymbol{x}) = \sum_{i=1}^{n_{\text{int}}} \alpha(\boldsymbol{x}, \boldsymbol{\xi}_i)\, [(\bullet)(\boldsymbol{\xi}_i)]\, \Omega_i = \frac{1}{\Omega_\alpha} \sum_{i=1}^{n_{\text{int}}} \alpha_0(r_i)\, [(\bullet)(\boldsymbol{\xi}_i)]\, \Omega_i \tag{88}$$

with n_{int} denoting the number of integration points located within the sphere spanned by the radius R from some material point $\boldsymbol{x}$. Consequently, Ω_i is the volume associated with the respective integration point. The weighted volume Ω_α in (88) is defined by the discrete counterpart of (86) as

$$\Omega_\alpha = \sum_{i=1}^{n_{\text{int}}} \alpha_0(r_i)\, \Omega_i \ . \tag{89}$$

The smoothing effect achieved by non-local averaging is utilized for an improved prediction of the direction of crack-propagation by applying non-local averaging to the strain tensor $\boldsymbol{\varepsilon}(\boldsymbol{x})$. The non-local strains $\bar{\boldsymbol{\varepsilon}}(\boldsymbol{x})$ following from (88) as

$$\bar{\boldsymbol{\varepsilon}}(\boldsymbol{x}) = \frac{1}{\Omega_\alpha} \sum_{i=1}^{n_{\text{int}}} \alpha_0(r_i)\, \boldsymbol{\varepsilon}(\boldsymbol{\xi}_i)\, \Omega_i \tag{90}$$

with $\alpha_0(r)$ according to (87) are used for predicting the direction of crack propagation.

A further improvement of the predicted crack direction is achieved, if the early stages of crack evolution, characterized by distributed microcracking, are described with a smeared crack model and a discontinuity is only embedded into a finite element, when the crack opening, predicted by the smeared crack model, has reached a prescribed threshold value [Jirasek(2001), Theiner(2009)]. In this way an initial misprediction of the crack direction can be corrected as the crack evolves, until the disconinuity in the displacement field is introduced.

3.5 Crack Tracking Algorithm

It can be shown by mesh studies that discontinuity segments $\Gamma^{(e)}$ should not be placed within the finite elements arbitrarily, because, depending on the employed layout of the finite element mesh, this may result in mesh-dependent results [Jirasek(2001), Samaniego(2003)]. E.g., the band of elements, enriched by a discontinuity may deviate from the correct crack path by following mesh lines in an erroneous manner.

As a remedy, continuity of the discontinuity segments across adjacent elements should be enforced. Thus, for the two-dimensional case discontinuity segments of neighboring elements have to share a common point at the common edge of neighboring elements. For three-dimensional discretizations the discontinuity segments have to share a common line at the common face of neighboring elements.

To this end local and global crack tracking algorithms have been developed. Local crack tracking algorithms are based on geometrical considerations, tracking a particular crack element by element, whereas global crack tracking algorithms [Oliver(2002)] are based on the representation of all existing and potential new discontinuities by isolines (for 2D-problems) or isosurfaces (for 3D-problems) of some scalar field, which is obtained as the solution of a boundary value problem for the whole domain under consideration. For the present embedded crack model a so-called partial domain crack tracking algorithm is used [Feist(2006)]. It is characterized by constructing a scalar field $\theta(\boldsymbol{x})$ with isolines or isosurfaces $\theta(\boldsymbol{x}) = $ const., representing existing and potential new discontinuities only for the subset of the domain occupied by those elements actually or potentially crossed by a discontinuity.

3.6 Validation

L-shaped panel test The tests on L-shaped panels, made of plain concrete, described in section 2.3, are used for the validation of the crack model with embedded discontinuities together with the partial domain crack tracking algorithm for ensuring a continuous crack path. The test set-up and the material parameters are provided in section 2.3. Hence, in this section only results are presented.

Figure 26 shows the predicted crack path determined on the basis of four different meshes. The first and second mesh are regular meshes, consisting of 600 and 2400 linear triangular elements, the third and fourth mesh are irregular meshes, consisting of 1151 and 2910 linear triangular elements. It follows from Figure 26 that irrespective of the employed FE-mesh the computed crack paths correspond well with the observed crack path, shown in Figure 10. In addition, also the computed load-displacement curves show good agreement with the experimental spectrum of the load-displacement curves determined from three tests on identical specimens (Figure 27).

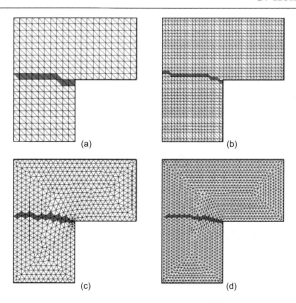

Figure 26. Elements enriched by a displacement jump (indicated grey shaded), computed on the basis of four different FE-meshes: regular meshes with (a) 600 and (b) 2400 elements; irregular meshes with (c) 1151 and (d) 2910 elements.

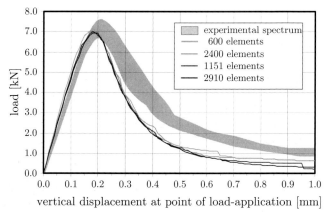

Figure 27. Comparison of measured load-displacement curves with predicted ones, computed for four different meshes.

Wedge splitting test An extensive series of wedge splitting tests on dam concrete for investigating the size dependence of fracture mechanics parameters, is documented in [Trunk(2000)]. A schematic diagram of the wedge splitting specimens is shown in Figure 28.

Here, the tests on specimens, made of dam concrete, with dimensions $H = B = 800, 1600$ and 3200 mm and a thickness of $t = 400$ mm are considered. The remaining dimensions are given as $k = 100$ mm, $s = 100$ mm and $a_0 = 375, 775$ and 1575 mm. In the vertical plane of symmetry the specimen has a notch of depth $(a_0 - s/2)$. The specimens rest on two supports, which are located below the centre of gravity of each half of the specimen. Young's modulus, Poisson's ratio, the uniaxial tensile strength and the specific fracture energy are given as $E = 28300$ MPa, $\nu = 0.18$, $f_t = 2.12, 2.11$ and 2.27 MPa and $G_f = 0.373, 0.482$ and 0.480 Nmm/mm^2.

In the tests, the relationship between the horizontal splitting force F, acting uniformly along the inner vertical faces of the specimen, and the change of the distance Δs of the points of load application, denoted as crack mouth opening displacement, was measured.

Again, in the numerical simulation continuous crack paths are achieved by means of the partial domain crack tracking algorithm. The irregular meshes for the three specimens consist of about 1500 linear triangular elements. Figure 29 depicts the predicted crack paths for the three specimens of different sizes and shows the displacements of the cracked specimens. The measured and the computed relationships between F and Δs for the three specimens of different sizes are shown in Figure 30.

Mixed mode fracture tests by Nooru-Mohamed Results from several series of mixed mode fracture tests with different load paths are reported in [Nooru-Mohamed(1992)] and [vanMier(1997)]. The square shaped, double edge notched specimens are characterized by dimensions of $200 \times 200 \times 50$ mm, a notch depth of 25 mm and a notch width of 5 mm. A schematic diagram of the specimens and of the testing arrangement, relevant to the results reported here, is shown in Figure 31(a).

Here, results from test series 4 and 6 are considered. In the three tests of test series 4, first a shear load was applied to the specimen in displacement control up to $P_s = 5, 10$ and 27.5 kN, with the axial load maintained at zero. Then an axial tensile load P_n was applied under displacement control, whilst the shear force was maintained at a constant level. In the three tests of test series 6 a shear load and an axial load were applied to the specimen simultaneously such that the ratio of the normal displacement u_n over the shear displacement u_s remains constant at values of $u_n/u_s = 1, 2$ and 3, respectively.

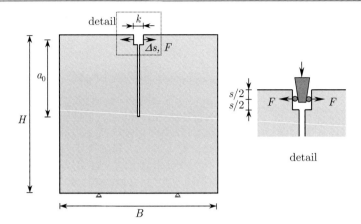

Figure 28. Test setup of the wedge splitting tests.

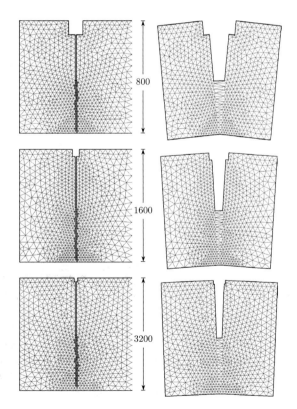

Figure 29. Predicted crack paths (left column) and displacements magnified by a factor of 50 (right column) for the three specimens of different sizes: (a) H=B=800 mm, (b) H=B=1600 mm, (c) H=B=3200 mm.

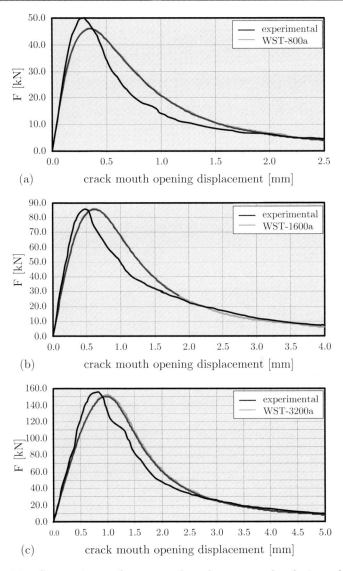

Figure 30. Comparison of measured and computed relations between the horiztonal splitting force F and the crack mouth opening displacement for the three specimens of different sizes: (a) H=B=800 mm, (b) H=B=1600 mm, (c) H=B=3200 mm.

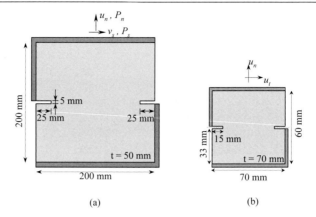

Figure 31. Test set up of mixed mode fracture tests: (a) tests by Nooru Mohamed, (b) tests by Hassanzadeh.

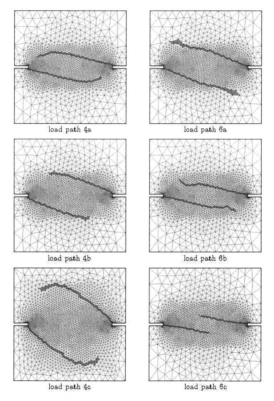

Figure 32. Computed crack paths for load paths 4a to 4c (left) and loads paths 6a to 6c (right) of the mixed mode fracture tests by Nooru-Mohamed.

Plasticity Based Crack Models and Applications   211

In [Nooru-Mohamed(1992)] no test data for the Young's modulus and Poisson's ratio are provided. However, in numerical simulations contained in [Nooru-Mohamed(1992)], these parameters are estimated as as $E = 30000$ MPa and $\nu = 0.2$. With respect to the fracture parameters only the splitting tensile strength is provided in [Nooru-Mohamed(1992)], data on the uniaxial tensile strength and the specific fracture energy are missing. They are estimated for the numerical simulations as $f_t = 3.0$ MPa and $G_f = 0.110$ Nmm/mm^2.

Figure 32 shows the predicted crack paths for the investigated load paths 4a to 4c and 6a to 6c. Again, the employed meshes consist of linear triangular elements and use of the partial domain crack tracking algorithm is made for ensuring continuous crack paths.

Whereas the computed crack paths show good agreement with the observed ones, the respective load displacement diagrams are characterized by a considerable overestimation of the ultimate load, as shown exemplarily in Figure 33. These discrepancies may partly be attributed to the estimated material parameters and partly to deviations of the test arrangement regarding the boundary conditions.

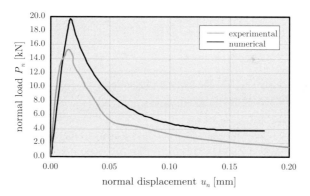

Figure 33. Experimentally and numerically obtained load displacement curves for load path 4a.

Mixed mode fracture tests by Hassanzadeh Specimens of the same geometric shape as used by Nooru-Mohamed [Nooru-Mohamed(1992)] were employed by Hassanzadeh [Hassanzadeh(1992)] for mixed mode fracture tests on plain concrete. However, the latter differ from the former by smaller dimensions (Figure 31b) and by different load paths.

In the tests by Hassanzadeh an axial displacement u_n is increased until the tensile strength is reached. Subsequently, a combination of axial and tangential displacements, u_n and u_t is applied. For the latter part of the load path two different displacement ratios were studied, i.e. $u_n = a\, u_t$ and $u_n = b\,\sqrt{u_t}$, where a and b represent constants. It is argued in [Hassanzadeh(1992)] that the latter better represent commonly encountered displacement ratios at cracks.

In [Hassanzadeh(1992)] only the compressive strength $f_c = 37.5$ MPa, the maximum aggregate size $D_{max} = 8$ mm and the water/cement ratio w/c = 0.5 are provided. However, [Hassanzadeh(1992)] also gives a summary of more extensive sets of material parameters for different concrete types with different compressive strengths, different maximum aggregate sizes and different w/c-ratios. From this summary the complete set of material parameters was taken for the numerical simulations, which corresponds to the compressive strength, the maximum aggregate size and w/c-ratio of the specimens used in the tests by Hassanzadeh. The respective material parameters are $E = 32000$ MPa, $f_t = 2.80$ MPa and $G_f = 0.108$ N/mm. Poisson's ratio is assumed as $\nu = 0.18$.

For the validation of the crack model with respect to the transfer of shear stresses across the crack faces, the test series, characterized by a displacement ratio of $b = u_n/\sqrt{u_t}$ is chosen.

For some of the latter tests Figure 34 shows comparisons of measured and computed evolutions of both, the tensile stress in terms of the normal crack opening, and the shear stress transferred across the crack faces in terms of the relative tangential displacement of the crack faces.

The comparison of computed and measured curves in Figure 34 confirms that the assumed traction-separation law allows describing (i) the normal stress transferred across an evolving crack in terms of the crack opening, (ii) the increase of the shear stress with increasing relative tangential displacement of the crack faces, (iii) the transfer of compressive stresses across crack faces due to aggregate interlock and (iv) the decrease of the transferred shear stress with increasing ratios of ζ_1/ζ_2 in a qualitatively correct manner.

3.7 Application

In the 2D anchor pull-out test, depicted in Figure 35, a vertical displacement u_n, resulting in the load P, was applied to the center of a steel head bolt embedded into a concrete panel. The material parameters are given as $E = 36630$ MPa, $f_t = 3.78$ MPa, and $G_f = 0.065$ N/mm. A detail of the concrete panel with the steel head bolt together with the observed crack

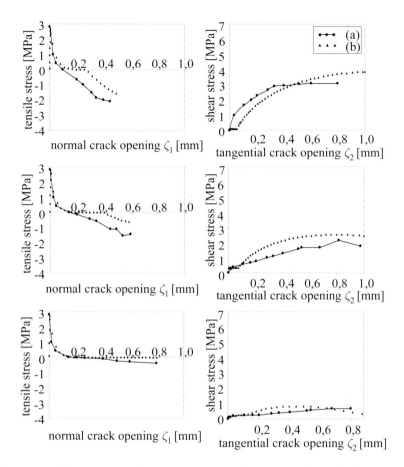

Figure 34. Comparison of experimental and computed results for the tensile stress in terms of crack opening (left column) and for the shear stress in terms of the relative tangential displacement of the crack faces (right column) for 3 different ratios of $b = u_n/\sqrt{u_t}$ (b = 0.5 (top), b = 0.6 (middle) and b = 0.8 (bottom)): (a) test results, (b) model predictions.

paths as well as the relationships between the applied load P and the crack opening ζ_1, measured at two measuring points during the test, are shown in Figure 36.

This test is simulated numerically using the delayed embedded discontinuity model. Thus, the early stages of crack development, characterized by smeared microcracking, are described by a smeared crack model with rotating crack directions and a discontinuity is embedded in the displacement field of the respective finite element only, if the crack opening, predicted by the smeared crack model, attains a prescribed threshold value. The partial domain crack tracking algorithm was employed in order to enforce continuity of the crack path, computed by the embedded crack model.

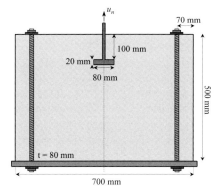

Figure 35. Test setup for the investigated anchor pull-out test.

For the early stages of cracking the cracked elements, predicted by the smeared crack model with rotating crack directions, are shown in the left column of Figure 37a, whereas the elements, into which discontinuities were embedded, are shown in the right column of Figure 37b. A comparison of the computed and the observed crack path shows good correspondence. However, the observed crack branching in the left part of the concrete panel, which is attributed to small deviations from symmetry in the test set-up, is not reproduced in the numerical simulation. Although the FE-mesh in Figure 37 seems to be symmetric at a first glance, it is not. It was obtained by free meshing of the concrete panel.

Figures 36b and c show a comparison of the computed relations between P and ζ_1 with the experimental data, recorded at two measurement points. It can be seen, that the computed load-crack opening curves agree quite well with the experimental ones.

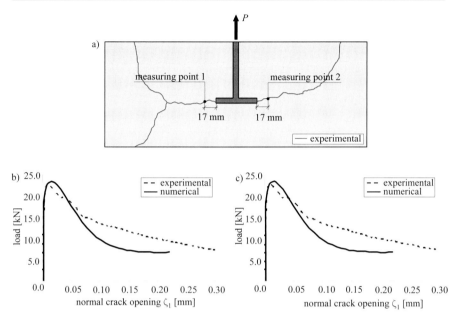

Figure 36. Anchor pull out test: (a) detail of the concrete panel with observed crack path, (b) and (c) load-crack opening relationships for the two measuring points.

Bibliography

[Kupfer(1969)] H. Kupfer, H.K. Hilsdorf and H. Rüsch, *Behavior of Concrete under Biaxial Stresses*, ACI Journal, Vol. 66, pages 656–666, 1969.

[Walraven(1981)] J. Walraven, H. Reinhardt, *Theory and experiments on the mechanical behavior of cracks in plain and reinforced concrete subjected to shear loading*, Heron 26:5–68, 1981.

[van Mier(1984)] J.G.M. van Mier, *Complete Stress-Strain Behavior and Damaging Status of Concrete under Multiaxial Conditions*, Proceedings of the International Conference on Concrete under Multiaxial Conditions, RILEM - CEB - CNRS, Presses de'l Université Paul Sabatier, Toulouse, 1984, pages 75–85, 1984.

[RILEM(1985)] *RILEM TC50-FMC: Determination of the fracture energy of mortar and concrete by means of three-point bend tests on notched beams*, RILEM Materials and Structures, Vol. 107, pages 407–413, 1985.

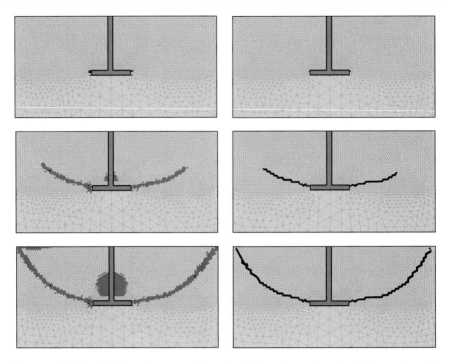

Figure 37. Evolution of smeared cracks (left column) and of elements with embedded cracks (right column).

[Oliver(1989)] J. Oliver, *A consistent characteristic length for smeared cracking models*, International Journal for Numerical Methods in Engineering 28, 431-474 1989.

[Crisfield(1989)] M.A. Crisfield and J. Wills, *Analysis of R/C panels using different concrete models*, Journal of Engineering Mechanics, Vol. 115, pages 578–597, 1989.

[Lubliner(1990)] J. Lubliner, *Plasticity Theory*, Macmillan, New York, 1990.

[CEB-FIP(1990)] *CEB-FIP Model Code 1990*. Bulletin d'information, Comité Euro-International du Béton (CEB), Lausanne, 1991.

[Hassanzadeh(1992)] Hassanzadeh M. *Behavior of fracture process zones in concrete influenced by simultaneously applied normal and shear displacements*. Dissertation, Lund Institute; 1992.

[Nooru-Mohamed(1992)] M.B. Nooru-Mohamed. *Mixed-mode fracture of concrete: An experimental approach.* Ph.D. Thesis, Delft University of Technology, Delft, 1992.

[Zienkiewicz(1992)] O.C. Zienkiewicz, J.Z. Zhu, *The superconvergent patch recovery (SPR) and adaptive finite element refinement.* Computer Methods in Applied Mechanics and Engineering, 101, 207–224, 1992.

[Simo(1993)] J. Simo, J. Oliver, F. Armero, *An analysis of strong discontinuities induced by strain-softening in rate-independent inelastic solids,* Computational Mechanics, 12, 277–296, 1993.

[Hofstetter(1995)] G. Hofstetter and H. Mang, *Computational Mechanics of Reinforced Concrete Structures,* Vieweg Verlag, 1995.

[Oliver(1996)] J. Oliver, *Modelling strong discontinuities in solid mechanics via strain softening constitutive equations. Part 1: Fundamentals,* International Journal for Numerical Methods in Engineering, 39:3575–3600, 1996.

[Feenstra(1996)] P.H. Feenstra and R. de Borst, *A composite plasticity model for concrete,* International Journal of Solids and Structures, Vol. 33, pages 707–730, 1996.

[vanMier(1997)] J.G.M. van Mier. *Fracture Processes of Concrete.* Series: New Directions in Civil Engineering, Vol. 12, CRC Press, 1997.

[Meiswinkel(1999)] R. Meiswinkel and H. Rahm, *Modelling tension stiffening in RC structures regarding non-linear design analyses,* European Conference on Computational Mechanics (ECCM 1999), CD-ROM, 20 pages, 1999.

[Borja(2000)] R. I. Borja, *A finite element model for strain localization analysis of strongly discontinuous fields based on standard Galerkin approximation,* Computer Methods in Applied Mechanics and Engineering, 190, 1529–1549, 2000.

[Vigl(2000)] A. Vigl, *Honeycomb segmental tunnel linings - simple, economical, successful,* Felsbau, Vol. 18, pages 24–31, 2000.

[Trunk(2000)] B. Trunk, *Einfluss der Bauteilgröße auf die Bruchenergie von Beton,* AEDIFICATIO Publishers, D-79104 Freiburg, 2000.

[Jirasek(2001)] Jirásek M, Zimmermann T, *Embedded crack model: Part II: Combination with smeared cracks,* International Journal for Numerical Methods in Engineering, 50(6):1291-1305, 2001.

[Winkler(2001a)] B. Winkler, *Traglastuntersuchungen von unbewehrten und bewehrten Betonstrukturen auf der Grundlage eines objektiven Werkstoffgesetzes für Beton,* Dissertation, University of Innsbruck, Austria, 2001.

[Winkler(2001b)] B. Winkler, G. Hofstetter and G. Niederwanger, *Experimental verification of a constitutive model for concrete cracking*, Proceedings of the Institution of Mechanical Engineers, Part L, Materials: Design and Applications, Vol. 215, pages 75–86, 2001.

[Jirasek(2002)] M. Jirásek, and T. Belytschko, *Computational resolution of strong discontinuities*, Proc.World Congress on Computational Mechanics, WCCM V, Eds.: Mang, H.A. and Rammerstorfer, F.G. and Eberhardsteiner, J.,Vienna University of Technology, Austria, http://wccm.tuwien.ac.at, 2002.

[Oliver(2002)] J. Oliver, A. E. Huespe, E. Samaniego, E. W. V. Chaves, *On strategies for tracking strong discontinuities in computational failure mechanics*, in: H. Mang, F. Rammerstorfer, J. Eberhardsteiner (Eds.), Proc.World Congress on Computational Mechanics, WCCM V, Vienna University of Technology, Austria, 2002, http://wccm.tuwien.ac.at.

[Samaniego(2003)] E. Samaniego, *Contributions to the Continuum Modelling of Strong Discontinuities in Two-dimensional Solids*, Ph.D. thesis, UPC Barcelona, 2003.

[Mosler(2003)] J. Mosler, G. Meschke, *3D modeling of strong discontinuities in elastoplastic solids: Fixed and rotating localization formulations*, International Journal for Numerical Methods in Engineering 57, 1553–1576, 2003.

[Rolshoven(2003)] S. Rolshoven, M. Jirásek. *Numerical aspects of nonlocal plasticity with strain softening*. In N. Bićanić, R. de Borst, H. Mang, and G. Meschke, editors, Computational Modelling of Concrete Structures (EURO-C 2003), St. Johann im Pongau, Austria, Lisse, The Netherlands, Swets & Zeitlinger B.V., 305–314, 2003.

[Winkler(2004)] B. Winkler, G. Hofstetter and H. Lehar, *Application of a constitutive model for concrete to the analysis of a precast segmental tunnel lining*, International Journal for Numerical and Analytical Methods in Geomechanics, Vol. 28, pages 797–819, 2004.

[Oliver(2006)] J. Oliver and A.E. Huespe and P.J. Sanchez, *A comparative study on finite elements for capturing strong discontinuities: E-FEM vs. X-FEM*, Computer Methods in Applied Mechanics and Engineering, 195, 4732-4752, 2006.

[Feist(2006)] C. Feist, G. Hofstetter, *An Embedded Strong Discontinuity Model for Cracking of Plain Concrete*, Computer Methods in Applied Mechanics and Engineering, 195, 7115-7138, 2006.

[ABAQUS] ABAQUS Inc., ABAQUS/Standard Users Manual, Version 6.6, Providence, RI, USA, 2006.

[Feist(2007)] C. Feist, G. Hofstetter, *Three-dimensional fracture simulations based on the SDA*, International Journal for Numerical and Analytical Methods in Geomechanics, 31, 189-212, 2007.

[Theiner(2009)] Y. Theiner, G. Hofstetter, *Numerical Prediction of Crack Propagation and Crack Widths in Concrete Structures*, Engineering Structures, 31, 1832-1840, 2009.

Crack models based on the extended finite element method

Nicolas Moës

GeM Institute, UMR CNRS 6183, Ecole Centrale de Nantes, France

1 Introduction

In spite of its decades of existence, the finite element method coupled with meshing tools does not yet manage to simulate efficiently the propagation of 3D cracks for geometries relevant to engineers in industry. Indeed, initial creation of the mesh and modification of this mesh during the propagation of a crack, remain extremely heavy and lack robustness. This fact was the motivation behind the design of the eXtended Finite Element Method (X-FEM).

Even if this operation were straightforward, the question of the projection of fields from one mesh to the next one would still be raised for history dependent problems (plasticity, dynamics, . . .). The possibility offered to preserve the mesh through the simulation is undoubtedly appealing.

The basic idea is to introduce inside the elements the proper discontinuities so as to relax the need for the mesh to conform to them. This introduction is done via the technique of the partition of unity (Melenk and Babuška, 1996; Babuška and Melenk, 1997). It should be noted that X-FEM is not the only method based on the partition of the unity (as in painting several schools exist). The GFEM approach (generalized finite element method) and PUFEM (partition of unity finite element method) are also based on the partition of unity.

The constant ambition which distinguishes the X-FEM approach since its beginnings is to use the partition of unity to release the mesh from constraints to conform to surfaces of discontinuity, while keeping the same performance as traditional finite element (optimality of convergence). Quickly also the X-FEM was coupled to the level set method to locate and evolve the position of surfaces of discontinuities.

Figure 1. Reference problem.

2 Background on discretization methods

2.1 Problem description and notations

The solid studied is depicted in Figure 1. It occupies a domain Ω whose boundary is denoted by S. This boundary is composed of the crack faces S_{c+} and S_{c+} assumed traction free, as well as a part S_u on which displacement $\boldsymbol{u}^\star$ are imposed and, finally, a part S_t on which tractions $\boldsymbol{t}^\star$ are imposed.

Stresses, strains and displacements are denoted by $\boldsymbol{\sigma}$, $\boldsymbol{\varepsilon}$ and $\boldsymbol{u}$, respectively. Small strains and displacements are assumed throughout the chapter. In the absence of volume forces, equilibrium equations read

$$\nabla \cdot \boldsymbol{\sigma} = 0 \text{ on } \Omega \tag{1}$$

$$\boldsymbol{\sigma} \cdot \boldsymbol{n} = \boldsymbol{t}^\star \text{ on } S_t \tag{2}$$

$$\boldsymbol{\sigma} \cdot \boldsymbol{n} = 0 \text{ on } S_{c+}, \quad \boldsymbol{\sigma} \cdot \boldsymbol{n} = 0 \text{ on } S_{c-} \tag{3}$$

where $\boldsymbol{n}$ is the outward normal. Kinematics equations read

$$\boldsymbol{\varepsilon} = \boldsymbol{\varepsilon}(\boldsymbol{u}) = \nabla_s \boldsymbol{u} \text{ on } \Omega \tag{4}$$

$$\boldsymbol{u} = \boldsymbol{u}^\star \text{ on } S_u \tag{5}$$

where ∇_s is the symmetrical part of the gradient operator. Finally, the constitutive law is assumed elastic: $\boldsymbol{\sigma} = \mathbb{E} : \boldsymbol{\varepsilon}$ where $\mathbb{E}$ is Hooke's tensor. The space of admissible displacement field is denoted U. whereas the space of admissible virtual displacements is denoted U_0:

$$U = \{\boldsymbol{v} \text{ regular} : \boldsymbol{v} = \boldsymbol{u}^\star \text{ on } S_u\} \tag{6}$$

$$U_0 = \{\delta\boldsymbol{v} \text{ regular} : \delta\boldsymbol{v} = 0 \text{ on } S_u\} \tag{7}$$

The regularity space to which the solution belongs is detailed in (Babuška and Rosenzweig, 1972) and (Grisvard, 1985). This space contains discontinuous fields of displacement across the crack faces S_c. The weak form of the equilibrium equations is written

$$\int_\Omega \boldsymbol{\sigma} : \varepsilon(\delta \boldsymbol{u}) \ \mathrm{d}\Omega = \int_{S_t} \boldsymbol{t}^\star \cdot \delta \boldsymbol{u} \ \mathrm{d}S \quad \forall \delta \boldsymbol{u} \in U_0 \qquad (8)$$

Let us note that the border S_c does not contribute to the weak form because it is traction free (this assumption will be released later). Combining (8) with the constitutive law and the kinematics equations, the displacement variational principle is obtained. Find $\boldsymbol{u} \in U$ such that

$$\int_\Omega \varepsilon(\boldsymbol{u}) : \mathbf{E} : \varepsilon(\delta \boldsymbol{u}) \ \mathrm{d}\Omega = \int_{S_t} \boldsymbol{t}^\star \cdot \delta \boldsymbol{u} \ \mathrm{d}S \quad \forall \delta \boldsymbol{u} \in U_0 \qquad (9)$$

2.2 Rayleigh-Ritz approximation

Within the Rayleigh-Ritz method, the approximation is written as a linear combination of displacement modes $\boldsymbol{\phi}_i(\boldsymbol{x}), i = 1, \ldots, N$ defined on the domain of interest:

$$\boldsymbol{u}(\boldsymbol{x}) = \sum_i^N a_i \boldsymbol{\phi}_i(\boldsymbol{x}) \qquad (10)$$

These modes must satisfy *a priori* the essential boundary conditions (imposed displacements are considered null to simplify the presentation). The introduction of this approximation into the variational principle (9) leads to the following system of equations

$$K_{ij} a_j = f_i, \quad j = 1, \ldots, N \qquad (11)$$

The summation rule over repeated indices is assumed.

$$K_{ij} = \int_\Omega \varepsilon(\boldsymbol{\phi}_i) : \mathbf{E} : \varepsilon(\boldsymbol{\phi}_j) \ \mathrm{d}\Omega \qquad (12)$$

$$f_i = \int_{S_t} \boldsymbol{t}^\star \cdot \boldsymbol{\phi}_i \ \mathrm{d}S \qquad (13)$$

The method of Rayleigh-Ritz offers a great freedom in the choice of the modes. These modes can for example be selected so as to satisfy the interior equations. However, this method has the disadvantage of leading to a linear system with dense matrix, on contrary to the finite element method which leads to a sparse system.

2.3 The finite element method

In the finite element method, the domain of interest, Ω, is broken up into geometrical subdomains of simple shape $\Omega_e, e = 1, \ldots, N_e$ called elements:

$$\Omega = \cup_{e=1}^{N_e} \Omega_e \qquad (14)$$

The set of elements constitutes the mesh. On each element, the unknown field is approximated using simple approximation functions, of polynomial type, as well as unknown coefficients called degrees of freedom. Degrees of freedom have a simple mechanical significance in general. For linear elements, the degrees of freedom are simply the displacement of the nodes along x and y directions. Let us indicate by u_i^α the displacement of node i in direction α ($\alpha = x$ or y) and by $\boldsymbol{\phi}_i^\alpha$ the corresponding approximation function. The finite element approximation on element Ω_e is written

$$\boldsymbol{u}(\boldsymbol{x})\,|_{\Omega_e} = \sum_{i \in N_n} \sum_\alpha a_i^\alpha \boldsymbol{\phi}_i^\alpha(\boldsymbol{x}) \tag{15}$$

where N_n denotes the set of nodes of element Ω_e. For instance, for a triangle, they are six approximation functions

$$\{\boldsymbol{\phi}_i^\alpha\} = \{\phi_1 \boldsymbol{e}_x, \phi_2 \boldsymbol{e}_x, \phi_3 \boldsymbol{e}_x, \phi_1 \boldsymbol{e}_y, \phi_2 \boldsymbol{e}_y, \phi_3 \boldsymbol{e}_y\} \tag{16}$$

where ϕ_1, ϕ_2 and ϕ_3 are scalar linear functions over the element with value of 0 or 1 at the nodes. Approximation (15) allows one to model any rigid mode or constant strain over the element. This condition must be fulfilled by the approximation for any type of elements. Continuity of the approximation over the domain is obtained by the use of nodal degrees of freedom shared by all elements connected to the node. The stiffness matrix, K_{ij}^e, and load vector, f_i^e, are given for a finite element by

$$K_{i\alpha,j\beta}^e = \int_{\Omega^e} \varepsilon(\boldsymbol{\phi}_i^\alpha) : \mathbf{E} : \varepsilon(\boldsymbol{\phi}_j^\beta)\; \mathrm{d}\Omega \tag{17}$$

$$f_{i\alpha}^e = \int_{S_t \cap \partial \Omega^e} \boldsymbol{t}^\star \cdot \boldsymbol{\phi}_i^\alpha\; \mathrm{d}S \tag{18}$$

The global system of equations is obtained by assembling the elementary matrices and forces in a global stiffness and force vector. In the assembly process, the equations related to degrees of freedom involved in Dirichlet boundary conditions are not built.

On the contrary to the Rayleigh-Ritz approximation, the local character of the finite element approximation leads to sparse matrices. Moreover, the finite element has a strong mechanical interpretation: kinematics is described by nodal displacements which are associated by duality to nodal forces. The behavior of the element is characterized by the elementary stiffness matrix which connects the nodal forces and displacements. The global system to solve enforces the equilibrium of the structure: the sum of the nodal forces at each node must be zero. Lastly, the finite element method did demonstrate a high level of robustness in industry which makes it a very appropriate approach for most applications.

However, the use of the finite element method for problems with complex geometry or evolution of internal surfaces is currently obstructed by meshing issues. This did yield a motivation to design the so called meshless methods.

2.4 Meshless methods

We give some insights on meshless methods because they are important to understand the history of the concept of enrichment. Within the framework of meshless methods, the support of the approximation function is more important than the elements (which actually do not exist any more). On these supports, enrichment functions may be introduced for example to model a crack tip (as in (Fleming et al., 1997)).

Years of active research on meshless methods did show the importance of the support. Many researches were undertaken in the nineties to develop methods in which the approximation does not rest on a mesh but rather on a set of points. Various methods exist to date: diffuse elements (Nayroles et al., 1992), Element Free Galerkin method (EFG) (Belytschko et al., 1994), Reproducing Kernel Particle Method (RKPM) (Liu et al., 1993), $h-p$ cloud method (Duarte and Oden, 1996).

Each point has a domain of influence (support) with a simple shape (circle or rectangle for example in 2D) on which approximations are built. These functions are zero on the boundary and outside the domain of influence. Abusively, we will speak about the support i for the support associated with node i. The approximation functions defined on the support i are denoted ϕ_i^α, $\alpha = 1, \ldots, N_f(i)$ where $N_f(i)$ is the number of functions defined over support i. The corresponding degrees of freedom are denoted a_i^α. The approximation at a given point x is written

$$\boldsymbol{u}(\boldsymbol{x}) = \sum_{i \in N_s(\boldsymbol{x})} \sum_{\alpha=1}^{N_f(i)} a_i^\alpha \phi_i^\alpha(\boldsymbol{x}) \qquad (19)$$

where $N_s(\boldsymbol{x})$ is the set of points whose support contains point $\boldsymbol{x}$. Figure 2 shows for example a point $\boldsymbol{x}$ covered by three supports. The approximations functions are built so that the approximation (19) can represent all rigid modes and constant strain modes on the domain. These conditions are necessary to prove the convergence of the method. Various approaches (diffuse element, EFG, RKPM, ...) are distinguished, among other things, by the techniques used for the construction of these approximation functions.

Once a set of approximation functions has been built, it is possible to add some by enrichment. Various manners of enriching exist and we will describe

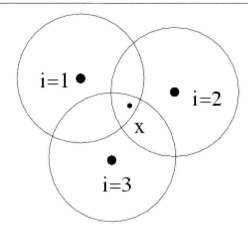

Figure 2. Three supports covering node x.

an enrichment type described as external by Belytschko and Fleming (1999) . The enrichment of the approximation makes it possible to represent a given displacement mode, for example $F(x)e_x$ on a subdomain denoted by $\Omega_F \subset \Omega$. Let us note N_F the set of supports which have a non empty intersection with Ω_f. The enriched approximation is written

$$u(x) = \sum_{i \in N_s(x)} \sum_{\alpha=1}^{N_f(i)} a_i^\alpha \phi_i^\alpha(x) + \sum_{i \in N_s(x) \cap N_F} \sum_{\alpha=1}^{N_f(i)} b_i^\alpha \phi_i^\alpha(x) F(x) \qquad (20)$$

where the new degrees of freedom, b_i^α, multiply the enriched approximation functions $\phi_i^\alpha(x) F(x)$. Let us show that the function $F(x)e_x$ may be represented on Ω_f. By setting to zero all degrees of freedom a_i^α and taking the function $F(x)$ out of the sum, the approximation at point $x \in \Omega_F$ reads

$$u(x) = \left(\sum_{i \in N_s(x) \cap N_F} \sum_{\alpha=1}^{N_f(i)} b_i^\alpha \phi_i^\alpha(x) \right) F(x) \qquad (21)$$

The degrees of freedom b_i^α can be selected so that the factor in front of $F(x)$ is the rigid mode e_x. That is possible since functions ϕ_i^α are able to represent any rigid mode. In conclusion, approximation (20) can represent $F(x)e_x$ on Ω_F. Enrichment made it possible within the framework of the Element Free Galerkin Method to solve problems of propagation of cracks in two and three dimensions without remeshing (Krysl and Belytschko, 1999): the crack is propagated through a set of points and is modeled by enrichment of the approximation with discontinuous functions $F(\boldsymbol{x})$ on the crack or representing the singularity on the crack front. Great flexibility in the writing of the approximation and its enrichment as well as the possibility of creating very regular fields of approximation are two important assets of meshless methods and the EFG approach in particular. The use of meshless methods however presents a certain number of difficulties compared to the finite element method:

- Within the finite element method the assembly of the stiffness matrix can be done by assembling the contributions of each element. In meshless methods, the assembly is done rather by covering the domain by points of integration and by adding the contribution of each one of them. The choice of the position and the number of integration points is tedious for an arbitrary set of approximation points;
- the approximation functions are to be built and are not explicit;
- the support size is a parameter in the method which the user must choose carefully;
- the boundary conditions of the Dirichlet type are delicate to impose.

Finally, it must be pointed out that due to the lack of the element concept, meshless methods are not at all trivial to implement in legacy finite element codes.

2.5 The partition of unity

Melenk et Babuška (1996) did show that the traditional finite element approximation could be enriched so as to represent a specified function on a given domain. Their point of view can be summarized as follows. Let us first us recall that the finite element approximation is written on an element as

$$\boldsymbol{u}(\boldsymbol{x})\,|_{\Omega_e} = \sum_{i \in N_n} \sum_{\alpha} a_i^{\alpha} \phi_i^{\alpha}(\boldsymbol{x}) \tag{22}$$

Since the degrees of freedom defined at a node have the same value for all the elements connected to it. The approximations on each element can be "assembled" to give a valid approximation in any point $\boldsymbol{x}$ of the domain:

$$\boldsymbol{u}(\boldsymbol{x}) = \sum_{i \in N_n(\boldsymbol{x})} \sum_{\alpha} a_i^{\alpha} \phi_i^{\alpha}(\boldsymbol{x}) \tag{23}$$

where $N_n(\boldsymbol{x})$ is the set of nodes belonging to the elements containing point $\boldsymbol{x}$. The domain of influence (support) of the approximation function ϕ_i^{α} is the set of elements connected to node i. The set $N_n(\boldsymbol{x})$ is thus also the set of

nodes whose support covers point x. The finite element approximation (23) can thus be interpreted as a particularization of the approximation (19) used in meshless methods:
- The set of points is the set of nodes in the mesh;
- The domain of influence of each node is the set of elements connected to it.

It is thus possible to enrich the finite element approximation by the same techniques as those used in meshless methods. Here is the enriched approximation which makes it possible to represent function $F(x)e_x$ on domain Ω_F:

$$u(x) = \sum_{i \in N_n(x)} \sum_\alpha \phi_i^\alpha a_i^\alpha + \sum_{i \in N_n(x) \cap N_F} \sum_\alpha b_i^\alpha \phi_i^\alpha(x) F(x) \qquad (24)$$

where N_F is the set of nodes whose support has an intersection with domain Ω_F. The proof is obtained by setting to zero coefficients a_i^α and by taking into account the fact that the finite element shape functions are able to represent all rigid modes and thus the e_x mode. We move now to the concrete use of the partition of unity for modeling discontinuities.

3 Discontinuity modeling with the X-FEM and level sets

The X-FEM introduces discontinuity inside elements using an enrichment based on the partition of unity technique. Proper enrichments for displacement discontinuities due to cracks will be discussed in this section. Proper enrichment for strain discontinuity maye also be found in the literature as in (Moës et al., 2003).

3.1 A simple 1D problem

We consider a bar shown in Figure 3. Two cases are considered : a crack is located at a node or between two nodes.

Case a : crack located at a node Using classical finite elements, this case is treated using double nodes. Node 2 is replaced by nodes 2^- and 2^+ sharing the same location but bearing different unknowns as shown in Figure 4. The approximation reads

$$u = u_1 N_1 + u_2^- N_2^- + u_2^+ N_2^+ + u_3 N_3 + u_4 N_4 \qquad (25)$$

where N_i indicates the approximation functions and u_i the corresponding degrees of freedom. Defining the displacement average, $<u>$, and (half)

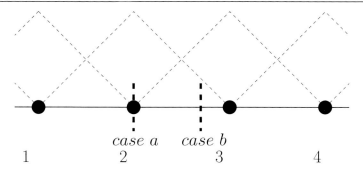

Figure 3. A bar with a crack located at a node (case a) or between two nodes (case b).

jump, $[u]$, at node 2

$$<u> = \frac{u_2^- + u_2^+}{2} \quad [u] = \frac{u_2^- - u_2^+}{2} \quad (26)$$

the approximation may be rewritten as

$$u = u_1 N_1 + <u> N_2 + [u] N_2 H(x) + u_3 N_3 \quad (27)$$

where

$$N_2 = N_2^- + N_2^+ \quad (28)$$

The generalized Heaviside function H (generalized because the original Heaviside function goes from 0 to 1) is represented in Figure 5. Abusively, we shall however call it Heaviside function. In the approximation (27), one distinguishes the continuous part modeled by functions N_1, N_2 and N_3 to which is added a discontinuous part given by the product of N_2 by the Heaviside function. Node 2 is called an enriched node because an additional degree of freedom is given to it.

Case b: crack located in between two nodes Let us study now case b, Figure 3, in which the crack is located between two nodes. As in case a, we wish to write the approximation as the sum of a continuous and a discontinuous part. Evolving on case a, we propose

$$u = u_1 N_1 + u_2 N_2 + u_3 N_3 + u_4 N_4 + a_2 N_2 H + a_3 N_3 H \quad (29)$$

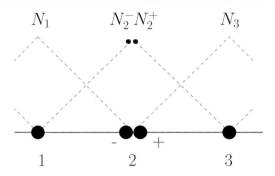

Figure 4. Double node to model a discontinuity located at a node.

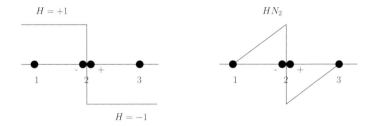

Figure 5. The generalized Heaviside function (left) as well as its product with function N_2 (right).

Nodes 2 and 3 are enriched by the Heaviside function. This enrichment was presented first (in 2D) by Moës et al. (1999). If the crack is located at a node, there is only one enriched node. In case b, two nodes are enriched because the support of nodes 2 and 3 are cut by the crack. A node is enriched by the Heaviside function if its support is cut into two by the crack. One can show that the approximation (29) makes it possible to represent two rigid modes (to the left of the crack and the other to the right). The fact that two (and not one) additional degrees of freedom are necessary may be surprising. Indeed, a crack implies a jump in displacement but can also imply a jump in strain. By linear combination of the various functions implied in (29), one notices that enrichment brings two functions on the element joining nodes 2 and 3. These two functions are shown in Figure 7. Note that for case a, only one additional degree of freedom is needed because finite element already exhibits strain jumps across element boundaries.

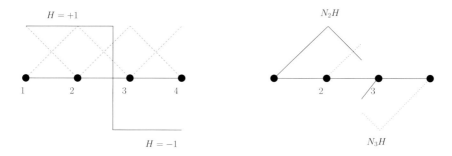

Figure 6. A crack located between two nodes. The Heaviside function (left) and enrichment functions (right).

Finally, it should be noted that proposed enrichment yields the same approximation space as if the cracked element is replaced by two elements and a double node. This observation is limited to 1D and will not carry over to 2D and 3D.

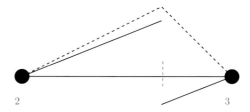

Figure 7. Two functions modeled by the X-FEM enrichment. We observe a continuous function with discontinuous slope (dashed line) and a discontinuous function with continuous slope (solid line).

Note that a set of variants to the Heaviside enrichment has been proposed in the literature: the "Hansbo" alternative (Hansbo and Hansbo, 2002), the use of virtual or phantom nodes (Molino et al., 2004), (Song et al., 2006) and, finally, the shifted basis from Zi and Belytschko (2003). All the variants listed above will lead to the same numerical solutions as the Heaviside enrichment described earlier. The choice is guided in general by the simplicity of implementation according to the target code. Also note that even if these various bases will lead to the same solution, the generated matrices will not be identical (and will not have the same condition number).

3.2 Extension to 2D and 3D

We consider now 2D and 3D meshes cut by a crack. Just like in the 1D case, we begin with the case of a crack inserted with double nodes.

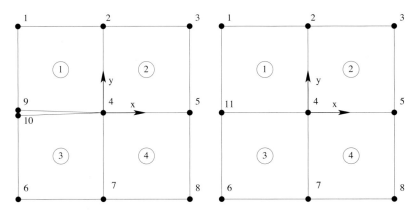

Figure 8. Finite element mesh near a crack tip, the circled numbers are element numbers

Figure 9. Regular mesh without a crack.

Figure 8 is taken from (Moës et al., 1999) and shows a four element mesh in which a crack has been introduced through double nodes (nodes 9 and 10). The finite element approximation associated with the mesh in Figure 8 is

$$u = \sum_{i=1}^{10} u_i N_i \quad (30)$$

where u_i is the (vectorial) displacement at node i and ϕ_i is the bilinear shape function associated with node i. Each shape function ϕ_i has a compact support ω_i given by the union of the elements connected to node i.

Let us rewrite (30) in such a way that we recover an approximation without crack corresponding to Figure 9 and a discontinuous additional displacement. Defining the average displacement a and the displacement jump b on the crack faces as

$$a = \frac{u_9 + u_{10}}{2} \quad b = \frac{u_9 - u_{10}}{2} \quad (31)$$

Crack Models, Based on the Extended Finite Element Method   233

we can express u_9 and u_{10} in terms of a and b

$$u_9 = a + b \quad u_{10} = a - b \tag{32}$$

Then replacing u_9 and u_{10} in terms of a and b in (30) yields

$$u = \sum_{i=1}^{8} u_i N_i + a(N_9 + N_{10}) + b(N_9 + N_{10})H(x) \tag{33}$$

where $H(x)$ is referred to here as a discontinuous, or 'jump' function. This is defined in the local crack coordinate system as

$$H(x,y) = \begin{cases} +1 & \text{for } y > 0 \\ -1 & \text{for } y < 0 \end{cases} \tag{34}$$

If we now consider the mesh in Figure 9, $N_9 + N_{10}$ can be replaced by N_{11}, and a by u_{11}. The finite element approximation now reads

$$u = \sum_{i=0}^{8} u_i N_i + u_{11} N_{11} + b N_{11} H(x) \tag{35}$$

First two terms on the right hand side represent the classical finite element approximation, whereas the last one represents the addition of a discontinuous enrichment. In other words, when a crack is modeled by a mesh as in Figure 8, we may interpret the finite element space as the sum of one which does not model the crack (such as Figure 9) and a discontinuous enrichment. The third term may be interpreted as an enrichment of the finite element function by a partition of unity technique.

Derivation that we have just carried out on a small grid of four elements may be reiterated on any 1D, 2D or 3D grid containing a discontinuity modeled by double nodes. This derivation will yield to the same conclusion: the modeling of a discontinuity by double nodes is equivalent to a traditional finite element modeling to which an enrichment by the partition of unity of the nodes located on the path of discontinuity is added. Let us note that the nodes which are enriched are characterized by the fact that their support is cut into two by the discontinuity.

Let us suppose now that one wishes to model a discontinuity which does not follow the edge of the elements. We propose to enrich all the nodes whose support is (completely) cut into two by the discontinuity (Moës et al., 1999). At these nodes, we add a degree of freedom (vectorial if the field is vectorial) acting on the traditional shape function at the node multiplied by a discontinuous function $H(x)$ being 1 on a side of the crack and -1 on the other. For example, in Figures 10 and 11, circled nodes are enriched. A node whose support is not completely cut by discontinuity must not be enriched by function H because that would result in enlarging the crack artificially. For example, for the mesh shown in Figure 11, if nodes C and D are enriched, the crack will be active up to the point R (since the displacement field will be discontinuous up to point R). However, if only nodes A and B are enriched by the discontinuity, the displacement field is discontinuous only up to point Q and the crack appears unfortunately shorter.

In order to represent the crack on its proper length, nodes whose support contains the crack tip (squared nodes shown in Figure 11) are enriched with discontinuous functions up to the crack tip but not beyond. Such functions are provided by the asymptotic modes of displacement (elastic if calculation is elastic) at the crack tip. This enrichment, already used by Belytschko and Black (1999) and Stroubolis et al. (2000) allows moreover precise calculations since the asymptotic characteristics of the displacement field are built-in. Let us note that if the solution is not singular at the crack tip (for example by the presence of a cohesive zone), other functions of enrichment can be selected (Moës and Belytschko, 2002; Zi and Belytschko, 2003).

We are now able to detail the complete modeling of a crack with X-FEM located arbitrarily on a mesh, Figure 12. The enriched finite element approximation is written:

$$\boldsymbol{u}^h(\boldsymbol{x}) = \sum_{i \in I} \boldsymbol{u}_i N_i(\boldsymbol{x}) + \sum_{i \in L} \boldsymbol{a}_i N_i(\boldsymbol{x}) H(\boldsymbol{x}) \qquad (36)$$
$$+ \sum_{i \in K_1} N_i(\boldsymbol{x}) (\sum_{l=1}^{4} \boldsymbol{b}_{i,1}^l F_1^l(\boldsymbol{x})) + \sum_{i \in K_2} N_i(\boldsymbol{x}) (\sum_{l=1}^{4} \boldsymbol{b}_{i,2}^l F_2^l(\boldsymbol{x}))$$

where:
- I is the set of nodes in the mesh;
- $\boldsymbol{u}_i$ is the classical (vectorial) degree of freedom at node i;
- N_i is the scalar shape function associated to node i;
- $L \subset I$ is the subset of nodes enriched by the Heaviside function. The corresponding (vectorial) degrees of freedom are denoted $\boldsymbol{a}_i$. A node belongs to L if its support is cut in two by the crack and does not contain the crack tip. Those nodes are circled on Figure 12;
- $K_1 \subset I$ et $K_2 \subset I$ are the set of nodes to enrich to model crack tips numbered 1 and 2, respectively. The corresponding degrees of freedom are $\boldsymbol{b}_{i,1}^l$ and $\boldsymbol{b}_{i,2}^l$, $l = 1, \ldots, 4$. A node belongs to K_1 (resp. K_2) if its

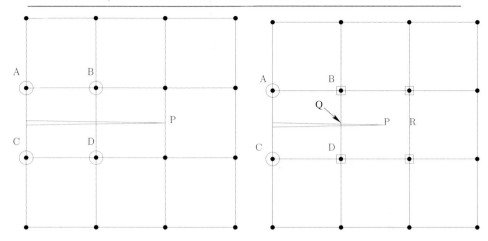

Figure 10. Crack not aligned with a mesh, the circled nodes are enriched with the discontinuous function $H(\boldsymbol{x})$.

Figure 11. Crack not aligned with a mesh, the circled nodes are enriched with the discontinuous $H(\boldsymbol{x})$ function and the squared nodes with the tip enrichment functions.

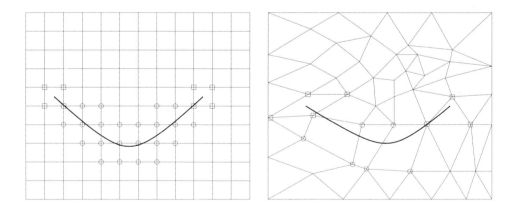

Figure 12. Crack located on a structured (left) and unstructured mesh (right). Circled nodes are enriched with the Heaviside function while squared nodes are enriched by tip functions.

support contains the first (resp. second) crack tip. Those nodes are squared in Figure 12.

Functions $F_1^l(\boldsymbol{x}), l = 1, \ldots, 4$ modeling the crack tip are given in elasticity by:

$$\{F_1^l(\boldsymbol{x})\} \equiv \{\sqrt{r}\sin(\tfrac{\theta}{2}), \sqrt{r}\cos(\tfrac{\theta}{2}), \sqrt{r}\sin(\tfrac{\theta}{2})\sin(\theta), \sqrt{r}\cos(\tfrac{\theta}{2})\sin(\theta)\} \quad (37)$$

where (r, θ) are the polar coordinates in local axis at the crack tip. It must be noted that the first function is discontinuous across the crack. The three others are able to model strain discontinuity across the crack fraces. Similarly, functions $F_2^l(\boldsymbol{x})$ are also given by (37); the local system of coordinates being now locate around the second crack tip.

The extension to the three-dimensional case of the modeling of cracks by X-FEM was carried out in (Sukumar et al., 2000). Just like in the two-dimensional case, the fact that a node is enriched or not and the type of enrichment depend on the relative position of the support associated with the node compared to the crack location. The support of a node is a volume, the crack front is a curve (or several disjoint curves) and the crack itself is a surface. Enrichment functions for the crack front remain given by (37). A node is enriched if its support is touched by the crack front. The evaluation of r and θ can be done by finding the nearest point on the crack front, then by establishing a local base there. The use of level sets dealt with in the following section makes it possible to avoid this operation.

3.3 Cracks located by level sets

To locate a curve in 2D, one can indicate all points located on this curve for example using a parametric equation. One can qualify this representation as explicit. Another manner, implicit, to represent the curve is to consider it as the iso-zero level of a signed distance function. The distance is counted positively if one is inside the curve and negatively in the contrary case (the curve is supposed to separate the space in two zones). On a finite element mesh the level set is interpolated between the nodes by traditional finite element shape functions. In short, the location of a surface in 3D (curve in 2D) is given by a finite element field defined near the surface (curve). The knowledge of the signed distance is indeed needed in a narrow band around the surface.

For instance, in the Figure 13 one can see the value of a level set locating a circle on a grid (negative inside the circle and positive outside). The iso-zero contour of the level set function indicates the position of the circle. The level set is defined likewise in 3D. Figure 14 gives for example the iso-zero for a level set defined on a fine grid. The level set locates the material interface between strands and matrix in a so-called 4D composite.

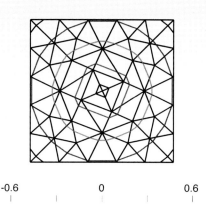

Figure 13. A level set locating a circle of radius 0.7 in a 2 by 2 plate. Five contours are plotted with level set values of -0.6, -0.3, 0.0, 0.3 and 0.6

As we indicated above, a level set separates space in two zones, a positive zone and a negative zone. A crack does not separate a domain into two (unless it is broken!). A unique level set is thus not enough to locate a crack. One needs two of them. The first one denoted ϕ_n separates space into two by considering a tangent extension from the crack whereas the function ϕ_t makes it possible to locate the front. These two level sets are represented in Figure 15. The set of points characterized by $\phi_n = 0$ and $\phi_t \leq 0$ defines the position of the crack whereas points for which $\phi_n = \phi_t = 0$ defines the front. The representation of a crack by two level set functions was for the first time introduced by Stolarska et al. (2001) in 2D and Moës et al. (2002) in 3D. Figure 15 gives the iso-zero contour of both level sets for a crack in 2D. Coordinates r and θ appearing in the enrichment functions (37) are computed from the equations given in (Stolarska et al., 2001)

$$r = (\phi_n^2 + \phi_t^2)^{1/2} \quad \theta = \arctan(\frac{\phi_n}{\phi_t}) \tag{38}$$

The implicit representation is particularly interesting when the curve (surface) evolves. Indeed, contrary to the explicit representation which does not make it possible to manage topological changes easily. These changes are taken into account very naturally in the implicit level set representation. By topological changes, one understands for example the fact that two bubbles meet to form a unique bubble or the fact that a drop can separate in two drops. Another example of topological change is the case of a crack initially inside a cube (Figure 16 left) which after some propagation cuts the four faces of the cube. The crack front initially circular is split into four independent curves (Figure 16 right).

Figure 14. The iso-zero of a level set function locating the interface between strands and a matrix in a 4D composite.

The article (Osher and Sethian, 1988) was one of the first to present robust algorithms for level sets propagation. The use of the level sets for computational science then very quickly developed as attested by a sequence of three books (Sethian, 1996, 1999; Osher and Fedkiw, 2002). These algorithms of propagation were initially mainly developed within the framework of finite differences. Indeed, level sets were initially used for fluid mechanics applications: for instance to follow free interfaces or interfaces between various phases. Some articles however did develop algorithms appropriate for unstructured finite element meshes as (Barth and Sethian, 1998).

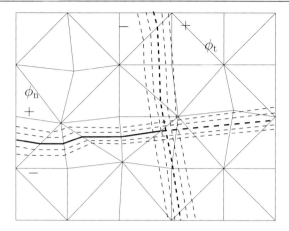

Figure 15. Two level set functions locating a crack on a 2D mesh.

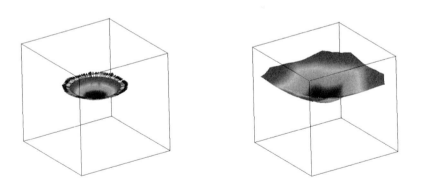

Figure 16. A lens shaped crack (left) propagating in a cube and finally cutting four cube faces (right).

4 Technical and mathematical aspects

4.1 Integration of the element stiffness

Integration on the elements cut by the crack is made separately on each side of the crack. The ϕ_n level set cuts a triangular (tetrahedral) element along a line (a plane). The possible cuts are indicated on Figures 17 and 18. For elements close to the crack tip, use of non polynomial enrichment functions requires special care (Béchet et al., 2005).

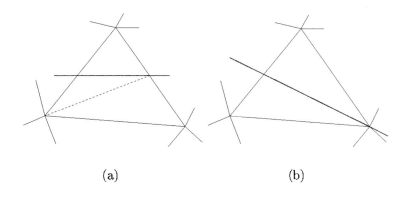

Figure 17. Two scenarios of the level set cut of a triangle.

4.2 Topological and geometrical enrichment strategies

The initial enrichment strategy for the crack tip consisted in enriching a set of nodes around the tip. A node is enriched if its support touches the crack tip (Moës et al., 1999). In 3D, nodes for which the support touches the crack front are enriched (Sukumar et al., 2000).

This type of enrichment may be called topological because it does not involve the distance from the node to the tip (front). As a matter of fact, the topological enrichment is active over an area which vanishes to zero as the mesh size goes to zero. Another enrichment, developed independently in (Béchet et al., 2005) and (Laborde et al., 2005) may be called geometrical because it consists in enriching all nodes located within a given distance to the crack tip. Both enrichment strategies are compared in Figure 19.

In order to study the influence of the enrichment on the convergence rate,

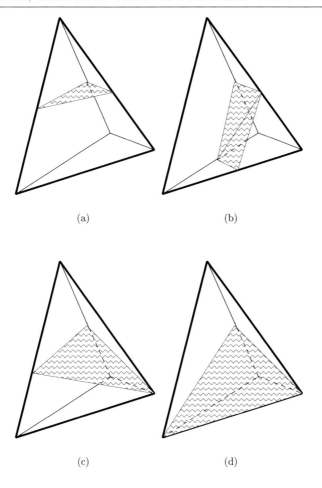

Figure 18. Four scenario of the level set cut of a tetrahedron.

a plane strain benchmark problem is set up. A square domain in plane strain is subjected to a pure mode I. The square boundary is subjected to the exact tractions corresponding to the mode I of infinite problem. Rigid modes are prevented. The exaggerated deformed shape of the benchmark problem is shown in Figure 20. To be precise the domain size is $\Omega = [0,1] \times [0,1]$ and tractions applied correspond to $K_\text{I} = 1$ and $K_\text{II} = 0$. The crack tip is located at the center of the square. Young modulus is 1 and Poisson ratio 0. A convergence analysis is performed for a uniform grid which is recursively refined. The energy norm error, ϵ, measuring the distance between the exact, $\boldsymbol{\sigma}, \boldsymbol{u}$, and approximated field $\boldsymbol{\sigma}^h, \boldsymbol{u}^h$

$$\epsilon = \left(\frac{\int_\Omega (\boldsymbol{\sigma}^h - \boldsymbol{\sigma}) : \mathbb{E}^{-1} : (\boldsymbol{\sigma}^h - \boldsymbol{\sigma}) \ \mathrm{d}\Omega}{\int_\Omega \boldsymbol{\sigma} : \mathbb{E}^{-1} : \boldsymbol{\sigma} \ \mathrm{d}\Omega} \right)^{1/2} \quad (39)$$

$$= \left(\frac{\int_\Omega \varepsilon(\boldsymbol{u}^h - \boldsymbol{u}) : \mathbb{E} : \varepsilon(\boldsymbol{u}^h - \boldsymbol{u}) \ \mathrm{d}\Omega}{\int_\Omega \varepsilon(\boldsymbol{u}) : \mathbb{E} : \varepsilon(\boldsymbol{u}) \ \mathrm{d}\Omega} \right)^{1/2} \quad (40)$$

is plotted in Figure 21. For the topological enrichment only the nodes whose support is touching the crack tip are enriched. In the case of the geometrical enrichment, nodes within a distance of $r_e = 0.05$ from the crack tip are enriched. It can be observed that the convergence rate is 0.5 when the topological or no enrichment is present. The topological enrichment yielding however a smaller error. On the contrary, the geometrical enrichment produces a order of 1 convergence. In order to analyze these convergence rates, we must recall the convergence rate result of the classical finite element method (see for instance (Bathe, 1996))

$$\epsilon = O(h^{\min(r-m, p+1-m)}) \quad (41)$$

The regularity of the solution is indicated by r ($\boldsymbol{u} \in H^r(\Omega)$) whereas p is the degree of the finite element interpolation and m is the error norm used. For our benchmark, $r = 3/2$, $p = 1$ and $m = 1$, so we indeed get a convergence rate of 0.5 The topological enrichment yields a lower error than a pure FEM analysis because the X-FEM approximation spans a larger space than the FEM one. However, it does not affect the convergence rate since the enrichment area goes to zero as the mesh size goes to zero. In the case of the geometrical enrichment, the enrichment is able to represent exactly (even as h to zero) the rough part of the solution. The classical part of the approximation is thus only in charge of the smooth part of the solution yielding the optimal order of 1 convergence rate. This was proved by Laborde et al. (2005). It was also shown in this paper that (for the benchmark problem) if the polynomial approximation is raised, higher (still

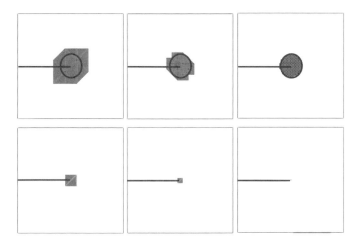

Figure 19. Extent of the enrichment zone as the mesh size decreases. Geometrical enrichment (top) and topological (bottom) enrichments are compared.

optimal) convergence rates are obtained. The polynomial degree needs only to be raised in the classical part and Heaviside parts of the approximation (first and second term term in the right hand side of (37)).

4.3 Solver and condition number

When solving a linear system of equation $Kx = f$, an important number to take into account is the condition number defined as the ratio between the maximum and minimum eigenvalue of the K matrix.

$$\kappa = \frac{\lambda_{\max}}{\lambda_{\min}} \qquad (42)$$

This condition number has a direct impact on the convergence rate for an iterative solver and on the propagation of round offs for a direct solver. For instance, for the conjugate gradient iterative solver, the error at iteration m reads (Saad, 2000):

$$\|x_- x_m\|_K \leq 2 \left[\frac{\sqrt{\kappa} - 1}{\sqrt{\kappa} + 1}\right]^m \|x_- x_0\|_K \qquad (43)$$

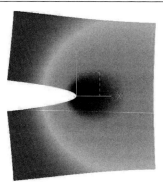

Figure 20. Mode I benchmark problem : Exaggerated deformed shape of a square slab under mode one loading.

where x_0 is the initial guess, x_m the iterate m and

$$\|a\|_K = \sqrt{a^T K a} \qquad (44)$$

Thus the higher the condition number, the slower is the convergence. To be precise, the bound (43) is in general pessimistic. Indeed, first of all κ can be calculated on the basis of the eigenvalues for which the corresponding eigenvector projected on the right hand side is not zero. Then, the κ can be reajusted progressively with the iterations while being based only on the eigenvectors remaining active through the iterations (see detail in (Saad, 2000)).

The conditioning of X-FEM was studied in (Béchet et al., 2005) and (Laborde et al., 2005) for the two types of enrichment: topological and geometrical. The evolution of the condition number according to the size of elements of the grid is given in Figure 22 for the stiffness and mass matrices. They are plotted for the benchmark problem already discussed in section 4.2.

We note that for the geometrical enrichment, the condition number grows dramatically with the mesh size. A specific preconditioner was designed in (Béchet et al., 2005) to circumvent the increase. The effect of the preconditioner is also given in Figure 22. This preconditioner could be called

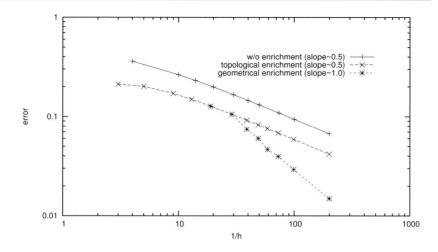

Figure 21. Relative error in the energy norm for the mode I benchmark problem. The top curve corresponds to no enrichment (slope 0.5). The middle curve is the result for the topological enrichment whereas the bottom curve (slope 1) is for the geometrical enrichment.

pre-preconditioner X-FEM preconditioner because it takes care of the specificity of X-FEM. After its application, regular FEM preconditioner may be used. The idea behind the X-FEM preconditioner is quite simple. On enriched nodes, the enriched shape functions are orthogonalized with respect to the classical shape function. The matrices related to a given node are thus diagonal.

4.4 Inf-sup condition for cracks under contact

We now consider a more complex scenario for which the crack faces may contact each other or may be loaded through hydraulic pressure for instance.

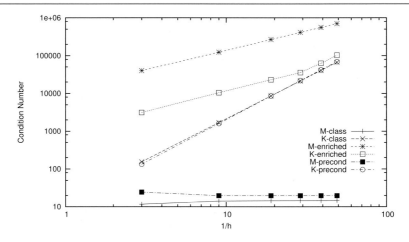

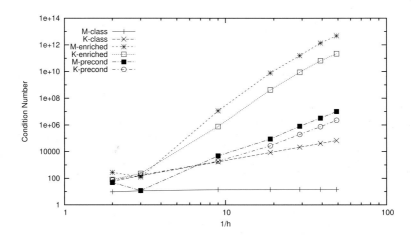

Figure 22. Condition number as a function of the mesh size for the mass and stiffness matrices. Topological (top) and geometrical (bottom) enrichments are considered as well as the influence of the preconditioner.

Crack Models, Based on the Extended Finite Element Method

To take into account this more general case, we need to reconsider the mathematical formulation of the problem. The new formulation will differ from the earlier formulation (9) for which the crack faces were assumed traction free. Let t^+ be the stress vector felt by the crack face S_{c^+} as shown in Figure 23.

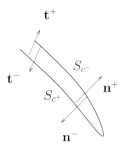

Figure 23. Notations to describe the stress vector on the cracks faces.

Due to the continuity of the stress vector across the crack, the crack S_{c^-} will feel the opposite action: $t^- = -t^+$. The traction free conditions (3) on the crack faces is now replaced by

$$\boldsymbol{\sigma}^+ \cdot \boldsymbol{n}^+ = \boldsymbol{t}^+ \text{ on } S_{c^+}, \quad \boldsymbol{\sigma}^- \cdot \boldsymbol{n}^- = -\boldsymbol{t}^+ \text{ on } S_{c^-} \qquad (45)$$

and the corresponding variational principle now reads

$$\int_\Omega \varepsilon(\boldsymbol{u}) : \mathbb{E} : \varepsilon(\delta\boldsymbol{u}) \, \mathrm{d}\Omega - \int_{S_{c^+}} \boldsymbol{t}^+ \cdot [\![\delta\boldsymbol{u}]\!] \, \mathrm{d}S = \int_{S_t} \boldsymbol{t}^\star \cdot \delta\boldsymbol{u} \, \mathrm{d}S \quad \forall \delta\boldsymbol{u} \in U_0 \qquad (46)$$

where $[\![\delta\boldsymbol{u}]\!]$ indicates the difference between the value of $\delta\boldsymbol{u}$ on S_{c^+} and S_{c^-}. To complete the formulation, we need to provide the relationship between the stress vector and the crack opening. In the case of an elastic joint gluing together both side of the crack, the relationship will simply read:

$$-\boldsymbol{t}^+ = k[\![\boldsymbol{u}]\!] \qquad (47)$$

where k is the "joint" stiffness. Note that the law above does not prevent the crack faces to penetrate each other, contact needs to be added.

In order to describe more complex laws on the crack faces, we shall introduce a couple of notations. The stress vector will be decomposed into its normal (scalar t_n) and tangential components (vector $\boldsymbol{t}_\tau$)

$$\boldsymbol{t}^+ = t_n \boldsymbol{n}^+ + \boldsymbol{t}_\tau, \quad t_n = \boldsymbol{t}^+ \cdot \boldsymbol{n}^+, \quad \boldsymbol{t}_\tau = \boldsymbol{t}^+ - t_n \boldsymbol{n}^+ \qquad (48)$$

Similarly the displacement jump is decomposed into a scalar normal jump (u_n) and tangential vectorial jump $\boldsymbol{u}_\tau$

$$-[\![\boldsymbol{u}]\!] = u_n \boldsymbol{n}^+ + \boldsymbol{u}_\tau \quad u_n = -[\![\boldsymbol{u}]\!] \cdot \boldsymbol{n}^+, \quad \boldsymbol{u}_\tau = -[\![\boldsymbol{u}]\!] - u_n \boldsymbol{n}^+ \qquad (49)$$

Regarding the normal part of the traction law on the interface, the most common choices are depicted in Figure 24. The cohesive type law is rather

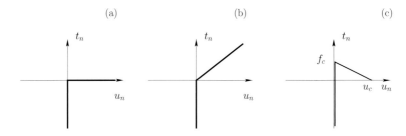

Figure 24. A set of classical normal law on the crack faces: basic contact (a), elastic interface (b), cohesive interface (c)

complex since it is non convex (t_n may not be expressed as the derivative of a convex potential in u_n). It is also irreversible in the sense that the unloading does not follow the loading curve as shown in Figure 25.

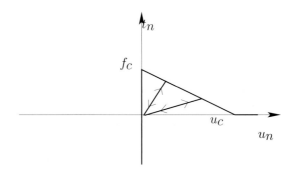

Figure 25. Cohesive law with the loading unloading depicting the gradual loss of stiffness of the interface

Regarding the tangential behavior of the interface, the case without friction leads to the nullity of the part of stress vector: $\boldsymbol{t}_\tau = 0$ whereas a

Coulomb fiction law is driven by the following relationship

$$u_\tau = \lambda t_\tau, \quad \lambda \geq 0, \quad f = \| t_\tau \| - \mu t_n \leq 0, \quad f\lambda = 0 \qquad (50)$$

An elegant way to formulate the Coulomb friction law is through the bipotential framework introduced by (de Saxcé, 1992). Note also that the cohesive law may be extremely rich in terms of mechanical phenomena: plasticity or speed effect may be added.

Using the decomposition (48) and (49), the variational principle (46) may be rewritten as

$$\int_\Omega \varepsilon(u) : \mathbf{E} : \varepsilon(\delta u) \, d\Omega + \int_{S_{c+}} t_n \delta u_n \, dS + \int_{S_{c+}} t_\tau \cdot \delta u_\tau \, dS$$

$$= \int_{S_t} t^\star \cdot \delta u \, dS \quad \forall \delta u \in U_0 \qquad (51)$$

The variational principle (51) gives the equilibrium condition to be met for given tractions on the crack faces. Since these are unknown they will be discretized. We first consider an interface law without friction and basic contact (case a in Figure 24). Mathematically, this law is expressed locally by

$$u_n \geq 0, \quad t_n \leq 0, \quad t_n u_n = 0 \qquad (52)$$

Let L be the space of regular function t_n defined on S_{c+}. The goal is to find the pair $(u, t_n) \in U_0 \times L$ such that

$$\int_\Omega \varepsilon(u) : \mathbf{E} : \varepsilon(\delta u) \, d\Omega + \int_{S_{c+}} t_n \delta u_n \, dS = \int_{S_t} t^\star \cdot \delta u \, dS \quad \forall \delta u \in U_0$$

$$\int_{S_{c+}} \delta t_n u_n \, dS = 0 \quad \forall \delta t_n \in L$$

The above does not in fact enforce correctly contact, it enforces in fact u_n to be zero on the crack (the crack cannot open). For the above to model contact, we need to impose a priori $u_n \geq 0$ and $t_n \leq 0$ in the approximation space which is of course very cumbersome. Fortunately, we do not need these a priori assumptions by using the work of Ben Dhia et al. (2000). The contact conditions (52) may be summarized by a single (highly nonlinear) equality

$$t_n = \chi(t_n, u_n)(t_n + \beta u_n) \qquad (53)$$

where β is a strictly positive parameter and

$$\chi = \chi(t_n, u_n) = 0 \quad \text{if } t_n + \beta u_n \geq 0 \qquad (54)$$
$$\chi = \chi(t_n, u_n) = 1 \quad \text{if } t_n + \beta u_n < 0 \qquad (55)$$

The variational principle now reads: find the pair $(\boldsymbol{u}, t_n) \in U_0 \times L$ such that

$$\int_\Omega \varepsilon(\boldsymbol{u}) : \mathbb{E} : \varepsilon(\delta \boldsymbol{u}) + \beta \chi u_n \delta u_n \ \mathrm{d}\Omega$$

$$+ \int_{S_{c+}} \chi t_n \delta u_n \ \mathrm{d}S = \int_{S_t} \boldsymbol{t}^\star \cdot \delta \boldsymbol{u} \ \mathrm{d}S \quad \forall \delta \boldsymbol{u} \in U_0$$

$$\int_{S_{c+}} \chi \delta t_n u_n + \frac{(\chi - 1)}{\beta} t_n \delta t_n \ \mathrm{d}S = 0 \quad \forall \delta t_n \in L$$

In the system above, spaces U and L no longer involve sign conditions and we may proceed to the space discretization. The X-FEM discrete displacement space U_0^h has already been described earlier. Regarding the discrete pressure space L, extra care needs to be taken in order to satisfy the so-called Inf-Sup Babuska-Brezzi condition.

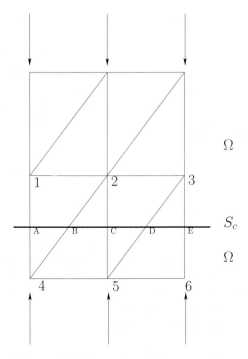

Figure 26. A crack cutting a mesh.

To illustrate the problem, consider the rectangular domain depicted in

Figure 26 (fully) cut by a crack. The two parts of the domain are pressed towards each other. It is tempting to discretize t_n based on the 1D mesh on S_c with 5 "nodes", labelled A to E. Unfortunately, this yield highly oscillatory pressure t_n over S_c (Ji and Dolbow, 2004; Moës et al., 2006). If the mesh is refined, the oscillatory behavior may even get worse. This oscillatory behavior is to be related to locking issue similar to what happens in incompressible formulation when the pressure is too rich and creates checkerboard type patterns. This problem is specific to the X-FEM because the crack lays inside the element. If the crack is meshed, this issue does not appear. A set of papers have been devoted to alleviate this locking issue, following different strategies

- following a Nitsche type approach (Nitsche, 1971; Hansbo and Hansbo, 2002, 2004)
- following a residual-free bubble stabilization approach (Mourad et al., 2007; Dolbow and Franca, 2008)
- following a Barbosa and Hugues type stabilization of the Lagrange multipliers (Haslinger and Renard, 2008)
- using a mortar based approach (Kim et al., 2007)
- proper choice of the Lagrange multiplier space (Moës et al., 2006), (Géniaut et al., 2007) and more recently (Béchet et al., 2009).

We will now detail the later paper (Béchet et al., 2009) which is particularly attractive because the pressure field is discretized using the same nodes as the displacement field. The nodes of all element cut by the crack will bear a pressure t_n degree of freedom (these nodes are numbered from 1 to 6 in Figure 26). It was proved in (Béchet et al., 2009) that by applying specific ties between the pressure degrees of freedom, the inf-sup was fulfilled.

The algorithm to create the ties goes as follows. Let E be the set of edges cut by the crack. We pick in E a set of independent edges. Two edges are said to be independent if they do not share a common nodes. Note that the choice of independent edges is not unique. In Figure 26, the set may be for instance (2,4),(3,5) or (1,4),(3,5) or even (1,4),(2,5),(3,6). On the more complex mesh depicted in Figure 27, a possible set of independent edges is indicated by dots (and square at the end nodes).

The pressure degrees of freedom at the end nodes of each independent edge are forced to be equal. These is illustrated by the numbers which are the same on Figure 27 (top), for the end nodes of each independent edge.

Once the independent edges have been selected, some nodes of edges in E may not have been taken care of. This is the case for the circled nodes in Figure 27 (bottom). The pressure field at these nodes is linked to the value at the squared nodes to which it is connected to through an edge in E (when the squared are multiple, either one is picked or a linear combination may be built with coefficients forming a partition of unity (Béchet et al., 2009)). The algorithm described above may also be applied in 3D.

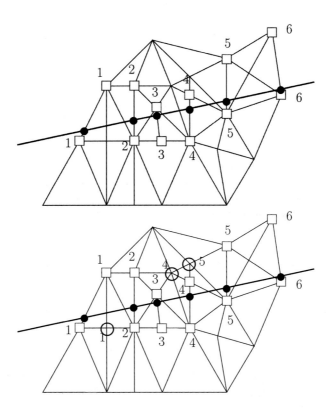

Figure 27. (Top) Selection of the independent edges. These edges are in between square nodes. The connected squared nodes will have the same pressure degree of freedom. (Bottom) Isolated nodes are located by circle nodes. Their pressure degree of freedom is forced equal to the one of a square they can connect to through a cut edge.

5 Configurational analysis of the crack front

5.1 The Eshelby tensor

Eshelby tensor is a second order tensor generally non symmetric defined by

$$P_{ij} = w\delta_{ij} - \sigma_{kj}u_{k,i}, \quad w = \frac{1}{2}\sigma_{kl}\epsilon_{kl} \tag{56}$$

in the case of small strain linear elasticity.

This tensor is a so called configurational tensor because it gives information on the change of energy in a system when its configuration is changed. Consider the clamped domain shown at the top in Figure 28. If a part of the domain is taken out (middle Figure), the energy int the system will change.

Let us assume that we keep removing material with a velocity of material retrieval v (bottom Figure 28). The rate of loss of potential energy is given by (in tensorial and indicial notations) by

$$\dot{\mathcal{U}} = \int_S \boldsymbol{v} \cdot \boldsymbol{P} \cdot \boldsymbol{n} \, \mathrm{d}S = \int_S v_i P_{ij} n_j \, \mathrm{d}S \tag{57}$$

where $\boldsymbol{n}$ is the outer normal.

Since the boundary of the material being retrieved is traction free ($\sigma_{ij}n_j = 0$) the integrand above reduces simply to $-w \parallel \boldsymbol{v} \parallel$ which is indeed negative implying a drop of potential energy in the system. The drop value is simply related to the elastic energy density present before the advance of the front. Again, we stress the fact that this simple expression was obtained on a traction free boundary. More complex expressions arises in the case of loaded boundaries and the general formula (57) must then be used.

Imagine now that the growing front depicted in Figure 28 (bottom) does not remove material but replaces a material with another one (phase change). We obtain then picture 29. The change in potential energy now reads

$$\dot{\mathcal{U}} = \int_S \boldsymbol{v} \cdot [\![\boldsymbol{P}]\!] \cdot \boldsymbol{n} \, \mathrm{d}S \tag{58}$$

The jump denoting the Eshelby tensor on right before the front minus the one right behind the front. This jump will be zero if the two material phases are identical. Indeed, replacing a material by the same one does not change at all the configuration and thus the potential energy.

The Eshelby tensor gives information on the change of energy in a system due to a change of configuration. On the other hand the Cauchy stress tensor gives information in the change of energy in system due to a change in spatial location. To understand this, consider now in Figure 28 (bottom) that the velocity $\boldsymbol{v}$ is an imposed velocity on the particles of the domain

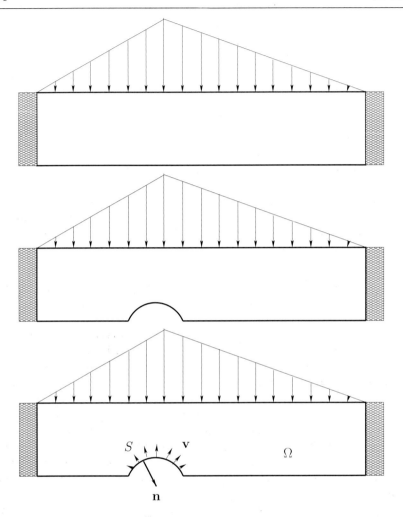

Figure 28. Top: a clamped domain subjected to some loadings, middle: the same loading applied to a different configuration, bottom: this Figure represents a spatial or configurational velocity on the boundary of the domain.

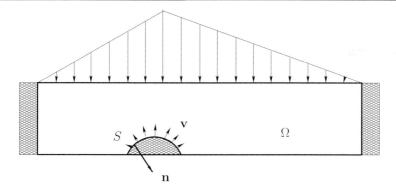

Figure 29. The clamped domain with a material interface.

(and no longer a rate at which material is being removed). This imposed velocity will generate stresses. The change in potential now reads

$$\dot{\mathcal{U}} = \int_S \boldsymbol{v} \cdot \boldsymbol{\sigma} \cdot \boldsymbol{n} \ \mathrm{d}S = \int_S v_i \sigma_{ij} n_j \ \mathrm{d}S \qquad (59)$$

The similarity with (57) is clear.

If we now consider Figure 29 a material interface being pushed (we are not talking about a phase change but the fact that the interface is being pushed at some velocity). The change in energy reads

$$\dot{\mathcal{U}} = \int_S \boldsymbol{v} \cdot [\![\boldsymbol{\sigma}]\!] \cdot \boldsymbol{n} \ \mathrm{d}S \qquad (60)$$

The jump term $[\![\boldsymbol{\sigma}]\!] \cdot \boldsymbol{n}$ is the reaction to the imposed velocity.

Assuming the stress $\boldsymbol{\sigma}$, strain $\boldsymbol{\varepsilon}$ and displacement $\boldsymbol{u}$ fields satisfy elasticity equations (equilibrium, compatibility and constitutive behavior):

$$\sigma_{ij,j} = 0 \quad \epsilon_{ij} = \frac{1}{2}(u_{i,j} + u_{j,i}) \quad \sigma_{ij} = E_{ijkl}\epsilon_{kl} \qquad (61)$$

the Eshelby tensor is divergence free

$$\nabla \cdot \boldsymbol{P} = 0, \quad P_{ij,j} = 0 \qquad (62)$$

5.2 Energy integrals

The fact that the Eshelby tensor is divergence free means that over any closed contour S the following integral is zero

$$\int_S \boldsymbol{P} \cdot \boldsymbol{n} \, \mathrm{d}S \tag{63}$$

where $\boldsymbol{n}$ is the outward normal to the contour. In fact, the assertion is true provided the domain described by the close contour has a smooth and continuous solution (and thus the divergence theorem may be applied). In other words the expression above is wrong if the contour S surrounds a crack tip. The integral (a related version to be precise) will be however very useful to characterize the strength of the singularity.

Next, we show that even though the integral around a crack tip is not zero, the result obtained will be the same whatever the contour chosen. More precisely, we have the following property introduced by Rice.

$$J = \int_{S_1} \boldsymbol{q} \cdot \boldsymbol{P} \cdot \boldsymbol{n}_1 \, \mathrm{d}S_1 = \int_{S_2} \boldsymbol{q} \cdot \boldsymbol{P} \cdot \boldsymbol{n}_2 \, \mathrm{d}S_2 \tag{64}$$

The contour S_1 and S_2 are depicted by dashed lines in Figure 30. The vector $\boldsymbol{q}$ is a vector indicating the direction of the crack (assumed straight at this point). To prove (64), we first define Ω_{12} as the domain bounded by S_1, S_2, S_{12c+} and S_{12c-}. Over the Ω_{12}, the mechanical field are smooth and we may apply the divergence theorem:

$$\int_{S_1 \cup S_2 \cup S_{12c+} \cup S_{12c-}} \boldsymbol{q} \cdot \boldsymbol{P} \cdot \boldsymbol{n} \, \mathrm{d}S = \boldsymbol{q} \cdot \int_{\Omega_{12}} \nabla \cdot \boldsymbol{P} \, \mathrm{d}\Omega = 0 \tag{65}$$

where $\boldsymbol{n}$ is the outward normal to Ω_{12}. Since $\boldsymbol{n}$ and $\boldsymbol{q}$ are orthogonal over $S_{12c+} \cup S_{12c-}$ and the traction is free over these segments, we have

$$\boldsymbol{q} \cdot \boldsymbol{P} \cdot \boldsymbol{n} = 0 \text{ on } S_{12c+} \cup S_{12c-} \tag{66}$$

yielding (64).

The physical meaning of the J integral is the power dissipated as the crack tip advances with the speed $\boldsymbol{q}$. Rice did show that J was related in linear elasticity to the stress intensity factor of the crack (for a unit crack tip velocity).

$$J = \frac{(1-\nu^2)}{E}(K_\mathrm{I}^2 + K_\mathrm{II}^2) + \frac{1}{2\mu} K_\mathrm{III}^2 \tag{67}$$

The contour integral (64) may be transformed into a so-called domain integral Destuynder et al. (1983). In order to perform the transition from

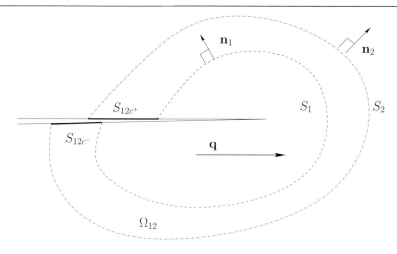

Figure 30. Notations to prove the J-rice contour independence property.

a contour to a domain integral, let us change slightly the definition of the virtual velocity q. It is a unit vector aligned with the crack within the contour S_0 and it drops to zero on the contour S, see Figure 31. We may write

$$J = \int_{S_0} q \cdot P \cdot n_0 \ dS + \int_S q \cdot P \cdot n \ dS = -\int_\Omega \nabla \cdot (q \cdot P) \ d\Omega = -\int_\Omega \nabla q : P \ d\Omega \tag{68}$$

Thus

$$J = -\int_\Omega \nabla q : P \ d\Omega \tag{69}$$

The domain Ω in the above is the domain enclosed by the contour S, since in the proof (68), the contour S_0 may be taken as small as one wishes around the tip. Compared to (64), the domain integral (69) is much more appropriate to finite element computations since the domain integral may be split as integral over elements. In fact, only one layer of elements do contribute to the integral as depicted in Figure 32 since the over the inner elements of the domain the q field is uniform.

5.3 Energetic information for cohesive cracks

So far, we discussed only straight traction free crack. Let us now consider that the crack faces are no longer traction free due to a contact or the

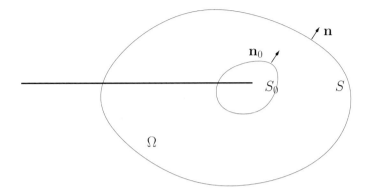

Figure 31. Notations for the domain integral proof.

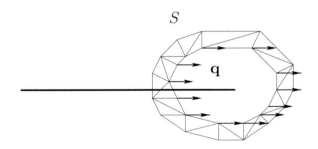

Figure 32. Elements contributing to the J domain integral.

presence of a cohesive zone. The J domain integral (69) (or its contour expression (64)) is then no longer contour independent. Independency may be regained by adding a boundary term to the J integral on the crack faces enclosed by the domain (contour). The more general expression is

$$J = \underbrace{-\int_\Omega \nabla q : P \ \mathrm{d}\Omega}_{J_\Omega} + \underbrace{\int_{S_{c+} \cup S_{c-}} q \cdot P \cdot n \ \mathrm{d}S}_{J_{\mathrm{coh}}} \qquad (70)$$

In order to show the domain independency, we use the notation depicted in Figure 33. We need to show that

$$-\int_{\Omega_1} \nabla \boldsymbol{q} : \boldsymbol{P} \, \mathrm{d}\Omega + \int_{S_{1c+} \cup S_{1c-}} \boldsymbol{q} \cdot \boldsymbol{P} \cdot \boldsymbol{n} \, \mathrm{d}S =$$
$$-\int_{\Omega_2} \nabla \boldsymbol{q} : \boldsymbol{P} \, \mathrm{d}\Omega + \int_{S_{2c+} \cup S_{2c-}} \boldsymbol{q} \cdot \boldsymbol{P} \cdot \boldsymbol{n} \, \mathrm{d}S \quad (71)$$

The notations are detailed in Figure 33. Removing the contribution from the domain Ω_0 from both sides, we get.

$$-\int_{\Omega_1 \setminus \Omega_0} \nabla \boldsymbol{q} : \boldsymbol{P} \, \mathrm{d}\Omega + \int_{S_{1c+} \cup S_{1c-} \setminus S_{0c+} \cup S_{0c-}} \boldsymbol{q} \cdot \boldsymbol{P} \cdot \boldsymbol{n} \, \mathrm{d}S =$$
$$-\int_{\Omega_2 \setminus \Omega_0} \nabla \boldsymbol{q} : \boldsymbol{P} \, \mathrm{d}\Omega + \int_{S_{2c+} \cup S_{2c-} \setminus S_{0c+} \cup S_{0c-}} \boldsymbol{q} \cdot \boldsymbol{P} \cdot \boldsymbol{n} \, \mathrm{d}S \quad (72)$$

Applying the divergence theorem on both sides and owing to the fact that the Eshelby tensor is divergence free, we get the proof.

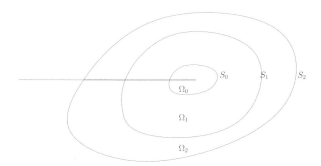

Figure 33. Notations to prove the J domain integral independency in the case of cohesive cracks. The domains Ω_i, $i = 1, \ldots, 3$ are enclosed by the contours S_i, $i = 1, \ldots, 3$. Similarly, the notation S_{ic+-} denotes the part of the crack enclosed by the contour S_i.

In the case of a cohesive crack it is interesting to detail the expression of the boundary term.

$$J_{\text{coh}} = \int_{S_{c+} \cup S_{c-}} \boldsymbol{q} \cdot \boldsymbol{P} \cdot \boldsymbol{n} \, \mathrm{d}S \tag{73}$$

$$= \int_{S_{c+}} (\boldsymbol{q} \cdot \nabla \boldsymbol{u}^+) \cdot (\boldsymbol{\sigma}^+ \cdot \boldsymbol{n}^+) \, \mathrm{d}S + \tag{74}$$

$$\int_{S_{c-}} (\boldsymbol{q} \cdot \nabla \boldsymbol{u}^-) \cdot (\boldsymbol{\sigma}^- \cdot \boldsymbol{n}^-) \, \mathrm{d}S \tag{75}$$

$$= -\int_0^l \boldsymbol{t}^+ \frac{\mathrm{d}[\![\boldsymbol{u}]\!]}{\mathrm{d}s} \, \mathrm{d}s \, \| \boldsymbol{q} \| \tag{76}$$

$$= -\int_0^{u_c} t_n \, \mathrm{d}u_n \, \| \boldsymbol{q} \| \tag{77}$$

The last inequality was obtained assuming no friction on the crack faces. The boundary term is thus negative and is the opposite of the area under the cohesive law (value $-0.5 f_c u_c$ is we consider the cohesive law depicted in Figure 25. The value of the J domain integral is zero for a cohesive crack since there is no singularity at the crack tip, we thus have

$$J = 0 = \underbrace{J_\Omega}_{\geq 0} + \underbrace{J_{\text{coh}}}_{\leq 0} \tag{78}$$

The J_Ω part is positive and represents the strength of the singularity at the crack tip if all cohesive forces were removed. The cohesive forces do create an opposite singularity. Even though no singularity exist at the tip of a cohesive crack, the stress field at some distance of the crack tip (larger that the cohesive zone length) behave as if there was a singularity (K dominance zone). The integral J_Ω is able to compute the strength of this K field. Note that the integral J_Ω is domain independant provided the domain always embeds fully the cohesive zone.

Finally, the condition $J = 0$ may be used as a robust way to find the proper load for a given extent of the cohesive zone as in (Moës and Belytschko, 2002). Since cohesive crack do not yield a singular field, the tip enrichment described in (37) is not appropriate. Non singular tip functions may be used as described in (Zi and Belytschko, 2003) or (Moës and Belytschko, 2002). Some authors prefer to consider only the Heaviside enrichment so that elements are either not cut or completely cut: (Wells and Sluys, 2001), (Mergheim et al., 2005) and (Meschke and Dumstorff, 2007).

Bibliography

I. Babuška and I. Melenk. Partition of unity method. *International Journal for Numerical Methods in Engineering*, 40(4):727–758, 1997.

I. Babuška and M. Rosenzweig. A finite element scheme for domains with corners. *Numer. Math.*, 20:1–21, 1972.

T. J. Barth and J. A. Sethian. Numerical schemes for the Hamilton-Jacobi and level set equations on triangulated domains. *Journal of Computational Physics*, 145(1):1–40, 1998.

K. J. Bathe. *Finite element procedures*. Prentice-Hall, 1996.

E. Béchet, H. Minnebo, N. Moës, and B. Burgardt. Improved implementation and robustness study of the x-fem method for stress analysis around cracks. *International Journal for Numerical Methods in Engineering*, 64: 1033–1056, 2005.

E. Béchet, N. Moës, and B. Wohlmuth. A stable lagrange mutliplier space for stiff interface condition within the extended finite element method. *International Journal for Numerical Methods in Engineering*, 78:931–954, 2009.

T. Belytschko and T. Black. Elastic crack growth in finite elements with minimal remeshing. *International Journal for Numerical Methods in Engineering*, 45(5):601–620, 1999.

T. Belytschko and M. Fleming. Smoothing, enrichment and contact in the element-free galerkin method. *Computers and Structures*, 71(2):173–195, 1999.

T. Belytschko, Y.Y. Lu, and L. Gu. Element-free galerkin methods. *International Journal for Numerical Methods in Engineering*, 37:229–256, 1994.

H. Ben Dhia, I. Vautier, and M. Zarroug. Problèmes de contact frottant en grandes transformations: du continu au discret. *Revue Européenne des éléments finis*, 9:243–261, 2000.

G. de Saxcé. A generalization of Fenchel's inequality and its applications to constitutive law. *Compte-Rendus Acad. Sci. Paris, Série II*, 314:125–129, 1992.

P. Destuynder, M. Djaoua, and S. Lescure. Some remarks on elastic fracture mechanics (quelques remarques sur la mécanique de la rupture élastique). *Journal de Mécanique théorique et appliquée*, 2(1):113–135, 1983.

J.E. Dolbow and L.P. Franca. Residual-free bubbles for embedded dirichlet problems. *Comp. Meth. in Applied Mech. and Engrg.*, 197:3751–3759, 2008.

C.A.M. Duarte and J.T. Oden. An hp meshless method. *Numerical methods for partial differential equations*, 12:673–705, 1996.

M. Fleming, Y. A. Chu, B. Moran, and T. Belytschko. Enriched element-free Galerkin methods for crack tip fields. *International Journal for Numerical Methods in Engineering*, 40(8):1483–1504, 1997.

S. Géniaut, P. Massin, and N. Moës. A stable 3D contact formulation for cracks using x-fem. *Revue européenne de mécanique numérique*, 16:259–276, 2007.

P. Grisvard. *Elliptic Problems in Nonsmooth Domains*. Pitman Publishing, Inc, Boston, 1985.

A. Hansbo and P. Hansbo. An unfitted finite element method, based on Nitsche's method, for elliptic interface problems. *Computer Methods In Appl. Mechanics Engineering*, 191:5537–5552, 2002.

A. Hansbo and P. Hansbo. A finite element method for the simulation of strong and weak discontinuities in solid mechanics. *Computer Methods In Applied Mechanics And Engineering*, 193:3523–3540, 2004.

J. Haslinger and Y. Renard. A new fictitious domain approach inspired by the extended finite element method. *SIAM Journal of numerical analysis*, 47:1474–1499, 2009.

H. Ji and J.E. Dolbow. On strategies for enforcing interfacial constraints and evaluating jump conditions with the extended finite element method. *International Journal for Numerical Methods in Engineering*, 61:2508–2535, 2004.

T. Y. Kim, J. Dolbow, and T. Laursen. A mortared finite element method for frictional contact on arbitrary interfaces. *Computational Mechanics*, 39(3):223–235, 2007.

P. Krysl and T. Belytschko. Element free Galerkin method for dynamic propagation of arbitrary 3-d cracks. *International Journal for Numerical Methods in Engineering*, 44(6):767–800, 1999.

P. Laborde, J. Pommier, Y. Renard, and M. Salaun. High-order extended finite element method for cracked domains. *International Journal for Numerical Methods in Engineering*, 64:354–381, 2005.

W.K. Liu, J. Adee, S. Jun, and T. Belytschko. Reproducing kernel particle methods for elastic and plastic problems. *Am Soc Mech Eng Appl Mech Div AMD*, 180:175–189, 1993.

J.M. Melenk and I. Babuška. The partition of unity finite element method: Basic theory and applications. *Comp. Meth. in Applied Mech. and Engrg.*, 39:289–314, 1996.

J. Mergheim, E. Kuhl, and P. Steinmann. A finite element method for the computational modelling of cohesive cracks. *Int. J. For Numerical Methods In Engineering*, 63(2):276–289, 2005.

G. Meschke and P. Dumstorff. Energy-based modeling of cohesive and cohesionless cracks via X-FEM. *Computer Methods In Appl. Mechanics Engineering*, 196(21-24):2338–2357, 2007.

N. Moës and T. Belytschko. Extended finite element method for cohesive crack growth. *Engineering Fracture Mechanics*, 69:813–834, 2002. URL http://dx.doi.org/10.1016/S0013-7944(01)00128-X.

N. Moës, J. Dolbow, and T. Belytschko. A finite element method for crack growth without remeshing. *International Journal for Numerical Methods in Engineering*, 46:131–150, 1999.

N. Moës, A. Gravouil, and T. Belytschko. Non-planar 3D crack growth by the extended finite element and level sets. part I: Mechanical model. *International Journal for Numerical Methods in Engineering*, 53:2549–2568, 2002.

N. Moës, M. Cloirec, P. Cartraud, and J.-F. Remacle. A computational approach to handle complex microstructure geometries. *Comp. Meth. in Applied Mech. and Engrg.*, 192:3163–3177, 2003. URL http://dx.doi.org/doi:10.1016/S0045-7825(03)00346-3.

N. Moës, E. Béchet, and M. Tourbier. Imposing essential boundary conditions in the extended finite element method. *International Journal for Numerical Methods in Engineering*, 67:1641–1669, 2006.

N. Molino, Z. Bao, and R. Fedkiw. A virtual node algorithm for changing mesh topology during simulation. *SIGGRAPH, ACM TOG 23*, pages 385–392, 2004.

H. M. Mourad, J. Dolbow, and I. Harari. A bubble-stabilized finite element method for Dirichlet constraints on embedded interfaces. *Int. J. For Numerical Methods In Engineering*, 69(4):772–793, 2007.

B. Nayroles, G. Touzot, and P. Villon. Generalizing the finite element method: Diffuse approximation and diffuse elements. *Computers and Structures*, 10(5):307–318, 1992.

J. Nitsche. Über ein Variationsprinzip zur lösung von Dirichlet-problemen bei Verwendung von Teilräumen, die keinen Randbedingungen unterworfen sind. *Abhandlungen aus dem Mathematischen Seminar des Universität Hamburg*, 36:9–15, 1971.

S. Osher and R. Fedkiw. *Level set methods and dynamic implicit surfaces.* Springer Verlag, 2002.

S. Osher and J. A. Sethian. Fronts propagating with curvature-dependent speed: Algorithms based on Hamilton-Jacobi formulations. *Journal of Computational Physics*, 79(1):12–49, November 1988.

Y. Saad. *Iterative methods for sparse linear systems.* PWS Publishing company, third edition edition, 2000.

J. A. Sethian. *Level Set Methods & Fast Marching Methods: Evolving Interfaces in Computational Geometry, Fluid Mechanics, Computer Vision, and Materials Science.* Cambridge University Press, Cambridge, UK, 1999.

J.A. Sethian. *Level Set Methods: Evolving Interfaces in Computational Geometry, Fluid Mechanics, Computer Vision, and Materials Science.* Cambridge Monographs on Applied and Computational Mathematics, 1996.

J.H. Song, P.M.A. Areais, and T. Belytschko. A method for dynamic crack and shear band propagation with phantom nodes. *International Journal for Numerical Methods in Engineering*, 67:868–893, 2006.

M. Stolarska, D. L. Chopp, N. Moës, and T. Belytschko. Modelling crack growth by level sets and the extended finite element method. *International Journal for Numerical Methods in Engineering*, 51(8):943–960, 2001.

T. Strouboulis, I. Babuška, and K. Copps. The design and analysis of the generalized finite element method. *Comp. Meth. in Applied Mech. and Engrg.*, 181:43–71, 2000.

N. Sukumar, N. Moës, T. Belytschko, and B. Moran. Extended Finite Element Method for three-dimensional crack modelling. *International Journal for Numerical Methods in Engineering*, 48(11):1549–1570, 2000.

G.N. Wells and L.J. Sluys. A new method for modelling cohesive cracks using finite elements. *International Journal for Numerical Methods in Engineering*, 50:2667–2682, 2001.

G. Zi and T. Belytschko. New crack-tip elements for XFEM and applications to cohesive cracks. *International Journal for Numerical Methods in Engineering*, 57:2221–2240, 2003.

Smeared Crack and X-FEM Models in the Context of Poromechanics

Günther Meschke[†], Stefan Grasberger[‡], Christian Becker[††], Stefan Jox[‡‡]

[†] Institute for Structural Mechanics, Ruhr University Bochum, Germany
[‡] Hilti Corporation, Schaan, Liechtenstein
[††] Institute for Continuum Mechanics, Ruhr University Bochum, Germany
[‡‡] ZERNA Ingenieure GmbH, Bochum, Germany

1 Elastoplastic-Damage Models for Concrete

1.1 Introductory Remarks

The process of cracking in quasi-brittle materials such as concrete is characterized by the formation and coalescence of micro-cracks which eventually form a propagating macro-crack. This process is connected with the gradual loss of load carrying capacity, generally denoted as material softening. As the most popular mode to represent cracking within numerical methods, such as the finite element method, is the representation of the cracking process in a continuum damage mechanics framework (see the reviews contained e.g. in Hofstetter and Mang (1995); de Borst et al. (1998); de Borst (2002); Mang et al. (2003)). Within this class of models, dissipation associated with softening is either formulated in a local (fracture-energy based) or in a non-local setting.

In this section, a coupled elastoplastic-damage model for concrete based upon the fracture energy concept, as proposed by Meschke et al. (1998a) is summarized in a continuum mechanics context first without considering hygromechanical couplings. This model is extended to poromechanics in Section 4. Coupling between plasticity theory and damage mechanics is motivated by the fact that cracking in concrete as well as compressive failure for low levels of confinement is associated with stiffness degradation as well as with inelastic deformations.

The simplest mode of coupling between damage and plasticity is a scalar damage elastoplastic model (Ju, 1989) based on the effective stress concept. It is characterized by the stress-strain relation

$$\boldsymbol{\sigma} = (1 - D(\kappa))\, \mathbf{C}^e : (\boldsymbol{\varepsilon} - \boldsymbol{\varepsilon}^p). \tag{1}$$

In (1), D is the isotropic damage parameter which depends on a measure of equivalent strains denoted as κ, ε^p is the tensor of plastic strains and $\mathbf{C}^e$ is the constant elasticity tensor. According to the effective stress concept, the yield function $f_p(\sigma_{eq}, \alpha_i)$ is formulated in terms of effective stresses $\hat{\boldsymbol{\sigma}}$ and a set of internal variables α_i:

$$f_p(\sigma_{eq}, \alpha_i) = \sigma_{eq}(\hat{\boldsymbol{\sigma}}) - q_i(\alpha_i) \quad \text{with} \quad \hat{\boldsymbol{\sigma}} = \frac{\boldsymbol{\sigma}}{1-D}. \qquad (2)$$

The evolution of the scalar damage parameter $D(\kappa)$ is controlled by a damage surface f_d formulated in the strain space

$$f_d = \eta(\boldsymbol{\varepsilon}) - \kappa \leq 0, \qquad (3)$$

with κ as the largest value of the equivalent strain $\eta(\boldsymbol{\varepsilon})$ experienced during the entire loading history.

In this section, an anisotropic elastoplastic-damage model for plain concrete is presented, in which stiffness degradation is taken into account by means of additional cracking strains and the growth of cracking and plastic strains is described by using only one damage function (Meschke et al. (1998a)). Conceptually, this model is an extension of the anisotropic damage model proposed by Govindjee et al. (1995). In contrast to this model, however, a rotating crack formulation is adopted in the present formulation.

1.2 Evolution Equations

The tensor of linearized strains is decomposed into an elastic and a plastic part

$$\boldsymbol{\varepsilon} = \boldsymbol{\varepsilon}^e + \boldsymbol{\varepsilon}^p. \qquad (4)$$

In analogy to classical plasticity, a region of admissible stress states is defined in the stress space by m active failure and yield surfaces f_k, respectively, that intersect in a non-smooth fashion:

$$\mathcal{E} = \{\boldsymbol{\sigma} | f_k(\boldsymbol{\sigma}, q_k) \leq 0, \quad k = 1, \ldots m\}, \qquad (5)$$

where q_k is a stress-like internal variable associated with the damage (yield) surface f_k related to a strain-like conjugate variable α_k by the relation

$$q_k(\alpha_k) = -\frac{1}{\rho} \partial_{\alpha_k} U(\alpha_k). \qquad (6)$$

ρ is the density of the material and $U(\alpha_k)$ is the part of the free energy associated with microstructural deterioration and slip processes in the material. The parameters $q_k(\alpha_k)$ determine the damage-dependent size of the damage surface f_k in the stress space. The degradation of the elastic moduli $\mathbf{C}$ and the growth of inelastic strains ε^p associated with the damage (yield) surface f_k are not regarded as independent processes. They are both controlled by a single scalar internal variable α_k. The function of free energy is defined as

$$\Psi(\varepsilon^e, \mathbf{C}, \alpha_k) = W(\varepsilon^e, \mathbf{C}) + U(\alpha_k) = \frac{1}{2\rho} \varepsilon^{e,T} \mathbf{C} \varepsilon^e + U(\alpha_k). \tag{7}$$

Restricting the present considerations to the purely mechanical theory, the *Clausius-Duhem* inequality requires

$$\mathcal{D} = \boldsymbol{\sigma} : \dot{\boldsymbol{\varepsilon}} - \rho \dot{\Psi} \geq 0. \tag{8}$$

From (6), (7) and (8), considering $\dot{\mathbf{C}}\mathbf{D} = -\mathbf{C}\dot{\mathbf{D}}$, follows

$$\mathcal{D} = \boldsymbol{\sigma}^T \dot{\boldsymbol{\varepsilon}}^p + \frac{1}{2} \boldsymbol{\sigma}^T \dot{\mathbf{D}} \boldsymbol{\sigma} + q_k(\alpha_k) \dot{\alpha}_k \geq 0. \tag{9}$$

In analogy to classical plasticity theory, the evolution of the compliance tensor $\mathbf{D}$, of the inelastic strains ε^p and of the internal variables α_k is obtained by exploiting the postulate of maximum dissipation (Simo and Hughes (1998)). For softening materials this hypothesis is replaced by the postulate of stationarity of the functional (9). Hence, for given admissible state variables $(\boldsymbol{\sigma}, q_k) \in \mathcal{E}$, the rates $\dot{\mathbf{D}}$, $\dot{\boldsymbol{\varepsilon}}^p$ and $\dot{\alpha}_k$ are those which yield a stationary point of the dissipation $\mathcal{D}$. To find the solution of this constrained optimization problem, the method of Lagrange multipliers is used, introducing the Lagrangean functional

$$\begin{aligned}
\mathcal{L}(\boldsymbol{\sigma}, q_k) &= -\mathcal{D} + \sum_{k=1}^{m} \dot{\gamma}_k f_k(\boldsymbol{\sigma}, q_k) \\
&= -\boldsymbol{\sigma}^T \dot{\boldsymbol{\varepsilon}}^p - \frac{1}{2} \boldsymbol{\sigma}^T \dot{\mathbf{D}} \boldsymbol{\sigma} - q_k(\alpha_k) \dot{\alpha}_k + \sum_{k=1}^{m} \dot{\gamma}_k f_k(\boldsymbol{\sigma}, q_k),
\end{aligned} \tag{10}$$

where $\dot{\gamma}_k \geq 0$ are m Lagrange multipliers. From the associated optimality conditions

$$\partial_{\boldsymbol{\sigma}} \mathcal{L} = \mathbf{0}, \qquad \partial_{q_k} \mathcal{L} = 0 \tag{11}$$

follows

$$\dot{\boldsymbol{\varepsilon}}^p + \dot{\mathbf{D}} \boldsymbol{\sigma} = \dot{\boldsymbol{\varepsilon}}^p + \dot{\boldsymbol{\varepsilon}}^d = \sum_{k=1}^{m} \dot{\gamma}_k \partial_{\boldsymbol{\sigma}} f_k(\boldsymbol{\sigma}, q_k), \quad \dot{\alpha}_k = \sum_{k=1}^{m} \dot{\gamma}_k \partial_{q_k} f_k(\boldsymbol{\sigma}, q_k), \tag{12}$$

where $\dot{\boldsymbol{\varepsilon}}^d$ are differential strains associated with the degradation of the compliance matrix and the loading/unloading conditions are given as

$$f_k \leq 0, \quad \dot{\gamma}_k \geq 0, \quad \dot{\gamma}_k f_k = 0. \tag{13}$$

Defining
$$\dot{\varepsilon}^{pd} = \dot{\varepsilon}^p + \dot{\varepsilon}^d, \tag{14}$$

(12$_1$) can be re-written in a form analogous to classical associative plasticity theory as

$$\dot{\varepsilon}^{pd} = \sum_{k=1}^{m} \dot{\gamma}_k \, \partial_{\boldsymbol{\sigma}} f_k(\boldsymbol{\sigma}, q_k). \tag{15}$$

Introducing a scalar parameter β_k, $0 \leq \beta \leq 1$, independently for each of the active damage surfaces, the plastic and the damage strains are given as

$$\begin{aligned}
\dot{\varepsilon}^p &= (1-\beta_k) \sum_{k=1}^{m} \dot{\gamma}_k \, \partial_{\boldsymbol{\sigma}} f_k(\boldsymbol{\sigma}, q_k), \\
\dot{\varepsilon}^d &= \dot{\mathbf{D}} \boldsymbol{\sigma} = \beta_k \sum_{k=1}^{m} \dot{\gamma}_k \, \partial_{\boldsymbol{\sigma}} f_k(\boldsymbol{\sigma}, q_k).
\end{aligned} \tag{16}$$

The parameter $0 \leq \beta_k \leq 1$ allows a partitioning of effects associated with inelastic deformations due to the crack-induced misalignment of the asperities of the crack surfaces, resulting in an increase of inelastic strains ε^p, and deterioration of the microstructure, resulting in a increase of the compliance moduli $\mathbf{D}$.

After some re-formulations (Meschke et al. (1998a)) the anisotropic evolution law for the compliance tensor $\mathbf{D}$ is obtained as

$$\dot{\mathbf{D}} = \sum_{k=1}^{m} \beta_k \, \dot{\gamma}_k \, \frac{\partial_{\boldsymbol{\sigma}} f_k(\boldsymbol{\sigma}, q_k) \otimes \partial_{\boldsymbol{\sigma}} f_k(\boldsymbol{\sigma}, q_k)}{\partial_{\boldsymbol{\sigma}} f_k(\boldsymbol{\sigma}, q_k) : \boldsymbol{\sigma}}. \tag{17}$$

To simplify the formulation of the plastic-damage evolution, the damage induced degradation may be assumed as isotropic. In this case, instead of the degradation of the stiffness tensor $\mathbf{C}$ the reduction of the isotropic integrity defined as

$$\psi = 1 - D \tag{18}$$

is used to describe evolving damage. Using, instead of the degrading stiffness tensor $\mathbf{C}$, the isotropic damage variable D or, equivalently, the integrity ψ as argument in the free energy (7) leads to the substitution of the anisotropic evolution according to (17) by the isotropic evolution equation for the reciprocal value of the integrity (ψ^{-1})

$$(\psi^{-1})\dot{\;} = \beta_k \sum_{k=1}^{4} \dot{\gamma}_k \frac{\dfrac{\partial f_k}{\partial \boldsymbol{\sigma}} : \mathbf{C} : \dfrac{\partial f_k}{\partial \boldsymbol{\sigma}}}{\dfrac{\partial f_k}{\partial \boldsymbol{\sigma}} : \boldsymbol{\sigma}}. \tag{19}$$

An elastoplastic model $((\psi^{-1})\dot{} = 0, \dot{\varepsilon}^p \neq \mathbf{0})$ and a damage model $((\psi^{-1})\dot{} \neq 0, \dot{\varepsilon}^p = \mathbf{0})$ are recovered as special cases by setting $\beta_k = 0$ and $\beta_k = 1$, respectively.

1.3 Damage and Yield Surfaces

To account for the brittle material behavior of concrete in tension, the maximum principle stress (*Rankine*) criterion is used. After crack initiation residual stresses are gradually decreasing. The damage surface can be formulated in terms of principal stresses as

$$f_R(\boldsymbol{\sigma}, q_R(\alpha_R)) = \sigma_A - (f_{tu} - q_R(\alpha_R)) \leq 0, \quad A = 1, \ldots 3, \quad (20)$$

with f_{tu} as the uniaxial tensile strength of concrete and the softening behavior controlling the post-cracking strength degradation is accounted for by the hyperbolic law

$$q_R(\alpha_R) = -\partial \mathcal{U}/\partial \alpha_R = f_{tu}\left[1 - (1 + \alpha_R/\alpha_{R,u})^{-2}\right]. \quad (21)$$

It should be noted, that the formulation in principal axes of the stress tensor does not account for damage induced anisotropy of the strength degradation after crack initiation. An extension to anisotropy has been proposed for 2D cases in Meschke et al. (1998b). For plane stress conditions, the damage surface formulated in terms of invariants reads

$$f_R(\boldsymbol{\sigma}, q_R) = \frac{\sigma_x + \sigma_y}{2} + \sqrt{\frac{1}{4}(\sigma_x - \sigma_y)^2 + \tau_{xy}^2} - [f_{tu} - q_R(\alpha_R)] \leq 0. \quad (22)$$

According to the fracture energy concept for softening materials (Willam (1984)), the parameter $\alpha_{R,u}$ is adjusted to the specific fracture energy of concrete in mode-I, G_f, and to the characteristic length l_c. In finite element analyses this length is associated to the length of the cracked finite element. From integrating

$$\int_0^\infty (f_{tu} - q_R(\alpha_R))\, d\alpha_R = G_f/l_c \quad (23)$$

follows

$$\alpha_{R,u} = G_f/(l_c f_{tu}). \quad (24)$$

For the determination of the characteristic length l_c an approach suggested by Oliver (1989) is used.

The ductile behavior of concrete subjected to compressive loading is described by a hardening/softening Drucker-Prager plasticity model. The yield function is defined as

$$f_D(\boldsymbol{\sigma}, q_D) = \sqrt{J_2} + \kappa_D I_1 - \frac{f_{cy} - q_D}{\beta_D} \leq 0, \qquad (25)$$

where f_{cy} is the uniaxial elastic limit, $q_D(\alpha_D)$, is the hardening/softening parameter dependent on the internal plastic variable α_D and I_1 and J_2 denote the first and second invariants of the stress tensor and stress deviator, respectively.

The material parameters κ_D and β_D are adjusted according to the ratio between the uniaxial and the biaxial strength of concrete f_{cb}/f_{cu} (Meschke et al. (1996)):

$$\kappa_D = \frac{1}{\sqrt{3}} \left(\frac{f_{cb}/f_{cu} - 1}{2 f_{cb}/f_{cu} - 1} \right), \quad \beta_D = \sqrt{3} \left(\frac{2 f_{cb}/f_{cu} - 1}{f_{cb}/f_{cu}} \right). \qquad (26)$$

The ratio f_{cb}/f_{cu} is assumed as 1.16. The nonlinear behavior of concrete in compression is governed by the hardening/softening law

$$q_D(\alpha_D) = \begin{cases} \frac{f_{cu} - f_{cy}}{\alpha_{D,u}^2} (\Delta \alpha_D)^2 - \Delta f_c & \text{for } \alpha_D < \alpha_{D,u}, \\ \frac{f_{cu}}{(\alpha_{D,c} - \alpha_{D,u})^2} (\Delta \alpha_D)^2 - \Delta f_c & \text{for } \alpha_{D,u} \leq \alpha_D < \alpha_{D,c}, \\ f_{cy} & \text{for } \alpha_{D,c} \leq \alpha_D, \end{cases} \qquad (27)$$

where $\alpha_{D,u}$ and $\alpha_{D,c}$ are material parameters, $\Delta f_c = f_{cu} - f_{cy}$ and $\Delta \alpha_D = \alpha_{D,u} - \alpha_D$. $\alpha_{D,u}$ is determined from the total strain level ε_u at $\sigma = f_{cu}$ in uniaxial loading as

$$\alpha_{D,u} = c (\varepsilon_u - f_{cu}/E_c), \quad \text{with} \quad c = \frac{1}{\beta_D (1/\sqrt{3} + \kappa_D)}. \qquad (28)$$

The critical value $\alpha_{D,c}$ is obtained from the uniaxial fracture energy of concrete G_c and the characteristic length l_c by setting

$$\int_0^{\alpha_{D,c}} (f_{cy} - q_D) \, d\alpha_D = G_c/l_c \qquad (29)$$

as

$$\alpha_{D,c} = \frac{3}{2} \frac{1}{f_{cu}} \left(c \frac{G_c}{l_c} - \frac{1}{3} f_{cy} \alpha_{D,u} \right). \qquad (30)$$

1.4 Numerical Analysis of a Notched Concrete Beam

A notched plain concrete beam subjected to cyclic quasi-static loading is analyzed numerically using the plastic-damage concrete model described previously. Respective test results are documented in Perdikaris and Romeo (1995). The geometrical specifications and the material data are taken from the test C2-D1-S3-R2. Figure 1 contains the geometry and the finite element discretization of the beam by means of 492 bilinear plane stress finite elements.

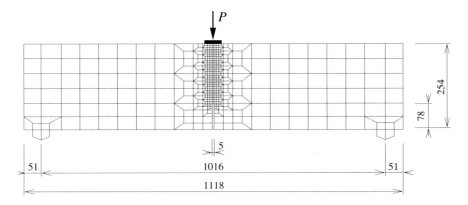

Figure 1. Notched concrete beam: dimensions and finite element discretization (units in [mm]).

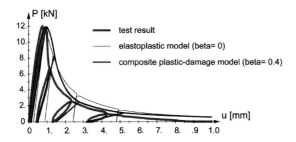

Figure 2. Re-analysis of a notched concrete beam: load-displacement diagrams obtained from the experiment and the plastic-damage model.

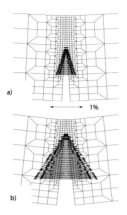

Figure 3. Re-analysis of a notched concrete beam: distribution of plastic strains in the vicinity of the notch at different loading stages: (a) u= 0.14 mm (P= 8.4 kN) and (b) u= 0.28 mm (P= 2.5 kN) (200-fold magnification of displacements).

The specimen was tested under displacement control according to the following procedure: as soon as the peak load was reached, the beam was unloaded, then reloaded up to a value of the crack mouth opening displacement (CMOD) twice the one previously attained at the peak load $CMOD_p$ and unloaded again. Subsequently, the beam was unloaded and reloaded at CMOD = 5 $CMOD_p$ and CMOD = 10 $CMOD_p$. The material parameters are chosen as follows: Young's modulus E= 43 600 N/mm^2, Poisson's ratio ν= 0.2, uniaxial tensile strength f_{tu}= 4.0 N/mm^2 and uniaxial compressive strength: f_{cu}= 63.4 N/mm^2. The model parameters $\alpha_{R,u}$, $\alpha_{D,c}$ and $\alpha_{D,u}$, have been calibrated from the fracture energy in uniaxial tension and compression, respectively, as $\alpha_{R,u}$ =8.365 10^{-3}, $\alpha_{D,u}$=7.459 10^{-4} and $\alpha_{D,c}$=5.637 10^{-2}. The fracture energy in mode I cracking G_f was reported as G_f= 0.1195 Nmm/mm^2 (Perdikaris and Romeo (1995)) and the fracture energy in uniaxial compression G_c was assumed as $G_c \tilde{} 50\ G_f$ =5.975 Nmm/mm^2. The plasticity-damage partitioning factor for cracking was chosen as β= 0.4. Figure 2 contains a comparison of the load-displacement diagrams obtained from the experiment (Perdikaris and Romeo (1995)) and from two finite element analyses based on the plastic-damage model (β= 0.4) and on an elastoplastic model (β= 0). The peak load and the post-peak regime of the curve are replicated fairly well by both analyses. In contrast to the results from the elastoplastic model, the un- and reloading paths obtained from the composite plastic-damage model also agree well with the test results. Figure 3 illustrates the distribution of the plastic tensile strains in the vicinity of the notch at different loading stages according to the plastic-damage model.

2 Hygro-mechanical Couplings in Concrete

2.1 Concrete Microstructure and Hygral Forces

Cementitious materials such as concrete are strongly hydrophyllic materials with a large internal surface, characterized by a large range of pore radii, ranging from nanometers to millimeters (Fig. 4). Due to molec-

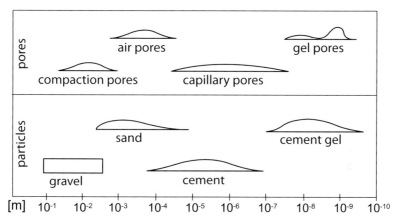

Figure 4. Comparison of pore sizes and solid components in concrete (Setzer (1975)).

ular adsorption and capillary condensation, saturation-dependent internal stresses, acting on the nano- and micro-scale, develop. Depending on the scale of observation, these hygral stresses result from different physical mechanisms: at the lowest level, surface forces (Van der Waals forces) are acting along the interfaces of CSH-gel pores. While in gel pores with a typical size of less then 1 μm disjoining pressures are induced by the repulsion between adsorbed water molecules within the gel pores, and capillary forces resulting from surface tension forces at the solid-liquid-vapor interface of capillary menisci are acting within capillary pores as shown in Figs. 5 and Fig.6 (Wittmann (1977); Bažant et al. (1997)).

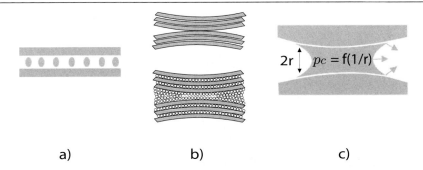

Figure 5. Illustration of hygral forces acting within the pore space of concrete: a) Van der Waals forces, b) gel pores of dry and disjoining pressure generated in wet cement paste as a consequence of hindered adsorption of water molecules, c) capillary forces (Visser (1998)).

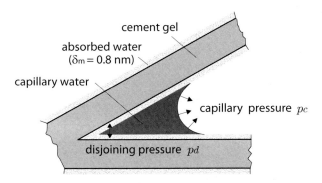

Figure 6. Illustration of disjoining and capillary pressure effects in the cement gel (after Powers (1965)).

The distribution of adsorbed and free water within the pore space depends on the degree of saturation. While for fully saturated cementitious materials more or less all pores, including also the larger capillary pores, are filled with water, for lower degrees of liquid saturation, the capillary pores are empty while the smaller pores remain to be filled with adsorbed water (Fig. 7).

Consequently, the level of these micro-stresses is highly dependent on the degree of saturation. On a macroscopic level, this dependence of hygral stresses, originating in fact from different sources, may be represented by a single macroscopic variable denoted as capillary pressure p_c (see Fig.8).

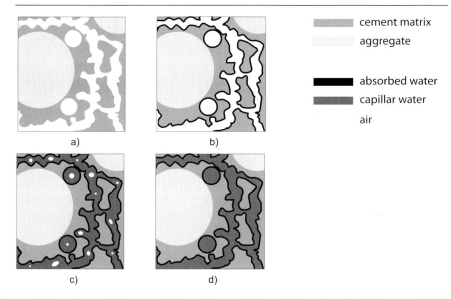

Figure 7. Schematic illustration of the moisture distribution within concrete for different degrees of saturation: a) dry material, b) generation of a continuous liquid phase of adsorbed water within the gel pores, c) larger capillary pores get filled with increasing water saturation, d) fully saturated conditions $S_l = 1.0$.

Due to the heterogeneity of the meso-structure of concrete, these internal stresses may lead to severe cracking in concrete structures when the material strength is exhausted (Alvaredo (1994); Colina and Acker (2000); Sadouki and Wittmann (2001)). Drying creep Pickett (1942); Bažant and Xi (1994), shrinkage and shrinkage-induced cracks Alvaredo (1994); Colina and Acker (2000) (Fig.9) are examples for such coupling effects between moisture transport and the constitutive behavior of cementitious materials. In turn, cracks promote the transport of moisture and render concrete structures even more vulnerable to external aggressive agents (Bažant and Raftshol (1982)).

2.2 Drying Creep

Long term creep depends on the moisture content. This causal relation is manifested in the observation that deformations measured on a loaded specimen simultaneously exposed to drying are appreciably larger than the sum of the deformations measured on a specimen loaded and dried separately

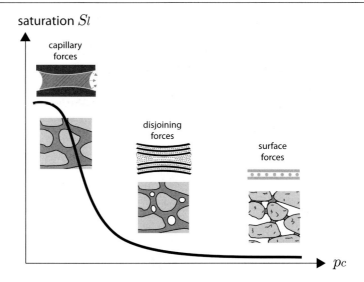

Figure 8. Saturation-dependence of hygral stresses.

Pickett (1942); Bažant and Xi (1994); Granger (1996); Bažant et al. (1997). This phenomenon is called *Pickett*-Effect and the excess of deformation is referred to as drying creep. It is widely accepted that drying creep may be explained by two mechanisms of different sources: an intrinsic process gives rise to creep when water molecules diffuse within the pore space. This contribution to creep is therefore called *intrinsic drying creep*. The second mechanism is caused by structural effects. Microcracks resulting from high tensile nonuniform stresses at the surface of a loaded specimen are getting compressed and partially closed compared to microcracks observed in a pure drying shrinkage experiment without a mechanical load. The difference in crack closure due to compressive load is referred to as cracking effect.

Cracks, irrespective of their origin, have a considerable influence on the moisture permeability of cementitious materials. As a consequence, the transport of aggressive substances may be promoted and the degradation process is further accelerated. The significant influence of fracture on the transport properties of porous materials was first recognized in the context of the coupled mechanical and hydraulic behavior of fractured rock masses. Experiments by Zoback and Byerlee (1975) indicate an increase of the permeability of granite caused by microcracking. Particularly for materials with very low moisture permeabilities, such as granite and shale, flow through the connected pore space was found to be insignificant com-

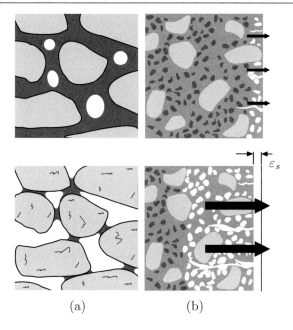

Figure 9. Effects of changing moisture conditions in cementitious materials: (a) Drying induces capillary forces which leads to shrinkage deformations indicated by shrinkage strains ε_s; (b) Cracks promote transport of moisture.

pared to flow through fracture zones. The role of cracks on the transport properties of cement-based materials has been investigated e.g. in Bažant et al. (1987); Brodersen and Nilsson (1992); Gérard et al. (1996); Aldea et al. (1999); Gérard and Marchand (2000); see Breysse and Gérard (1997) for a state-of-the-art survey. It has been shown, that the permeability is increased by several orders of magnitude when cracking is considered.

3 Fundamentals of Poromechanics

The Theory of Porous Media (TPM) provides a suitable framework for the homogenization of microscopic or submicroscopic quantities to describe coupled hygro-mechanical mechanisms and damage on a macroscopic level. In this section, basic concepts of the TPM relevant for the macroscopic formulation of hygro-thermo-mechanical processes in partially saturated concrete are summarized.

The macroscopic description of multi-phase materials is based upon the concept of poromechanics according to Coussy (1989a, 2004). Closely related approaches are the theory of mixtures extended by the concept of volume fractions (Ehlers and Bluhm (2000)) and the averaging theory (Lewis and Schrefler (1998); Hassanizadeh and Gray (1979)).The Biot-Coussy-Theory of porous media provides a thermodynamical framework according to Coleman and Noll (1963) for open continua, consisting of different, mechanically interacting phases moving with different kinematics.

According to the TPM, concrete may be idealized as a partially saturated porous material, with pores partially filled with air and water. The macroscopic representation of microstructural quantities is based on an averaging procedure, in which the individual phases are characterized by their respective fraction of volume (Fig. 10). During such an homogenization procedure, no information regarding the local topology is transferred from the micro- to the macro-level. The model representation can either be based upon a

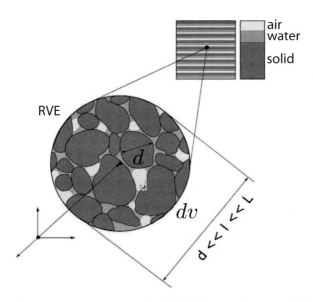

Figure 10. Averaging process in the TPM for a 3-phase model of concrete, characterized by a solid, liquid and gaseous phase at the microstructure.

three phase model, in which the liquid and gaseous phases are considered to move independently through the skeleton, or as a two phase model, in which the liquid phase is idealized as an ideal mixture of water and air. The latter representation is generally used in concrete, if the air pressure within the pores may be assumed to be more or less at atmospheric level. This is, in general, the case in durability oriented analyses. If phase changes in the pore space of concrete occur, e.g. when subjected to very high temperatures or to freezing, a three phase model, however, is mandatory (Gawin et al. (2003)).

3.1 Concept of Volume Fractions

Concrete is described as a porous material consisting of n constituents (a connected pore space partially filled with $n_f = n - 1$ fluids and gases, respectively, and a solid matrix). In the context of a macroscopic description, a statistical view of the microstructure is taken, where neither the exact location nor the topology of the individual constituents are known and the only macroscopic variable describing the individual constituents is the fraction of the total volume that is occupied by the phase α:

$$\phi^\alpha = \frac{\mathrm{d}v^\alpha}{\mathrm{d}v}. \tag{31}$$

ϕ^α is denoted as the volume fraction of the constituent ϕ_α. The fraction of the volume of the effective connected pore space volume, which enables the transport of fluids and gases will be denoted as

$$\phi = \sum_{\beta=1}^{n_f} \phi^\beta, \tag{32}$$

with n_f as the number of fluid phases filling the pore space ϕ^1. Consequently, the volume fraction of the solid matrix is $1 - \phi$.

The ratio between the volume fraction of the individual fluid phases β with respect to the effective pore space is denoted as the degree of saturation of the liquid or gaseous phase β:

$$S_\beta = \frac{\phi_\beta}{\phi}, \tag{33}$$

with the saturation condition

$$\sum_{\beta=1}^{n_f} S_\beta = 1. \tag{34}$$

According to the definition of the volume fraction ϕ_α two definitions for the densities of each phase φ^α can be used. The effective density ϱ^α relates the

---
[1] The notation of effective porosity ϕ is related to the connected pore space and is not necessarily identical to the total pore space

local mass $\mathrm{d}m^\alpha$ of the constituent φ^α to the volume $\mathrm{d}v^a$ actually occupied by φ^α:

$$\varrho^\alpha = \frac{\mathrm{d}m^\alpha}{\mathrm{d}v^\alpha}. \tag{35}$$

The partial density ρ^a relates the local mass $\mathrm{d}m^\alpha$ to the volume $\mathrm{d}v$ of the mixture φ:

$$\rho^\alpha = \frac{\mathrm{d}m^\alpha}{\mathrm{d}v}. \tag{36}$$

With Eq.(31) the relation between the effective density ϱ^α and the partial density ρ^α is obtained as:

$$\rho^\alpha = \frac{\mathrm{d}v^\alpha}{\mathrm{d}v}\frac{\mathrm{d}m^\alpha}{\mathrm{d}v^\alpha} = \phi^\alpha \varrho^\alpha. \tag{37}$$

Obviously, changes of the partial density ρ^α may result from changes of the volume fraction ϕ^α as well as from changes of the effective density ϱ^α.

3.2 Kinematics

In the framework of the geometrically linear theory, the linearized strain measure $\boldsymbol{\varepsilon}$ is linearly related to the symmetric part of the displacement gradient grad $\boldsymbol{u}$ of the skeleton as

$$\boldsymbol{\varepsilon} = \frac{1}{2}\left(\mathrm{grad}\ \boldsymbol{u} + \mathrm{grad}^T\ \boldsymbol{u}\right). \tag{38}$$

3.3 Balance of Momentum

The Cauchy stress tensor $\boldsymbol{\sigma}$ is related to the stress vector $\boldsymbol{t}$ acting on a surface with the normal $\boldsymbol{n}$ according to the Cauchy theorem:

$$\boldsymbol{t} = \boldsymbol{\sigma}\cdot\boldsymbol{n}. \tag{39}$$

$\boldsymbol{\sigma}$ represents the sum of the partial stresses $\boldsymbol{\sigma}^\alpha$ of the different phases:

$$\boldsymbol{\sigma} = \sum_{\alpha=1}^{n}\boldsymbol{\sigma}^\alpha = \sum_{\alpha=1}^{n}\phi^\alpha\boldsymbol{\sigma}_\alpha = (1-\phi)\boldsymbol{\sigma}_s + \sum_{\beta=1}^{n_f}\phi S^\beta \boldsymbol{\sigma}_\beta \tag{40}$$

In concrete, the pore space is usually filled with a mixture of water and air and the respective intrinsic stresses are represented by the liquid and air pressures p_l and p_a. Hence, the total stresses according to (40) are formulated as

$$\boldsymbol{\sigma} = (1-\phi)\boldsymbol{\sigma}_s - \phi\left(\sum_{\beta=1}^{n_f} S_\beta p_\beta\right)\boldsymbol{I} = (1-\phi)\boldsymbol{\sigma}_s + \phi(S_l p_l + S_a p_a)\boldsymbol{I}, \tag{41}$$

where $\boldsymbol{\sigma}_s$ are the intrinsic matrix stresses, p_β represents the fluid pressures of the liquid and gaseous phases filling the pore space and $\boldsymbol{I}$ denotes the second order unity tensor.

Assuming quasi-static conditions, the balance of momentum for the mixture is given as

$$\text{div } \boldsymbol{\sigma} + \rho \mathbf{b} = \mathbf{0} \tag{42}$$

with the volume forces acting on the mixture given as

$$\rho \mathbf{b} = \sum_{\alpha=1}^{n} \rho_\alpha \mathbf{b}_\alpha. \tag{43}$$

The intergranular stresses resulting from the deformation of the skeleton are expressed by means of effective stresses. Often, the *Bishop* effective stress concept is employed (Schrefler (2001)):

$$\boldsymbol{\sigma}' = \boldsymbol{\sigma} - (S_l p_l + S_a p_a) \boldsymbol{I}, \tag{44}$$

(see also the discussion on effective stresses used in concrete models in Section 4).

3.4 Mass Balance Equations

Conservation of mass within a body $\mathcal{B}$ requires, that

$$m = \int_{\mathcal{B}} dm = \int_{\mathcal{B}} \rho \, dv \tag{45}$$

remains constant, i.e.

$$\frac{d}{dt} \int_{\mathcal{B}} \rho \, dv = 0. \tag{46}$$

The local format of the mass balance can be written as

$$\dot{\rho} + \rho \, \text{div}(\dot{\mathbf{x}}) = 0, \tag{47}$$

with $\dot{\mathbf{x}}$ as the velocity of the mixture. In formulating the local balance equation for the individual phases β, mass exchanges between the phases may have to be taken into account:

$$(\rho^\beta)'_\beta + \rho_\beta \, \text{div}(\mathbf{x}'_\beta) = \hat{\rho}_\beta. \tag{48}$$

In (48) the symbol $(\bullet)'_\beta$ denotes the material time derivative following the motion of phase β and $\hat{\rho}_\beta$ represents the mass exchanged between phase β and other phases.

In the formulation of the mass balance equations for the liquid and gaseous phase in concrete phase transformations are disregarded. The pore water is assumed as incompressible and air is described as a compressible, ideal gas. The local mass balance of a phase β with respect to its current configuration is given by

$$(m_\beta)' = (\phi_\beta \varrho_\beta)'_\beta + \phi_\beta \varrho_\beta \text{div} \mathbf{x}'_\beta = 0. \tag{49}$$

3.5 Constitutive Equations

The starting point of the formulation of constitutive equations is, according to Truesdell and Toupin (1960), the assumption of a unique dependence of the internal energy $\mathcal{E}$ from a set of thermodynamic state variables which characterize the state of the open system:

$$\mathcal{E} = \mathcal{E}(S, \boldsymbol{\varepsilon}, m_\beta, \alpha_1, \alpha_2, ..., \alpha_N). \tag{50}$$

In (50), several external variables are used to describe, on a macroscopic level, the state of the material: S denotes the entropy, $\boldsymbol{\varepsilon}$ the strain tensor related to the skeleton and m_β the mass of fluid phase β. The internal variables $\alpha_1 \ldots \alpha_N$ are related to dissipative processes. The free *Helmholtz* free energy can be written as (Coussy (1995))

$$\Psi = \mathcal{E} - TS = \Psi(T, \boldsymbol{\varepsilon}, m_\beta, \alpha_1, \alpha_2, ..., \alpha_N), \tag{51}$$

with $T > 0$ as the absolute temperature. From (51) follows the expression for the material time derivative

$$\dot{\Psi} = \frac{\partial \Psi}{\partial T}\dot{T} + \frac{\partial \Psi}{\partial \boldsymbol{\varepsilon}}:\dot{\boldsymbol{\varepsilon}} + \frac{\partial \Psi}{\partial m_\beta}\dot{m}_\beta + \frac{\partial \Psi}{\partial \alpha_i}\dot{\alpha}_i. \tag{52}$$

Non-negativeness of the intrinsic dissipation leads to

$$\left(\boldsymbol{\sigma} - \frac{\partial \Psi}{\partial \boldsymbol{\varepsilon}}\right):\dot{\boldsymbol{\varepsilon}} + \left(g_\beta - \frac{\partial \Psi}{\partial m_\beta}\right)\dot{m}_\beta - \left(S + \frac{\partial \Psi}{\partial T}\right)\dot{T} - \frac{\partial \Psi}{\partial \alpha_i}\dot{\alpha}_i \geq 0, \tag{53}$$

where g_β is the *Gibbs* energy per unit of liquid mass of the fluid phase β as the thermodynamically conjugate variable to the variation of mass of phase β. Since Eq.(53) must be fulfilled for independent variations of the external variables (Coleman and Noll (1963)) the state equations are obtained as

$$\boldsymbol{\sigma} = \frac{\partial \Psi}{\partial \boldsymbol{\varepsilon}}, \qquad g_\beta = \frac{\partial \Psi}{\partial m_\beta}, \qquad S = -\frac{\partial \Psi}{\partial T}, \tag{54}$$

together with the dissipation inequality

$$\mathcal{D}_{\text{int}} = \zeta_i \dot{\alpha}_i \geq 0 \qquad \text{with} \quad \zeta_i = -\frac{\partial \Psi}{\partial \alpha_i}, \quad i = 1....N. \tag{55}$$

3.6 Transport of Water and Heat

For the description of the fluid flux $\mathbf{q}_\beta = \phi_\beta \dot{\mathbf{x}}_\beta$ of the fluid phase φ_β within the connected pore space of concrete *Darcy*'s law is applied:

$$\mathbf{q}_\beta = \phi_\beta \boldsymbol{v}_\beta = \phi_\beta \boldsymbol{v}_\beta = -\frac{\boldsymbol{k}_\beta}{\gamma_\beta} \left(\mathrm{grad} p_\beta - \varrho_\beta \boldsymbol{g}\right), \tag{56}$$

with γ_β as the dynamic viscosity of the fluid phase ϕ_β. The permeability $\mathbf{k}^\beta$ is given by the product of the intrinsic permeability $k_\beta(\phi, T)$, which depends on the porosity and the temperature, and the saturation dependent, relative permeability $k_{\beta,r}(S_l)$. In cracked concrete, the (considerable) effect of cracks on the fluid transport has to be considered by adding a term $\mathbf{k}_c(w_c)$ related to the additional permeability of crack channels with crack width w_c. A detailed description of transport models for cracked concrete is contained in Subsection 4.7.

Heat transport is generally formulated by means of *Fourier*'s law as

$$\mathbf{q}_T = -\mathbf{k}_T \,\mathrm{grad} T, \tag{57}$$

with $\mathbf{q}_T$ as the heat flux and $\mathbf{k}_T$ as the thermal conductivity tensor. Coupling between liquid and heat transport may be formulated within a thermomechanical framework by considering non-negativeness of the sum of dissipations associated with transport of liquid phases and heat, respectively, in the form (Coussy (1995))

$$\begin{aligned}\mathbf{q}_\beta &= -\frac{k_\beta}{\gamma_\beta}\left(\mathrm{grad} p_\beta - \varrho_\beta \mathbf{g}\right) - \boldsymbol{\kappa}_\beta \frac{\mathbf{q}_T}{T}, \\ \mathbf{q}_T &= -\mathbf{k}_T \mathrm{grad} T - \boldsymbol{\kappa}_\beta \mathbf{q}_\beta, \end{aligned} \tag{58}$$

with $\boldsymbol{\kappa}_\beta$ denoting a tensor associated with coupled heat and fluid transport.

4 Poroplastic-Damage Model for Concrete

In this section, the plastic damage model described in Section 1 is extended to a coupled thermo-hygro-mechanical elastoplastic damage model for durability-oriented finite element analyses of concrete structures (Grasberger and Meschke (2004); Meschke and Grasberger (2003); Grasberger and Meschke (2003)) in the framework of poromechanics (Coussy (1995)).

In accordance with the hygral processes acting at the various levels of the nano-porous skeleton (nano, micro and capillary pores) (see Section 2), the effect of shrinkage is taken into consideration by means of a macroscopic capillary pressure which represents, on a macroscopic level, hygrally induced stresses (Meschke and Grasberger (2003)). In addition to deformations and cracking resulting from drying shrinkage, the effect of cracks on the moisture transport, the moisture-dependence of the strength and stiffness of concrete and deformations resulting from long-term creep are considered in this model. Since long-term creep is associated with dislocation-like processes in the nano-pores of the cement gel which are prestressed by disjoining pressure (Bažant et al. (1997)), creep is strongly dependent on the moisture content. This coupling between moisture transport and creep deformations is also considered in the model.

The coupled thermo-hygro-mechanical material behavior of concrete is formulated within the context of thermodynamics of deformable porous media (Biot (1941); Coussy (1995)). In the model, concrete is assumed to consist of the matrix material (subscript s) – a mixture of cement paste and the aggregates – and the pores, which are partially saturated by liquid water (subscript l) and an ideal mixture of water vapor and dry air (subscript g). Provided that there is thermodynamic equilibrium between the mixture of water vapor and dry air and the external atmosphere, it is assumed that the gaseous phase is at constant atmospheric pressure, taken as zero (Bear and Bachmat (1991)). Therefore, for the sake of simplicity, the capillary pressure is expressed as $p_c = -p_l$ in what follows.

4.1 State Equations

Coupled phenomena at the micro-level of cementitious materials are described within the macroscopic framework using state variables as summarized in Section 3, based upon the formulation of the function of free energy

$$\Psi = \mathcal{W}(\varepsilon - \varepsilon^p - \varepsilon^f, m_l - \rho_l \phi_l^p, \psi, T, \gamma_f) + \mathcal{U}(\alpha_R, \alpha_D), \qquad (59)$$

which depends on three external variables (ε, m_l, T) and internal variables $\varepsilon^p, \varepsilon^f, \phi_l^p, \psi, \alpha_R, \alpha_D, \gamma_f$ (Coussy (1995); Coussy and Ulm (1996)). The quantities used in Equation (59) are summarized below:

- ε denotes the linearized strain tensor,
- ε^p is the tensor of plastic strains,
- ε^f are the flow strains corresponding to long-term creep effects,
- m_l denotes the liquid mass content variation,
- ρ_l is the mass density of the liquid phase,
- ϕ_l^p stands for the non-recoverable portion of the porosity occupied by the liquid phase, $\psi = 1 - d$ is the integrity with d denoting the isotropic damage parameter $0 \leq d \leq 1$,
- T denotes the absolute temperature,

- γ_f denotes the viscous slip associated with relative motions within the micro-pores and
- α_R and α_D characterize the inelastic pre- and postfailure behavior of concrete in tension (subscript R) and compression (subscript D).

From the entropy inequality, the state equations are obtained:

$$\boldsymbol{\sigma} = \frac{\partial \mathcal{W}}{\partial(\boldsymbol{\varepsilon} - \boldsymbol{\varepsilon}^p - \boldsymbol{\varepsilon}^f)}, \quad p_l = \frac{\partial \mathcal{W}}{\partial(\frac{m_l}{\rho_l} - \phi_l^p)}, \quad (60)$$
$$S = -\frac{\partial \mathcal{W}}{\partial T}, \quad q_R = -\frac{\partial \mathcal{U}}{\partial \alpha_R}, \quad q_D = -\frac{\partial \mathcal{U}}{\partial \alpha_D},$$

where $\boldsymbol{\sigma}$ is the tensor of total stresses, p_l is the liquid pressure and S is the entropy. The thermodynamic forces conjugate to α_R and α_D are q_R and q_D. They determine the damage-dependent size of the damage (f_R) and loading (f_D) surface in the stress space. The differential form of (60) is obtained as

$$\begin{aligned}
\mathrm{d}\boldsymbol{\sigma} &= \mathbb{C}_u^{\mathrm{ed}} : (\mathrm{d}\boldsymbol{\varepsilon} - \mathrm{d}\boldsymbol{\varepsilon}^p - \mathrm{d}\boldsymbol{\varepsilon}^f) - \psi M \mathbb{B}\left(\frac{\mathrm{d}m_l}{\rho_l} - \mathrm{d}\phi_l^p\right) \quad (61) \\
&\quad + \boldsymbol{\Lambda}_u \mathrm{d}\psi - \mathbf{A}_u \mathrm{d}T - \boldsymbol{\Delta}\gamma_f, \\
\mathrm{d}p_l &= \psi M \left(\frac{\mathrm{d}m_l}{\rho_l} - \mathrm{d}\phi_l^p\right) - \psi M \mathbb{B} : (\mathrm{d}\boldsymbol{\varepsilon} - \mathrm{d}\boldsymbol{\varepsilon}^p - \mathrm{d}\boldsymbol{\varepsilon}^f) \quad (62) \\
&\quad + \Xi \mathrm{d}\psi + L \mathrm{d}T + \Lambda \gamma_f, \\
\mathrm{d}S &= \frac{C_u}{T_0} \mathrm{d}T + \mathbf{A}_u : (\mathrm{d}\boldsymbol{\varepsilon} - \mathrm{d}\boldsymbol{\varepsilon}^p - \mathrm{d}\boldsymbol{\varepsilon}^f) + s_l (\mathrm{d}m_l - \rho_l \mathrm{d}\phi_l^p) \quad (63) \\
&\quad - L\left(\frac{\mathrm{d}m_l}{\rho_l} - \mathrm{d}\phi_l^p\right) + \Pi \mathrm{d}\psi + \Theta \gamma_f.
\end{aligned}$$

The coefficients introduced in (61 - 64) represent the mixed partial derivatives of the free energy and can be interpreted as follows:

- $\mathbb{C}_u^{\mathrm{ed}} = \psi \mathbb{C}_u$ denotes the degraded (undrained) fourth-order stiffness tensor,
- the term $\psi M \mathbb{B}$ represents the hygro-mechanical couplings with M as the isotropic *Biot* modulus and $\mathbb{B} = b\mathbf{1}$ as the second-order tensor of tangential *Biot* coefficients b,
- $\boldsymbol{\Lambda}_u$ is the undrained second-order tensor describing the coupling mechanisms between damage evolution and the total stress increment,
- $\mathbf{A}_u = \mathbb{C}_u^{\mathrm{ed}} : \mathbf{1}\alpha_{t,u}$ denotes the undrained second-order thermo-mechanical coupling tensor with $\alpha_{T,u}$ as the undrained thermal dilatation coefficient,

- Ξ is a coupling coefficient connected with the change of the liquid pressure due to damage evolution,
- $L = 3\psi M \alpha_{t,u}$ characterizes the thermo-hygral coupling mechanisms,
- C_u denotes the undrained volume heat capacity and T_0 the reference temperature,
- s_l is the internal entropy of the liquid phase and
- $T_0 \Pi$ represents the latent heat due to damage evolution.
- Λ, Δ and Θ are couplings with viscous slip processes leading to creep deformations. As can be shown from considering Maxwell symmetries, these couplings can be neglected and the coefficients may be set to zero.

Inserting (62) into (61) yields an alternative drained formulation for the differential stress tensor as (Grasberger and Meschke (2004))

$$d\boldsymbol{\sigma} = \mathbb{C}^{\text{ed}} : (d\boldsymbol{\varepsilon} - d\boldsymbol{\varepsilon}^p - d\boldsymbol{\varepsilon}^f) - \mathbb{B} dp_l + \boldsymbol{\Lambda} d\psi - \mathbb{A} dT, \tag{64}$$

with the drained stiffness tensor

$$\mathbb{C}^{\text{ed}} = \psi \left[\mathbb{C}_u - Mb^2 (\mathbf{1} \otimes \mathbf{1})\right] = \psi \mathbb{C}, \tag{65}$$

the drained thermo-mechanical coupling tensor

$$\mathbb{A} = \mathbb{A}_u - 3\alpha_{t,u} M \mathbb{B} = \mathbb{C}^{\text{ed}} : \mathbf{1}\alpha_t, \tag{66}$$

and the drained tensor

$$\boldsymbol{\Lambda} = \boldsymbol{\Lambda}_u + \Xi \mathbb{B}, \tag{67}$$

respectively.

4.2 Coupling Coefficients

According to Grasberger and Meschke (2004) the poroelastic hygro-mechanical coefficients b and M can be determined by relating differential stress and differential strain quantities defined on the meso-level to respective homogenised quantities on the macro-level. The so-obtained tangential *Biot* coefficient is determined as

$$b = S_l \left[1 - \psi \frac{K}{K_s}\right], \tag{68}$$

which includes the expression $b = S_l$ suggested by Coussy (1995) for the special case of poroelastic materials with incompressible matrix behavior.

K and K_s denote the bulk moduli of the skeleton and of the solid (matrix) phase, respectively. An expression for the *Biot* modulus $\overline{M} = \psi M$ is obtained as

$$\overline{M} = \left[\phi\left(1 - \frac{S_l p_l}{K_s}\right)\frac{\partial S_l}{\partial p_l} + \frac{\phi S_l}{K_l} + \frac{S_l(b - \phi S_l)}{K_s}\right]^{-1}, \qquad (69)$$

with K_l as the bulk modulus of the liquid phase, see Schrefler and Zhan (1993) and Lewis and Schrefler (1998) for a similar formulation. For cementitious materials, expression (69) can be replaced by

$$\overline{M} \approx \left[\phi \frac{\partial S_l}{\partial p_l}\right]^{-1}. \qquad (70)$$

In the special case of a fully saturated material ($S_l = 1$), Eq.(69) yields the classical relation (Coussy (1995); Lewis and Schrefler (1998))

$$\overline{M}_{S_l=1} = \left[\frac{\phi}{K_l} + \frac{(b-\phi)}{K_s}\right]^{-1}. \qquad (71)$$

The coefficients $\mathbf{\Lambda}$ and Ξ related to damage processes are identified by exploiting the symmetry relations that are connected to the existence of a macroscopic potential. Using the Maxwell symmetries, the drained tensor $\mathbf{\Lambda}$ can be expressed as

$$\mathbf{\Lambda} = \mathbb{C} : (\varepsilon - \varepsilon^p - \varepsilon^f) + \frac{K}{K_s}\int_{p_l} S_l \mathrm{d}p_l \mathbf{1} - \mathbb{C} : \mathbf{1}\alpha_T T. \qquad (72)$$

The coupling coefficient Ξ is obtained as

$$\Xi = \frac{\overline{M}S_l K}{K_s^2}\int_{p_l} S_l \mathrm{d}p_l \approx 0. \qquad (73)$$

4.3 Effective stresses

The concept of effective stress is a generally accepted approach in soil mechanics for the determination of stresses in the skeleton of fully saturated soils. In addition to the original proposal of von Terzaghi (1936), several alternative suggestions for the definition of effective stresses exist, taking the compressibility of the matrix material or the porosity into account (see e.g. Biot and Willis (1957); Nur and Byerlee (1971); Bishop (1973)).

Based on the relevance of the concept of effective stress for the analysis of fully saturated soils, this concept has also been adapted for the description of partially saturated soils. Early formulations introduced the capillary pressure in the (elastic) effective stress definition (Bishop (1959)). However, difficulties to obtain satisfactory agreements with experimental results have motivated the use of two independent stress fields for the constitutive modelling of unsaturated soils (see e.g. Bishop and Blight (1963); Alonso et al. (1990)). As far as the numerical modelling of partially saturated cement-based materials is concerned, the assumption of (elastic) effective stresses seems not well suited for the description of shrinkage-induced cracks using stress-based crack-models. However, the concept of plastic effective stress first introduced at a macroscopic level by Coussy (1989b) for saturated porous media allows to overcome these difficulties in the framework of poroplasticity – porodamage models. The proposed form of the plastic effective stress is the same as the classical *Biot*-type, however, a plastic effective stress coefficient is used. A similar form has been derived from micromechanical considerations by Lydzba and Shao (2000). This concept has been extended to partially saturated materials e.g. by Burlion et al. (2000); Meschke and Grasberger (2003), and is also adopted in the present formulation. In attempts to describe drying creep processes in concrete structures, the coupling between spatially and temporarily varying moisture distributions and the extent of long term creep have been considered by Benboudjema et al. (2003) and Grasberger and Meschke (2003). Starting point for the derivation of the elastic effective stress tensor $\boldsymbol{\sigma}'^e$ for isothermal conditions is the integrated form of (64). Inserting (65) for the drained stiffness tensor $\mathbb{C}^{\mathrm{ed}}$, (68) for the tangential *Biot* coefficient b and the relation $p_c = -p_l$, the drained formulation of the increment of the total stress tensor is obtained as

$$d\boldsymbol{\sigma} = \psi \mathbb{C} : (d\boldsymbol{\varepsilon} - d\boldsymbol{\varepsilon}^p - d\boldsymbol{\varepsilon}^f) + \mathbb{C} : (\boldsymbol{\varepsilon} - \boldsymbol{\varepsilon}^p - \boldsymbol{\varepsilon}^f) d\psi \\ + S_l \left[1 - \psi \frac{K}{K_s}\right] \mathbf{1} dp_c - \frac{K}{K_s} \int_{p_c} S_l(p_c) dp_c \mathbf{1} d\psi. \tag{74}$$

Integration of (74) results in

$$\boldsymbol{\sigma} = \psi \mathbb{C} : (\boldsymbol{\varepsilon} - \boldsymbol{\varepsilon}^p - \boldsymbol{\varepsilon}^f) + \left[1 - \psi \frac{K}{K_s}\right] \int_{p_c} S_l(p_c) dp_c \mathbf{1}, \tag{75}$$

which gives rise to the following definition of the elastic effective stress tensor representing intergranular forces acting on the skeleton:

$$\boldsymbol{\sigma}'^e = \psi \mathbb{C} : (\boldsymbol{\varepsilon} - \boldsymbol{\varepsilon}^p - \boldsymbol{\varepsilon}^f). \tag{76}$$

Using (76), Eq.(75) can be written as

$$\boldsymbol{\sigma} = \boldsymbol{\sigma}'^e + \left[1 - \psi \frac{K}{K_s}\right] \int_{p_c} S_l(p_c) dp_c \mathbf{1}. \tag{77}$$

The plastic effective stress tensor $\boldsymbol{\sigma}'^p = \boldsymbol{\sigma}'$, defined as

$$\boldsymbol{\sigma}' = \boldsymbol{\sigma} - b^p p_c \mathbf{1}, \tag{78}$$

characterizes the thermodynamic force associated with the plastic strain rate (Coussy (1995)). In contrast to the elastic effective stress tensor, $\boldsymbol{\sigma}'$ represents the macroscopic counterpart to matrix-related micro-stresses with the coefficient b^p as the plastic counterpart of the *Biot* coefficient b. By relating stress quantities on the meso-scale to respective macroscopic quantities, a possible identification of b^p as a function of the integrity ψ, the porosity ϕ and the liquid saturation S_l can be accomplished as

$$b^p = \psi \phi S_l(p_c). \tag{79}$$

See Grasberger and Meschke (2004) for details.

4.4 Multisurface Poroplastic-Damage Model

According to the concept of multisurface damage-plasticity theory as described in Section 1, mechanisms characterized by the degradation of stiffness and inelastic deformations are controlled by four threshold functions defining a region of admissible stress states in the space of plastic effective stresses $\boldsymbol{\sigma}'$

$$\mathcal{E} = \{(\boldsymbol{\sigma}', q_k) | \ f_k(\boldsymbol{\sigma}', q_k(\alpha_k)) \leq 0, \quad k = 1, ..., 4\}. \tag{80}$$

In (80), the index $k = 1, 2, 3$ stands for an active cracking mechanism associated with the damage function $f_{R,k}(\boldsymbol{\sigma}', q_R)$ (Eq.20) and $k = 4$ represents an active hardening/softening mechanism in compression associated with the loading function $f_D(\boldsymbol{\sigma}', q_D)$ (Eq.25). Cracking of concrete is accounted for by means of the *Rankine* criterion, employing three failure surfaces $f_{R,A}(\boldsymbol{\sigma}', q_R)$ perpendicular to the axes of principal stresses. The softening parameter $q_R(\alpha_R) = -\partial \mathcal{U}/\partial \alpha_R$ is given by Eq.(21). For the description of concrete subjected to (predominantly biaxial) states of compressive stresses a hardening/softening *Drucker-Prager* plasticity model $f_D(\boldsymbol{\sigma}', q_D)$ (see Section 1), with the hardening/softening parameter $q_D(\alpha_D) = -\partial \mathcal{U}/\partial \alpha_D$ given by Eq.(27).

Adopting the same arguments as described in Subsection 1.2, the evolution equations of the tensor of plastic strains $\dot{\boldsymbol{\varepsilon}}^p$, of the elastic compliance moduli $\dot{\boldsymbol{\mathcal{D}}}_u$ of the plastic porosity occupied by the liquid phase $\dot{\phi}_l^p$ and of the internal variables $\dot{\alpha}_R$ and $\dot{\alpha}_D$ are obtained from the postulate of stationarity of the dissipation functional (Govindjee et al. (1995)) as

$$\dot{\varepsilon}^p = (1-\beta_k)\sum_{k=1}^{4}\dot{\gamma}_k\frac{\partial f_k}{\partial \boldsymbol{\sigma}'}, \tag{81}$$

$$(\psi^{-1})^{\cdot} = \beta_k\sum_{k=1}^{4}\dot{\gamma}_k\frac{\frac{\partial f_k}{\partial \boldsymbol{\sigma}'}:\mathbf{C}_u:\frac{\partial f_k}{\partial \boldsymbol{\sigma}'}}{\frac{\partial f_k}{\partial \boldsymbol{\sigma}'}:\boldsymbol{\sigma}'}, \tag{82}$$

$$\dot{\phi}_l^p = \sum_{k=1}^{4}\dot{\gamma}_k\frac{\partial f_k}{\partial \boldsymbol{\sigma}'}:\mathbf{1}b^p, \tag{83}$$

$$\dot{\alpha}_R = \sum_{A=1}^{3}\dot{\gamma}_{R,A}\frac{\partial f_{R,A}}{\partial q_R}, \quad \dot{\alpha}_D = \dot{\gamma}_D\frac{\partial f_D}{\partial q_D}, \tag{84}$$

together with the loading/unloading conditions

$$f_k(\boldsymbol{\sigma}',q_k)\leq 0; \quad \dot{\gamma}_k\geq 0; \quad \dot{\gamma}_k\, f_k(\boldsymbol{\sigma}',q_k)=0. \tag{85}$$

4.5 Long-term Creep

The adopted creep law embedded in the present model is based on the microprestress-solidification theory (Bažant et al. (1997)). Within this theory, creep is attributed to relaxation of the microprestress S_f acting in the micropores. Tensile microprestress generated by disjoining pressure of hindered adsorbed water in the pore space and by local volume changes is carried by bonds and bridges crossing the micro-pores in the hardened cement gel. As bond breakage is happening, a relaxation of the microprestress occurs, leading to viscous shear slip γ_f between the opposite walls of the micro-pores. These microstructural movements are attributed to long-term creep $\boldsymbol{\varepsilon}^f$. Consequently, variations of the internal pore humidity h due to drying, which entail a changing disjoining pressure, lead to drynig creep (*Pickett* effect), see Section 2.

Consideration of long-term or flow creep effects is accomplished in the framework of the Microprestress-Solidification theory (Bažant et al. (1997)). The evolution law of the flow strains is based on a linear relation between the rate $\dot{\boldsymbol{\varepsilon}}^f$ and the stress tensor $\boldsymbol{\sigma}$ as

$$\dot{\boldsymbol{\varepsilon}}^f = \frac{1}{\eta_f(S_f)}\mathbf{G}^{\mathrm{ed}}:\boldsymbol{\sigma}' = \frac{1}{\eta_f(S_f)}\mathbf{G}^{\mathrm{ed}}:\boldsymbol{\sigma} - \frac{1}{\eta_f(S_f)}\mathbf{G}^{\mathrm{ed}}:\mathbf{1}p_c, \tag{86}$$

with the fourth-order tensor $\mathbf{G}^{\mathrm{ed}} = E(\mathbf{C}^{\mathrm{ed}})^{-1}$ and Young's modulus E. The first part in (86) accounts for basic creep while the second term is connected to hygral stresses and accounts for the intrinsic part of drying creep.

In contrast to the original formulation of the Microprestress-Solidification Theory, where the microprestress S_f is considered to be affected by changes of the pore humidity due to moisture diffusion, the viscosity η_f is assumed as a decreasing function of the microprestress S_f independent of the pore humidity

$$\frac{1}{\eta_f(S_f)} = cp S_f^{p-1}, \tag{87}$$

where c and $p > 1$ are positive constants. From the entropy inequality, the state equation for the microprestress is obtained from Eq.(61) as

$$S_f = -\frac{\partial \psi}{\partial \gamma_f} = S_f(\gamma_f) = S_{f,0} - H_f \gamma_f, \tag{88}$$

assuming a linear dependence between S_f and the viscous slip γ_f. The parameters S_{f0}, H and c have to be calibrated according to experimental data. From Eq.(88) it becomes evident, that the irreversible viscous slip γ_f is energetically conjugated to the microprestress S_f (Sercombe et al. (2000); Lackner et al. (2002)). The evolution law of the viscous slip is given by

$$\dot{\gamma}_f = c S_f^2 \tag{89}$$

which coincides with the adopted formula given in Bažant et al. (1997) when the influence of moisture on the microprestress is neglected.

4.6 Capillary-Pressure Relation

Recalling the identification of the hygro-mechanical coupling coefficients in Subsection 4.2, the derivative of the liquid saturation S_l with respect to p_c remains to be defined. Since a direct experimental determination is difficult (Helmig (1997)), an indirect determination of the $S_l - p_c$-relation using the desorption isotherm is generally used. In the sorption experiment, the relative humidity h is controlled while the moisture content within the sample is measured. Applying *Kelvin's* law to relate changes in relative humidity to changes of the capillary pressure yields an expression for the missing capillary pressure curve (see Coussy et al. (1998); Carmeliet and Abeele (2000)). Based upon the pore network model of Mualem (1976), a relation between the capillary pressure and the liquid saturation is proposed by van Genuchten (1980) in the form

$$S_l(p_c) = \left[1 + \left(\frac{p_c}{p_r}\right)^{1/(1-m)}\right]^{-m} \quad \text{for} \quad p_c > 0. \tag{90}$$

For concrete, the reference pressure p_r and the coefficient m have been specified in Baroghel-Bouny et al. (1999) on the basis of experiments as $p_r = 18.6237$ N/mm^2 and $m = 0.4396$.

4.7 Moisture Transport

Starting with a simplified nonlinear diffusion approach, in which the different moisture transport mechanisms in liquid and in vapor form are represented by means of a single macroscopic moisture-dependent diffusivity (Bažant and Najjar (1972)), the relation between the moisture flux $\boldsymbol{q}_l$ and the spatial gradient of the capillary pressure ∇p_c is given by

$$\boldsymbol{q}_l = \frac{\boldsymbol{k}}{\mu_l} \cdot \nabla p_c. \tag{91}$$

In (91), $\boldsymbol{k}$ denotes the intrinsic liquid permeability tensor and μ_l is the water viscosity. According to the hypothesis of dissipation decoupling (Coussy et al. (1998)), possible couplings between heat and moisture transport are disregarded in the present formulation. In order to account for the dependence of the moisture transport properties on the nonlinear material behavior of concrete, $\boldsymbol{k}$ is additively decomposed into two portions as

$$\boldsymbol{k} = k_r(S_l) k_\phi(\phi) \, \boldsymbol{k}_0 + k_{rc}(S_l) \boldsymbol{k}_c(\alpha_\mathrm{R}), \tag{92}$$

one related to the moisture flow through the partially saturated pore space and one related to the flow within a crack, respectively (Snow (1969)). This approach is consistent with the smeared crack concept. In (92), $\boldsymbol{k}_0$ denotes the initial isothermal permeability tensor, k_r is the relative permeability of the bulk material, k_t accounts for the dependence of the isothermal moisture transport properties on the temperature, k_ϕ describes the relationship between the permeability and the porosity, k_{rc} is the relative permeability of the cracked material and $\boldsymbol{k}_c$ is the permeability tensor relating plane *Poiseuille* flow through discrete fracture zones to the degree of damage in the continuum model, see Meschke and Grasberger (2003); Grasberger (2002) for details.

Influence of Moisture Content on the Permeability Starting from the statistical pore size distribution, an analytical expression for the relative permeability depending on the liquid saturation was suggested by van Genuchten (1980) for soils as

$$k_r(S_l) = \sqrt{S_l} \left[1 - (1 - S_l^\delta)^{\frac{1}{\delta}} \right]^2. \tag{93}$$

Relation (93) is also used for concrete in the present model with $\delta = 2.275$ as suggested by Baroghel-Bouny et al. (1999).

Influence of Porosity on the Permeability The relationship between permeability and the pore structure of hardening cement paste at early ages was investigated experimentally in Nyame and Illston (1981). From the presented data, the permeability k_ϕ can be related to the porosity ϕ as follows (Grasberger (2002))

$$k_\phi(\phi) = 10^\delta \quad \text{with} \quad \delta = \frac{6(\phi - \phi_0)}{0.3 - 0.4\phi_0}, \tag{94}$$

where $\phi - \phi_0$ denotes the change of porosity due to elastic as well as plastic deformations.

Influence of Cracking on the Permeability The increase of permeability due to the change of porosity is insufficient when cracking of concrete is to be considered. The influence of cracks on the moisture transport is significantly larger compared to the effect of elastic and inelastic change of porosity. In addition to the dependence of the crack width w_c also the saturation of the crack has an influence, which is capture by the relative permeability k_{rc}. From water flow in rock joints the exponential function

$$k_{rc}(S_l) = a_0 \exp(a_1 S_l), \tag{95}$$

with $a_0 = 8 \cdot 10^{-6}$ and $a_1 = 11.7$ (Zoback and Byerlee (1975); Helmig (1997)).

Starting from an idealized crack formation, assumed to be planar, parallel and of constant opening width w_c, it is possible to estimate the moisture flux along one single crack from the solution of the Navier-Stokes equation for plane *Poiseuille* flow as (Witherspoon et al. (1979))

$$q_c = \frac{w_c^3}{12\mu_l} \nabla p_c. \tag{96}$$

The cubic flow law (96) represents an upper bound, since the roughness of the cracks as well as aperture variation and tortuosity effects have not been taken into account (Bažant and Raftshol (1982)). Following the approach of Barton et al. (1985), the mechanical crack width w_c is replaced by the equivalent hydraulic width w_h

$$w_h = \frac{w_c^2}{R^{2.5}} \quad \text{for} \quad w_c \geq w_h, \tag{97}$$

where w_c and w_h are the nominal and the equivalent hydraulic crack width in μm and the parameter R describes the roughness of the crack. This modified cubic flow law is based on the evaluation of test data from jointed rocks. From evaluation of two different experimental series performed by Aldea et al. (2000) and Oshita and Tanabe (2000) concerning the relation between the normalized permeability k/k_0 and the mechanical crack width w_c, R was determined as $R = 15.0$ (Grasberger (2002)). Inserting (97) into (96) gives the enhanced permeability along cracks as

$$k_c = \frac{w_h^3}{12\mu_l} \qquad (98)$$

The permeability $\boldsymbol{k}'_c = w_h^2/12$ of a single crack is related to a local coor-

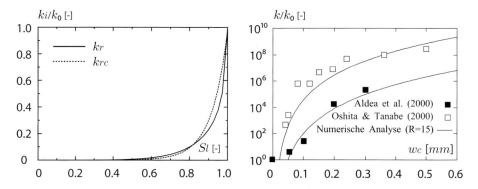

Figure 11. Moisture transport in cracks. Left: relative permeability according to (93), right: influence of the crack width on the permeability (Grasberger and Meschke (2004)).

dinate system $x' - y'$ defined by the current crack direction with the crack normal $\mathbf{n} \equiv \mathbf{e}_{x'}$. In finite element analyses, in analogy to the smeared crack concept, in which the energy dissipated within discrete fracture planes is distributed within the respective finite element, the moisture flow along a single crack is accordingly distributed within the cracked element characterized by the element length l_c. For 3D cases, the anisotropic damage permeability tensor $\boldsymbol{k}_c(w_c)$ is obtained from transformation to global axes by means of the rotation tensor $\mathbb{T}$:

$$\boldsymbol{k}_c = \mathbb{T}^T \cdot \boldsymbol{k}'_c \cdot \mathbb{T}, \qquad \text{with} \qquad \boldsymbol{k}'_c = \begin{bmatrix} 0 & 0 & 0 \\ 0 & w_h^3/12l_c & 0 \\ 0 & 0 & w_h^3/12l_c \end{bmatrix}. \tag{99}$$

It should be noted, that moisture flow perpendicular to the crack, i.e. drying of the surrounding matrix via open cracks, is implicitly represented (within the scale set by the finite element discretization) in the model, since the moisture gradient in this direction increases with continuing drying of the crack. The relationship between the crack width w_c and the damage state, represented by the internal variable α_R is established in the context of the smeared crack concept. A uniaxially stretched bar of length l_c with a crack is taken as a conceptional model. The crack width w_c is obtained as the difference between the total elongation w of the bar and the change of length w_0 of the intact (unloading) parts of the bar:

$$w_c = w - w_0 = l_c \left[\sigma(\alpha_R) \left(D(\alpha_R) - D_0 \right) + \varepsilon^p(\alpha_R) \right], \tag{100}$$

with $D(\alpha_R)$ (D_0) as the macroscopic compliance modulus of the damaged (intact) material. According to Meschke et al. (1998a) the compliance modulus $D(\alpha_R)$ can be obtained in terms of α_R as

$$D(\alpha_R) = \frac{\beta \alpha_{R,u}}{3 f_{tu}} \left[(1+\kappa)^3 - 1 \right] + 1/E_0, \tag{101}$$

where $\kappa = \alpha_R/\alpha_{R,u}$. From inserting (101) into (100), an expression of the crack width as a function of the internal variable α_R is obtained:

$$w_c(\alpha_R) = \frac{l_c \beta \alpha_{R,u}}{3} \frac{\left[(1+\kappa)^3 - 1 \right]}{(1+\kappa)^2} + l_c(1-\beta)\alpha_R. \tag{102}$$

4.8 Finite Element Formulation

The coupled system of mechanical deformation and moisture transport is characterized by the displacement field $\boldsymbol{u}$ and the capillary pressure p_c together with a set of internal variables. The primary variables in the domain Ω are controlled by the balance of linear momentum and balance of liquid mass as

$$\text{div}\,\boldsymbol{\sigma} = \boldsymbol{0}, \quad \text{div}(\rho_l \, \boldsymbol{q}) + \dot{m}_l = 0. \tag{103}$$

The related boundary conditions at the boundary Γ are prescribed as

$$\boldsymbol{\sigma}\cdot\boldsymbol{n} = \boldsymbol{t}^\star \text{ on } \Gamma_\sigma, \quad \boldsymbol{q}\cdot\boldsymbol{n} = q^\star \text{ on } \Gamma_q, \quad \boldsymbol{u} = \boldsymbol{u}^\star \text{ on } \Gamma_\mathbf{u}, \quad p_c = p_c^\star \text{ on } \Gamma_{p_c}, \quad (104)$$

where $\mathbf{n}$ is the normal vector at the boundary surface, $q^\star$ is the liquid mass flux across the boundary, $\boldsymbol{u}^\star$ are prescribed displacements and $p_c^\star$ is the prescribed capillary pressure. The initial conditions are defined in the domain Ω

$$\boldsymbol{u}(t=0) = \boldsymbol{u}_0, \quad p_c(t=0) = p_{c,0}. \tag{105}$$

The numerical solution is based on the weak formulations of the mechanical and the hygral field problem. The weak form of the balance of linear momentum $(103)_1$ together with the Neumann boundary condition $(104)_1$ yields

$$\int_\Omega \delta\boldsymbol{\varepsilon} : \boldsymbol{\sigma} \, d\Omega - \int_{\Gamma_\sigma} \delta\boldsymbol{u}\cdot\boldsymbol{t}^\star \, d\Gamma_\sigma = 0 \, . \tag{106}$$

In the present implementation of the model, $\boldsymbol{\sigma}_{n+1}$ at the end of a time interval $[t_n, t_{n+1}]$ is obtained from semi-explicit integration of (61), computing b at t_n.

The weak form of the balance of liquid mass $(103)_2$ together with the respective Neumann boundary conditions $(104)_2$ is given as

$$\int_\Omega \delta p_c \, \frac{\dot{m}_l}{\rho_l} \, d\Omega - \int_\Omega \delta\nabla p_c \cdot \boldsymbol{q}_l \, d\Omega - \int_{\Gamma_q} \delta p_c \, q^\star \, d\Gamma_q = 0, \tag{107}$$

with the liquid mass flux vector $\boldsymbol{q}_l$ according to (91).

The finite element method is used for the solution of the governing differential equations of the 2-field problem. Discretization in space results in the following set of equations (Lewis and Schrefler (1998)):

$$\begin{bmatrix} \mathbf{K}_u & \mathbf{Q} \\ \mathbf{0} & \mathbf{H}_p \end{bmatrix} \begin{Bmatrix} \overline{\mathbf{u}} \\ \overline{\mathbf{p}}_c \end{Bmatrix} + \begin{bmatrix} \mathbf{0} & \mathbf{0} \\ \mathbf{Q}^T & \mathbf{S}_p \end{bmatrix} \frac{d}{dt} \begin{Bmatrix} \overline{\mathbf{u}} \\ \overline{\mathbf{p}}_c \end{Bmatrix} = \begin{Bmatrix} \mathbf{f}_u \\ \mathbf{f}_{p_c} \end{Bmatrix}, \tag{108}$$

with $\overline{\mathbf{u}}, \overline{\mathbf{p}}_c$ denoting the nodal values of the displacements and the capillary pressure, respectively. The coefficients of (108) are defined as follows:

- $\mathbf{K}_u = \int_\Omega \mathbf{B}_u^T \mathbf{C} \mathbf{B}_u \, d\Omega$ is the secant stiffness matrix,
- $\mathbf{Q} = \int_\Omega \mathbf{B}_u^T \mathbf{m} \, b \mathbf{N}_{p_c} \, d\Omega$ is the hygro-mechanical coupling matrix with $\mathbf{m} = [1, 1, 1, 0, 0, 0]^T$,

- $\mathbf{H}_p = -\int_\Omega \mathbf{B}_{p_c}^T \mathbf{D}_l \mathbf{B}_{p_c} \, d\Omega$ is the isothermal liquid permeability matrix, and
- $\mathbf{S}_p = -\int_\Omega \mathbf{N}_{p_c}^T \frac{1}{M} \mathbf{N}_{p_c} \, d\Omega$ is the moisture capacitance matrix.

$\mathbf{N}_{pc}$ contains the interpolation polynomials for the capillary pressure, $\mathbf{B}_u$ and $\mathbf{B}_{pc}$ are the respective gradient matrices. Rewriting (108) in a more concise format

$$\mathbf{K}(\mathbf{x})\mathbf{x} + \mathbf{S}(\mathbf{x})\frac{d\mathbf{x}}{dt} = \mathbf{f}(\mathbf{x}), \tag{109}$$

with $\mathbf{x} = \{\overline{\mathbf{u}}, \overline{\mathbf{p}}_c\}$ as the vector of unknown nodal values, discretization in time in conjunction with a fully implicit integration scheme yields the following nonlinear system of coupled algebraic equations

$$[\mathbf{S} + \Delta t \mathbf{K}]_{n+1} \mathbf{x}_{n+1} = \mathbf{S}\, \mathbf{x}_n + \Delta t\, \mathbf{f}_{n+1}. \tag{110}$$

(110) is solved by means of a *Newton-Raphson* procedure, using a consistently linearized algorithmic tangent operator.

4.9 Application: Drying and Re-wetting of a Base-Restrained Concrete Wall

Restrained shrinkage is one of the major causes of damage in concrete structures. In order to demonstrate the capability of the hygro-mechanical model to reproduce shrinkage-induced cracks, a numerical simulation of a concrete wall subjected to drying, but restrained at the base by a non-shrinking foundation, is described in this section. Figure 12 shows the geometry and the finite element discretization of the investigated concrete wall. The following material parameters were assumed: Young's modulus $E = 43600$ N/mm^2, Poisson's ratio $\nu = 0.2$, uniaxial tensile strength $f_{tu} = 4.0$ N/mm^2, uniaxial compressive strength $f_{cu} = 63.4$ N/mm^2, fracture energy in tension $G_f = 0.120$ N/mm, fracture energy in compression $G_c = 5.975$ N/mm and $\beta = 0.2$. The following relationship between the capillary pressure and the liquid saturation was used (Baroghel-Bouny et al. (1999)):

$$p_c(S_l) = 18.62 \left[S_l^{-2.27} - 1 \right]^{1-1/2.27}. \tag{111}$$

Furthermore, the initial pore humidity and the initial porosity were specified as $h_0 = 0.93$ and $\phi_0 = 0.20$.

A 2D analysis of one half of the concrete wall, using 8910 four-noded plane stress elements, was performed. A uniform humidity was assumed within the wall. This simplified assumption is justified by the large surface exposed to the surrounding air and the small thickness of the wall ($d=$

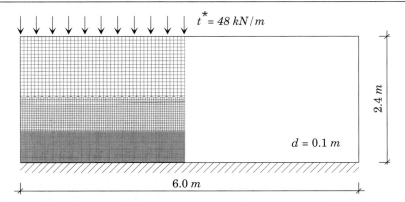

Figure 12. Numerical analysis of a concrete wall: Geometry and finite element discretization.

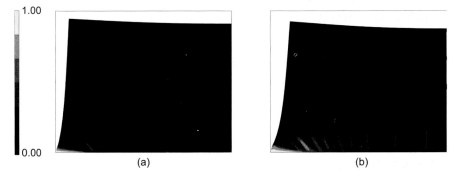

Figure 13. Numerical analysis of a drying concrete wall: Computed distribution of damage ($0 \leq d \leq 1$) at different states of drying: a) $h = 0.80$ b) $h = 0.70$ (400-fold magnification of displacements).

0.1 m), respectively. After application of an in-plane load taken as eight times of the self-weight, the relative humidity was decreased from $h_0 = 0.93$ to $h = 0.70$.

Figure 13 illustrates the crack distribution due to the uniform drying of the concrete wall at $h = 0.80$ and $h = 0.70$. A horizontal crack opens already at $h = 0.90$. At $h = 0.80$, skew cracks start to open in the vicinity of the crack process zone of the horizontal crack with an inclination of 45° with respect to the base line. As the drying proceeds, these skew cracks are further opening, while additional cracks start to open along the bottom of the slab, characterized by a larger angle approaching an angle of 90° as the cracks approach the center of the slab. Figure 14 shows the moisture

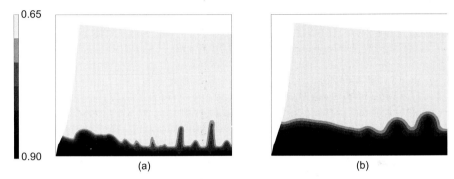

Figure 14. Numerical analysis of the re-wetting of a concrete wall: Computed moisture distribution h after a) $t = 10$ years, b) $t = 50$ years (400-fold magnification of displacements).

distribution after 10 and 50 years, respectively, after the moisture boundary condition at the bottom face of the slab has been increased up to $h^\star = 0.90$. The considerable influence of existing cracks on the moisture transport is clearly illustrated.

5 Hygromechanical Extended Finite Element Model for Concrete

5.1 Introduction

Up to the mid of the 1990's research in computational failure analysis of concrete and reinforced concrete structures was focused on two alternative approaches: continuum-based models (see the reviews contained e.g. in de Borst (2002); Mang et al. (2003)) and discrete representations of fracture along mesh boundaries using concepts of Linear Elastic Fracture Mechanics (LEFM) and cohesive crack models (see, e.g., Ingraffea and Saouma (1985); Xie and Gerstle (1995)). Since the mid of the 1990's, the goal of a discrete representation of cracks within the framework of the Finite Element Method has resulted in approaches to represent cracks as embedded discontinuities within finite elements, circumventing the need for re-meshing as cracks evolve. From a conceptual point of view, this strategy has the character of multi-scale methods, characterized by superimposing the small scale resolution (the displacement jump across cracks) onto a large scale resolution of the (smooth) displacement field.

These formulations can generally be categorized into element-based formulations, generally denoted as Embedded Crack Models (see Simo et al. (1993),Oliver (1996),Jirásek and Zimmermann (2001),Armero and Garikipati (1996),Mosler and Meschke (2003), among others), and nodal-based formulations, i.e. the Extended Finite Element Method (X-FEM) (see Moës et al. (1999); Moës and Belytschko (2002); Wells and Sluys (2001b)). For a comparative assessment of both approaches we refer to Jirasek and Belytschko (2002); Dumstorff et al. (2003). The Extended Finite Element Method has been successfully applied to model traction-free cracks in the context of the LEFM in two- and three-dimensional settings (see. e.g. Moës et al. (1999); Belytschko et al. (2001); Sukumar et al. (2000)) as well as for cohesive cracks (e.g. Wells and Sluys (2001a); Moës and Belytschko (2002); Mariani and Perego (2003); Zi and Belytschko (2003); Mergheim et al. (2005); Dumstorff and Meschke (2007)). Since the topology of cracks is held fixed once they are signaled to open, the determination of the crack propagation direction and of the crack path has received considerable attention (see Oliver and Huespe (2002); Feist and Hofstetter (2006); Dumstorff and Meschke (2007); Meschke and Dumstorff (2007b); Jaeger et al. (2008)). In recent years the Extended Finite Element Method has been successfully applied to fully three-dimensional problems (Sukumar et al., 2000; Moës et al., 2000; Sukumar et al., 2003; Gasser and Holzapfel, 2005).

In this section, a 3D X-FEM model is formulated in a generalized, hygromechanical framework as recently proposed in Becker et al. (2010) considering the influence of cracks on the liquid permeability of concrete (Barton et al. (1985); Meschke and Grasberger (2003)). Using the transport model for the liquid phase in cracks according to Subsection 4.7, the effect of cracks on the moisture transport is considered using the information on the crack width and crack topology from the X-FEM model.

5.2 X-FEM Resolution of Cracks

This section contains a summary of the Extended Finite Element Method in a 3D setting, restricted to the geometrically linear theory. It is assumed that cracks fully penetrate through elements. Consequently, no crack tip enhancements as proposed e.g. in Zi and Belytschko (2003); Asferg et al. (2005) are considered. For a more elaborate discussion on the background of the X-FEM we refer to Chapter 5 and to Moës et al. (1999); Sukumar et al. (2000), among others.

To incorporate a small scale resolution of the displacement jump across cracks into the finite element the displacement field is decomposed into a continuous part $\bar{\boldsymbol{u}}$ and discontinuous part $\check{\boldsymbol{u}}$ (see Figure (15)):

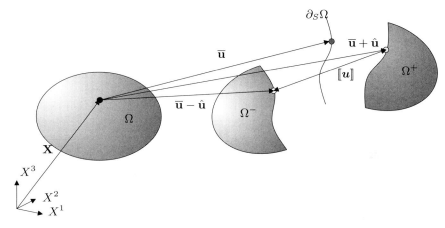

Figure 15. Separation of a body Ω by a discontinuity $\partial_S \Omega$ into sub-domains Ω^- and Ω^+.

$$\boldsymbol{u}(\mathbf{X}) = \bar{\boldsymbol{u}}(\mathbf{X}) + \check{\boldsymbol{u}}(\mathbf{X}). \tag{112}$$

The discontinuous part is described with the help of the Signum function which is multiplied with the enhanced regular part of the displacements $\hat{\boldsymbol{u}}$:

$$\check{\boldsymbol{u}}(\mathbf{X}) = S_S(\mathbf{X})\, \hat{\boldsymbol{u}}(\mathbf{X}). \tag{113}$$

The displacement jump at the discontinuity can be expressed as

$$[\![\boldsymbol{u}]\!] = 2\, \hat{\boldsymbol{u}}(\mathbf{X}) \Big|_{\mathbf{X} \in \partial_S \Omega}. \tag{114}$$

5.3 Weak Form of Balance of Momentum

Incorporating the discontinuous displacement field (112) in the weak form of the balance of momentum

$$\int_\Omega \delta \nabla \boldsymbol{u} : \boldsymbol{\sigma}\, dV = \int_{\Gamma_\sigma} \delta \bar{\boldsymbol{u}} \cdot \boldsymbol{t}^\star\, d\Gamma_\sigma. \tag{115}$$

leads to

$$\int_\Omega [\delta\nabla\bar{u} + S_S\,\delta\nabla\hat{u}] : \sigma\,dV + 2\int_\Omega \delta_S\,[\delta\hat{u} \otimes n] : \sigma\,dV = \int_{\Gamma_\sigma} \delta\bar{u} \cdot t^\star\,d\Gamma_\sigma. \tag{116}$$

With the definition of an enhanced strain tensor

$$\bar{\varepsilon} = \mathbf{B}\,u = \mathbf{B}\,\bar{u} + S_S\,\mathbf{B}\,\hat{u}, \tag{117}$$

the weak form of balance of momentum including a surface of discontinuity is obtained as

$$\int_\Omega \delta\bar{\varepsilon} : \sigma\,dV + 2\int_{\partial_S\Omega} \delta\hat{u} \cdot \underbrace{\sigma \cdot n}_{t_S}\,dA = \int_{\Gamma_\sigma} \delta\bar{u} \cdot t^\star\,d\Gamma_\sigma. \tag{118}$$

In Eq.(118) t_S is the traction vector acting on the surface of discontinuity which satisfies the balance of tractions.

In the present 3D implementation of the X-FEM, a new crack is assumed to open if the average principal stress in an element exceeds the tensile strength (see Subsection 5.5). The transition from a gradual opening of a zone of micro-cracks to a fully developed macro-crack in concrete is represented by a cohesive zone model. For simplicity, the focus is laid on mode-I crack opening in this section. Consequently, the adopted softening interface law relates crack openings and tractions t_S in the direction of the crack normal $\mathbf{n}$ (see Figure(16)):

$$\mathbf{t}_S = t_n(\llbracket u \rrbracket)\,\mathbf{n}. \tag{119}$$

This mode-I cohesive interface law is obtained as a special case of the mixed mode interface law proposed by Camacho and Ortiz (1996) which is used in the 2D X-FEM model using a more sophisticated crack propagation criteria (Meschke and Dumstorff (2007a)).

Adopting concepts of damage theory, the relation between normal tractions and crack openings is given as Wells and Sluys (2001a):

$$t_n = \left[T - T^{da}(\llbracket u^{eq} \rrbracket)\right]\,\llbracket u^{eq} \rrbracket, \tag{120}$$

with the initial stiffness T and the damaging (softening) portion T^{da} depending on the displacement jump:

$$T^{da} = T\left[1 - \frac{\alpha}{\llbracket u^{eq} \rrbracket}\exp\{-\frac{f_{tu}}{G_f}[\llbracket u^{eq} \rrbracket - \alpha_0]\}\right]. \tag{121}$$

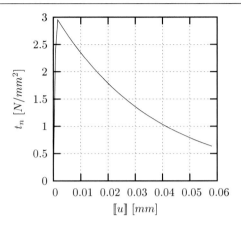

Figure 16. Interface law: Softening relation between normal tractions and crack separation.

In Eq.(121), α is the largest value of $[\![u^{eq}]\!]$ experienced in the loading history according to the damage criterion

$$\Phi([\![u^{eq}]\!], \alpha) = [\![u^{eq}]\!] - \alpha \leq 0, \tag{122}$$

f_{tu} is the tensile strength and G_f the fracture energy of concrete.

5.4 Extension to Coupled Hygro-Mechanical Analyses

The extension to hygro-mechanical model is formulated in the framework of poromechanics (see Section 3). In other words, for the uncracked bulk material a two-phase poromechanical formulation is considered as described in Section 4, without taking into account continuum damage. Assuming Navier-Stokes flow with crack channels, the enhanced transport capacity of a discrete crack with crack width w_c is formulated in analogy to Eq.(92) as

$$k_c(S_l, w_c) = k_{r,c}(S_l) k_{c,0}(w_c), \quad \text{with} \quad k_{c,0}(w_c) = \frac{w_c^2}{12}. \tag{123}$$

5.5 Finite Element Implementation

For the finite element implementation of the X-FEM concept in a three-dimensional setting the following assumptions are made:
- Cracks are assumed to fully propagate through elements after the opening of a crack is signaled in the respective element. Hence, no

crack-tip functions are used and the crack tip is represented by homogeneous boundary conditions for the enhanced displacement field at the crack front.
- Cracks are approximated as element-wise plane surfaces in the isoparametric domain. Consequently, in physical coordinates, non-plane surfaces according to the *Jacobi*-transformation are generated.
- An average weighted principal stress criterion characterized by the evaluation of the stresses within the integration points of the elements adjacent to the existing crack front and taking the principal values and axes of the averaged tensor is used to determine crack propagation and direction.
- The enhanced displacement field is restricted to linear approximations whereas the regular part may be approximated arbitrarily high taking into account the kinematics associated with the specific geometry of the investigated structure, e.g. shell-like or slab-like structures (Becker et al. (2010)):

$$u(\mathbf{X}) \approx \sum_{i=1}^{NN_p} \mathbf{N}_{p1,p2,p3} \bar{u}^{ei} + \sum_{j=1}^{8} \mathbf{N}_{1,1,1} S(\mathbf{X}) \tilde{u}^{ej}. \qquad (124)$$

5.6 Crack Tracking Algorithm

The representation of cracks as plane surfaces fully penetrating through finite elements requires special considerations as far as tracking of cracks is concerned. If a strictly C_0-continuous evolution of cracks would be followed (see e.g. Areias and Belytschko (2005)), in case that two intersections of an existing crack front with an uncracked element exist, the topology of the new crack plane opening in the uncracked element would be already defined by the two lines characterizing the two crack fronts, without taking into account any information regarding the direction of the new evolving crack from the cracking criterion. Therefore, a less strict algorithm is adopted, following suggestions made by Gasser and Holzapfel (2005), see Becker et al. (2010).

5.7 Numerical Solution

In the context of a two-field hygro-mechanical model the primary variables $\mathbf{u}(\mathbf{x},t)$ and $p_c(\mathbf{x},t)$ in the domain Ω are governed by the balance of linear momentum and balance of liquid mass

$$\text{div}\,\boldsymbol{\sigma} = 0, \quad \text{div}(\rho_l \mathbf{q}_l) + \dot{m} = 0. \qquad (125)$$

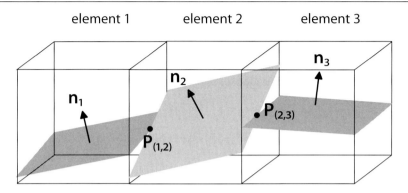

Figure 17. 3D crack tracking algorithm using 3D quadrilateral finite elements.

In Eq.(125) body forces are neglected. The system of governing equations (125) is completed by the boundary conditions

$$\begin{aligned}
\boldsymbol{\sigma}(\mathbf{x},t) \cdot \mathbf{n}_\Gamma &= \mathbf{t}^*(t) \quad \forall \mathbf{x} \in \Gamma_\sigma, & \mathbf{q}_l(\mathbf{x},t) \cdot \mathbf{n}_\Gamma &= q_l^*(t) \quad \forall \mathbf{x} \in \Gamma_{q_l}, \\
\mathbf{u}(\mathbf{x},t) &= \mathbf{u}^*(t) \quad \forall \mathbf{x} \in \Gamma_u, & p_c(\mathbf{x},t) &= p_c^*(t) \quad \forall \mathbf{x} \in \Gamma_{p_c}
\end{aligned} \quad (126)$$

where $\mathbf{n}_\Gamma$ is the normal vector on the boundary of the structure, q_l^* is the liquid flux across the boundary, p_c^* is the prescribed capillary pressure, $\mathbf{u}^*$ is the prescribed displacement and $\mathbf{t}^*$ the traction vector on the boundary. Finally, the initial conditions are given by

$$\mathbf{u}(\mathbf{x}, t=0) = \mathbf{u}_0, \quad p_c(t=0) = p_{c,0}. \tag{127}$$

Using, for simplicity, instead of Eq.(77), the standard formulation for the effective stresses according to *Bishop*, the total stresses $\boldsymbol{\sigma}$ are expressed in terms of the effective skeleton stresses $\boldsymbol{\sigma}'$, the *Biot* coefficient b and the capillary pressure p_c as

$$\boldsymbol{\sigma} = \boldsymbol{\sigma}' + b p_c \mathbf{1}. \tag{128}$$

The weak form of the balance of momentum is obtained according to Eq.(116)

after inserting Eq.(128) as

$$\begin{aligned} \delta W_m &= \int_{\Omega^e} \delta\varepsilon : \boldsymbol{\sigma}\, dV - \int_{\Gamma_\sigma} \delta \boldsymbol{u} \cdot \boldsymbol{t}^* \, dA \\ &= \int_{\Omega^e} \nabla \delta\bar{\mathbf{u}} : (\boldsymbol{\sigma}' + b p_c \mathbf{1})\, dV + \int_{\Omega^{+-}} S_S \nabla \delta\hat{\mathbf{u}} : (\boldsymbol{\sigma}' + b p_c \mathbf{1})\, dV \\ &\quad + 2 \int_{\partial_S \Omega^e} \delta\hat{\mathbf{u}} \cdot \underbrace{(\mathbf{t}'_S + b p_c \mathbf{n}_S)}_{\mathbf{t}_S}\, dA - \int_{\Gamma_\sigma} \delta \boldsymbol{u} \cdot \boldsymbol{t}^* \, dA, \end{aligned} \quad (129)$$

with the total and effective traction vectors $(\mathbf{t}_S(\llbracket \mathbf{u} \rrbracket, p_c))$ and $(\mathbf{t}'_S(\llbracket \mathbf{u} \rrbracket))$ at the surface of discontinuity $\partial_S \Omega^e$. The weak form of balance of liquid mass and the hygral *Neumann* boundary conditions remains unchanged compared to the continuum formulation in Section 4:

$$\delta W_h = \int_{\Omega^e} \delta p_c \frac{\dot{m}_l}{\rho_l}\, dV - \int_{\Omega^e} \delta \nabla p_c \cdot \mathbf{q}_l \, dV - \int_{\Gamma_q} \delta p_c\, q_l^* \, dA = 0. \quad (130)$$

Spatial discretization of Eqs.(129) and (130) yields the semi-discrete coupled set of equations

$$\mathbf{S}^e(\mathbf{x}^e)\,\dot{\mathbf{x}}^e + \mathbf{K}^e(\mathbf{x}^e)\mathbf{x}^e = \mathbf{r}^e(\mathbf{x}^e) \quad (131)$$

with the storage and stiffness matrices

$$\mathbf{S}^e = \begin{bmatrix} 0 & 0 & 0 \\ 0 & 0 & 0 \\ \mathbf{Q}_{p\bar{u}} & \mathbf{Q}_{p\hat{u}} & \mathbf{S}_{pp} \end{bmatrix}, \quad \mathbf{K}^e = \begin{bmatrix} \mathbf{K}_{\bar{u}\bar{u}} & \mathbf{K}_{\bar{u}\hat{u}} & \mathbf{Q}_{\bar{u}p} \\ \mathbf{K}_{\hat{u}\bar{u}} & \mathbf{K}_{\hat{u}\hat{u}} & \mathbf{Q}_{\hat{u}p} \\ 0 & 0 & \mathbf{H}_{pp} \end{bmatrix} \quad (132)$$

and the nodal degrees of freedom and nodal forces

$$\dot{\mathbf{x}}^e = \begin{bmatrix} \dot{\bar{\boldsymbol{u}}} \\ \dot{\hat{\boldsymbol{u}}} \\ \dot{\boldsymbol{p}}_c \end{bmatrix} \quad \mathbf{x}^e = \begin{bmatrix} \bar{\boldsymbol{u}} \\ \hat{\boldsymbol{u}} \\ \boldsymbol{p} \end{bmatrix} \quad \mathbf{r}^e = \begin{bmatrix} \boldsymbol{r}_{\bar{u}} \\ \boldsymbol{r}_{\hat{u}} \\ \boldsymbol{r}_p \end{bmatrix}. \quad (133)$$

After assembly, applying time discretization using finite differences together with a fully implicit time integration scheme and consistent linearization by means of the *Newton-Raphson* algorithm (Lewis and Schrefler (1998)) the resulting algebraic system of equations is solved within the time interval $[t_n, t_{n+1}]$:

$$[\mathbf{S} + \Delta t\,\mathbf{K}]_{n+1}\, \mathbf{x}_{n+1} = \mathbf{S}\,\mathbf{x}_n + \Delta t\, \mathbf{r}_{n+1}. \quad (134)$$

Numerical integration For the numerical integration of the weak form (129) and (130) at the element level the discontinuous character of the enhanced strain field has to be taken into account. Consequently, both parts of elements separated by a crack plane have to be integrated separately. As a more efficient alternative to a standard subdivision using *Delaunay*-triangularization, cracked elements are subdivided into a fixed set of six sub-tetrahedrons (Figure 18). For each of these tetrahedrons it is checked,

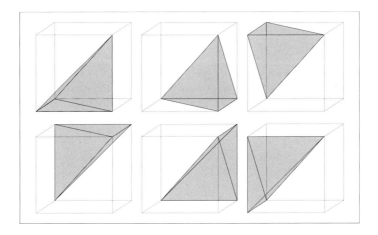

Figure 18. Numerical intgeration of cracked elements: Subdivision of an element into six sub-tetrahedrons.

whether it contains the crack plane segment. The crack plane cutting the element may either have a triangular or a quadrilateral shape. Consequently, a sub-tetrahedron can be split into two tetrahedrons, two pentahedrons, a tetrahedron and a pentahedron or into a tetrahedron and a pyramid. For the integration of the traction-separation law and the moisture flow the respective sub-domains are either triangles or quadrilaterals. The volume integral of a cracked finite element is computed, after transformation to natural element coordinates ξ^i with the help of the Jacobian $\boldsymbol{J}$, by summing all contributions of $nsub$ sub-continua that are integrated numerically over the natural coordinates η^j of the sub-domains:

$$\int_{\Omega^e} (\bullet)\, dV = \int_{V_\xi} (\bullet) \underbrace{\left\| \frac{\partial \mathbf{X}}{\partial \boldsymbol{\xi}} \right\|}_{\|\boldsymbol{J}\|} dV_\xi = \sum_{i=1}^{nsub} \int_{V_\eta} (\bullet) \left\| \frac{\partial \mathbf{X}}{\partial \boldsymbol{\xi}} \frac{\partial \boldsymbol{\xi}}{\partial \boldsymbol{\eta}} \right\| dV_\eta. \qquad (135)$$

The relation between both natural coordinates $\boldsymbol{\xi} = \boldsymbol{\xi}(\boldsymbol{\eta})$ is obtained from determining the ξ^i coordinates of the nodes of the sub-domains. These correspond either to the vertices of the element or to points on the crack surface. The latter ones are identified with the help of the ϕ-level sets.

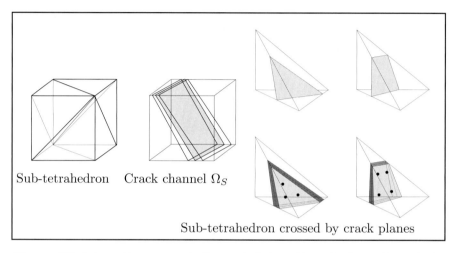

Figure 19. Integration concept: Separate integration of sub-continua (sub-tetrahedrons) and the crack channel Ω_S.

Numerical integration of the crack channel Intact parts and the crack plane have to be considered as different domains in the integration procedure. Hence, the stiffness matrix $\mathbf{K}_{\hat{u}\hat{u}}$ and the liquid permeability matrix $\mathbf{H}_{pp}$ are evaluated as

$$\mathbf{K}_{\hat{u}\hat{u}} = \int_{\Omega^{+-}} S_S^2 \, \mathbf{B}_u^T \, \mathbf{C} \, \mathbf{B}_u \, dV + \int_{\partial_S \Omega} \mathbf{N}_u^T \, \mathbf{T} \, \mathbf{N}_u \, dA,$$

$$\mathbf{H}_{pp} = -\int_{\Omega} \mathbf{B}_p^T \, \mathbf{k}_f / \mu_l \, \mathbf{B}_p \, dV - \int_{\Omega_S} \mathbf{B}_p^T \, \mathbf{A}^T \, k_c^t(w) / \mu_l \, \mathbf{A} \, \mathbf{B}_p \, dV.$$

(136)

$\mathbf{N}_u$ contains the hierarchical shape functions of the displacement field, $\mathbf{B}_u$ and $\mathbf{B}_p$ are the gradient matrices of the displacement and the capillary pressure field and $k_c^t(w)$ is the liquid permeability in the crack channel according to Eq.(98). The matrix $\mathbf{A}(\mathbf{n})$ is the projection of ∇p_c onto the crack channel characterized by the normal unit vector $\mathbf{n}$.

The permeability matrix $\mathbf{H}_{pp}$ is computed as

$$\mathbf{H}_{pp} \sim -\sum_{n=1}^{GP_p} \mathbf{B}_p^T \mathbf{k}_f/\mu_l \, \mathbf{B}_p |\mathbf{J}_n| \alpha_n - \sum_{j=1}^{GP_c} \mathbf{B}_p^T \mathbf{A}^T k_c^t(w)/\mu_l \, \mathbf{A} \, \mathbf{B}_p \, w_j \, |\mathbf{J}_{\partial_S \Omega, j}| \, \alpha_j \,, \tag{137}$$

using GP_p integration points for the continuum and GP_c integration points for the crack channel (Figure 19). The integration is performed either over triangles or quadrilaterals according to the integration concept described above.

5.8 Numerical Applications

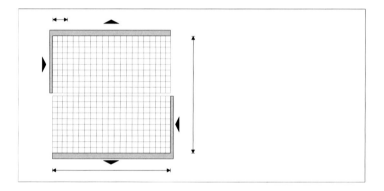

Figure 20. 3D X-FEM analysis of a double notched slab: Geometry and finite element mesh.

Double Notched Panel As a first benchmark example a double notched concrete panel subjected to a combined loading of shear and normal forces which has been investigated experimentally by (Nooru-Mohamed, 1992) is analyzed using the described 3D X-FEM model (see also Becker et al. (2010)). The geometry, material parameters and the finite element discretization are contained in Figure 20. The thickness of the slab is 50 mm. The material parameters are chosen as follows: Young's modulus $E = 30.000\,[N/mm^2]$, Poisson's ratio $\nu = 0.2\,[-]$, fracture energy $G_f = 0.11\,[N/mm]$, tensile strength $f_{tu} = 3.0\,[N/mm^2]$. According to the geometry of the slab the approximation of the displacement field is chosen as $vecu \approx \boldsymbol{u}_{3,3,1}$. In a first loading stage, the slab is subjected to a shear force $F_s = 10\,kN$. Subsequently, keeping the load level of the shear force

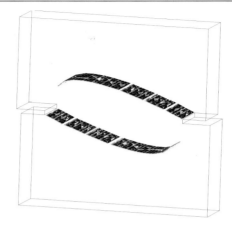

Figure 21. 3D X-FEM analysis of a double notched slab: Visualization of the crack topology by the $\phi = 0$-level set.

constant, a normal force F_n is applied incrementally by controlling the respective displacement.

Figure 21 illustrates the curved crack paths obtained from the proposed crack model for the combined loading scenario. The visualization of the crack path is accomplished by plotting the $\phi = 0$-level set. The left hand side of Figure 22 shows a comparison of the crack topologies obtained from the present 3D X-FEM model using a *slab-like* higher order solid FEM formulation, with a variational 2D X-FEM model proposed by (Meschke and Dumstorff, 2007b) and with the experimentally determined range of cracks (in grey color). The crack topology obtained from the proposed 3D X-FEM model correlates well with the experimental results and the results from 2D analyses using an energy-based X-FEM model with crack tip enhancements (Meschke and Dumstorff, 2007b). Figure 22 (right) shows a comparison of the corresponding load displacement curves. In this figure, also computational results from (Feist, 2004) are included. All three numerical results are within a relatively small range while they differ significantly with respect to the experimental results. It should be emphasized, that, due to the lack of reported material parameters the set of parameters taken by (Feist, 2004; Meschke and Dumstorff, 2007b) has been adopted.

Drying of a Three-Point Bending Beam The applicability of the proposed finite element formulation for prognoses of moisture transport in

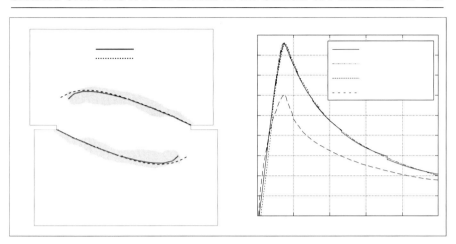

Figure 22. 3D X-FEM analysis of a double notched slab: Comparison of results from the proposed 3D X-FEM model with a variational 2D X-FEM model (Meschke and Dumstorff, 2007b) and the experimental results. left: crack topologies, right: load-displacement curves.

cracks is shown by the numerical analysis of a notched three point bending beam subjected to drying (Becker et al. (2010)). The material parameters are chosen as follows: Young's modulus $E = 2000\,[N/mm^2]$, Poisson's ratio $\nu = 0.2[N/mm^2]$, thickness $t = 10\,[mm]$ and liquid permeability $k_0 = 2.77 \cdot 10^{-21}\,[m^2]$. Figure 23 contains the geometry, the mechanical

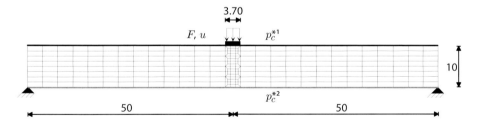

Figure 23. Hygro-mechanical simulation of a concrete beam: Geometry and finite element mesh.

and hygral boundary conditions, the material parameters and the spatial discretization by means of 272 3D-p-elements. For the approximation of the

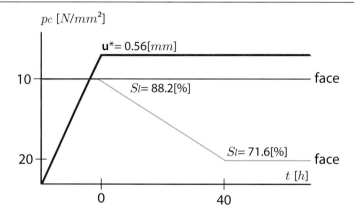

Figure 24. Hygro-mechanical simulation of a concrete beam: mechanical and hygral loading history.

regular displacement field, the anisotropic Ansatz $u \approx u_{2,2,1}$ is chosen while trilinear shape functions are used for the enhanced displacements and the capillary pressure, respectively. Figure 24 illustrates the mechanical and hygral loading history of the beam structure. A fixed crack located in the center at the bottom side of the beam with a length of $3.25\,mm$ is generated by applying a displacement $u^* = 0.56\,mm$ at the center of the top face of the beam. After applying the mechanical loading a drying process at the lower face, starting from a liquid saturation of $S_l^* = 88.2\,\%$ ($p_c = 10.0\,N/mm^2$) to a final saturation of $S_l^* = 58.8\,\%$ ($p_c = 20.0\,N/mm^2$) is prescribed while the liquid saturation at the upper face remains constant. Due to the drying process, the existing crack opens (Figure 25).

Figure 26 illustrates the distribution of capillary pressure p_c in the vicinity of the crack at different stages of the drying process. Due to the hygral environmental conditions, the moisture front penetrates from the bottom upwards. In the vicinity of the crack, an accelerated drying process is observed. This is also illustrated on the left hand side of Figure 26 showing the distribution of the capillary pressure p_c along the height of the beam at the position of the initial crack at different stages of the drying process.

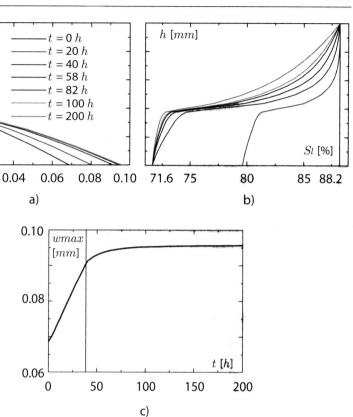

Figure 25. Hygro-mechanical simulation of a concrete beam: a) temporal evolution of the crack opening profie along crack depth, b) temporal evolution of the saturation within the crack channel, c) temporal evolution of maximum crack width at lower face of the beam.

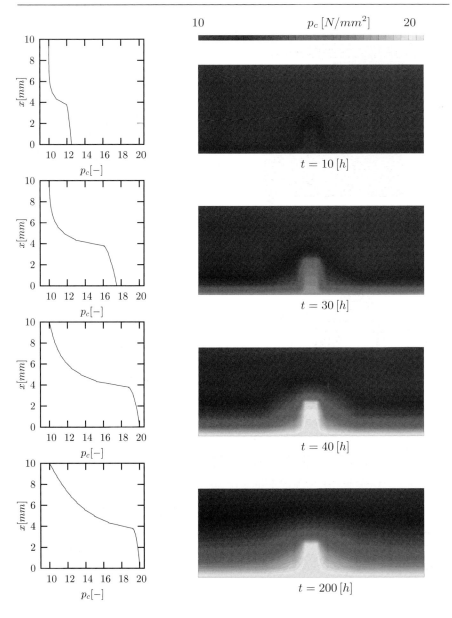

Figure 26. Hygro-mechanical simulation of a concrete beam: temporal evolution of the capillary pressure in the vicinity of the initial crack.

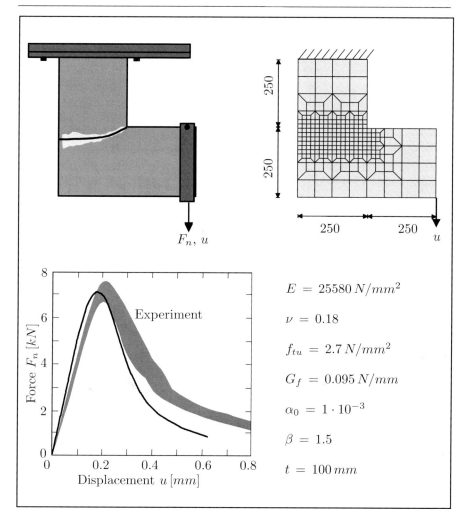

Figure 27. Analysis of a L-shaped panel (Dumstorff (2005)). Geometry, loading, finite element discretization and load-displacement curves obtained in the experiment Winkler (2001) and from the analysis (Units in $[mm]$).

Hygromechanical Analysis of a L-Shaped Panel A L-shaped panel is investigated as an example for a sequential hygromechanical analysis with crack propagation. The experimental investigation under purely mechanical loading conditions is described in Winkler (2001). The simulation procedure and the material and geometry parameters are illustrated in Figure 27. The structure is loaded with a prescribed displacement at the right boundary until failure. Various crack propagation criteria (global energy criterion, principal stress criterion) and the mesh dependency of the crack paths have been investigated in Dumstorff and Meschke (2007). In this work the work-equivalent anisotropic damage model described in Meschke and Dumstorff (2007a) with a shear parameter of $\beta = 1.5$, i.e. an average influence of the shear strength, is used as traction-displacement-relation. In comparison to the experimental results, the choice of this parameter ensures a good approximation in terms of crack path topology. In this work the structure is analyzed with regard to the crack path and the crack width by applying the global energy criterion proposed in Meschke and Dumstorff (2007a). For this purpose, a finite element mesh consisting of 261 square elements (see Figure 27) is used, in which a quadratic approximation is chosen for the continuous part and a linear approximation is chosen for the discontinuous part of the displacement field. According to the energy-based X-FEM model Meschke and Dumstorff (2007a), the crack propagation angle contributes to the global system of equations as a degree of freedom.

A progressing crack is induced by the point load acting on the structure as illustrated in Figure 27. The panel is analyzed until failure of the structure. When using the global energy criterion, the corresponding load-displacement curve and the crack path correspond well with the experimental results of Winkler (2001). Like in several other numerical analyses, see, e.g. Feist (2004), the load bearing capacity is over-estimated - a fact that results from the missing data regarding the fracture energy. In a second loading sequence the drying of the surface is modeled by applying hygral boundary conditions at the interior boundary of the L-shaped panel. The hygral analysis is not coupled to the mechanical analysis. The crack width and the topology of the crack path are available as input parameters, and transferred to the hygral analysis through a preprocessing step. In contrast to the purely mechanical load in the proximity of the crack and the hygral loading boundary where the boundary condition p_c^{*2} is applied, a refined mesh with 1006 elements is used. As described in Subsection 4.7, a dependency on the crack width w is assumed for the in-crack permeability, which has its maximum at the crack origin at the nook of the panel and allows for an accelerated permeation of moisture into the structure through the crack. The change of the hygral loading (see Figure 28) is prescribed along the complete interior boundary of the cracked L-shaped panel, and hence in the proximity of the maximum crack opening width of the induced crack. On the exterior a constant capillary pressure p_c^{*1} is assumed, which corresponds to the initial capillary pressure of the overall structure. The top face and the right face of the panel, however, are isolated

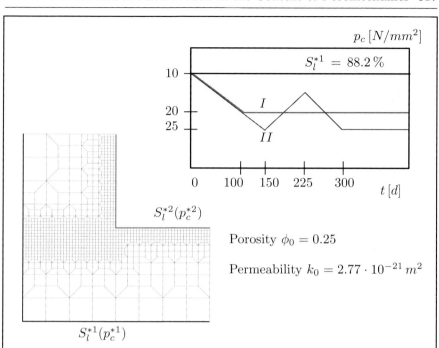

Figure 28. Hygral loading of a cracked L-shaped panel - Finite Element discretization, boundary conditions and loading history.

against drying in the model using by means of *Neumann* boundary conditions $q^* = 0\,m/s$. Two load cases are investigated for the hygral loading sequence. For load case I a drying of the interior surface of the L-shaped panel within a time period of $\Delta t = 100\,d$, corresponding to a reduction of the saturation from $S_l^{*2} = 88.2\,\%$ to $S_l^{*2} = 71.6\,\%$ and an increase of the capillary pressure from $p_c^{*2} = 10\,N/mm^2$ to $p_c^{*2} = 20\,N/mm^2$. Load case II is characterized by a cyclic hygral loading. First, a drying process, characterized by an increase of p_c from $p_c^{*2} = 10\,N/mm^2$ to $p_c^{*2} = 25\,N/mm^2$ ($S_l^{*2} = 64.7\,\%$) within $150\,d$. Subsequently, a wetting to $p_c^{*2} = 17.5\,N/mm^2$ ($S_l^{*2} = 75.5\,\%$) within $75\,d$, followed again by a drying to $p_c^{*2} = 25\,N/mm^2$ within $75\,d$ is applied. In the last sequence the capillary pressure and hence the saturation is, as illustrated in figure 28, kept constant on the interior surface of the L-shaped panel.

Figure 30 shows the distribution of the capillary pressure at different points in time of the hygral loading sequence for load case I. At three points the evolution of the saturation over time t is plotted. Point A represents the

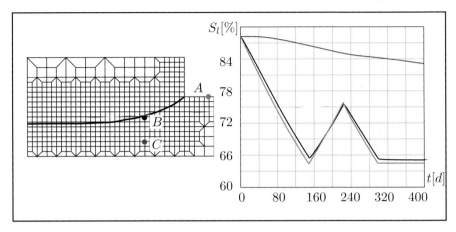

Figure 29. Results from hygro-mechanical X-FEM analysis of a L-shaped panel (Load case II).

boundary condition at the hygral loading boundary of the panel. For point B - in direct proximity of the crack in the interior of the structure - a rapid decrease of saturation is observed due to the highly accelerated moisture transport within the crack. In the following loading phase, characterized by constant capillary pressures at the interior boundary, only a slight decrease of saturation can be noticed. For point C, which is situated within the structure, a significantly decelerated decrease of saturation can be noticed; an attenuation of the saturation gradient can only be noticed after a longer period of time. At this point, the evaluation shows a considerably decelerated diffusion into the matrix material.

For load case II, which represents one cycle of partial drying and wetting after completion of the primary drying process, a deceleration of the saturation process can be recognized as well. For this purpose the evolution of the saturation $S_l(p_c)$ is again plotted over the time t at the three evaluation points of the panel.

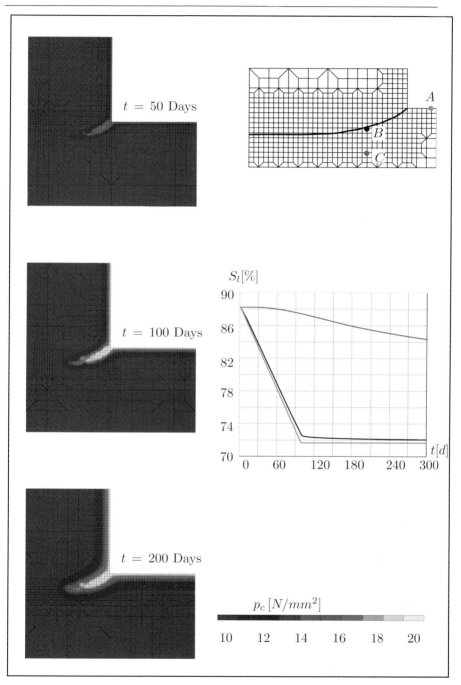

Figure 30. Results of hygro-mechanical X-FEM analysis of a L-shaped panel (Load case I).

Bibliography

C.-M. Aldea, S.P. Shah, and A. Karr. Permeability of cracked concrete. *Materials and Structures*, 32:370–376, 1999.

C.-M. Aldea, M. Ghandehari, S.P. Shah, and A. Karr. Estimation of water flow through cracked concrete under load. *ACI Materials Journal*, 97: 567–575, 2000.

E. Alonso, A. Gens, and A. Josa. A constitutive model for partially saturated soils. *Gétotechnique*, 40(2):405–430, 1990.

A.M. Alvaredo. Drying shrinkage and crack formation. Technical Report 5, Building Materials Reports, Aedificatio, Freiburg, 1994.

P.M.A. Areias and T. Belytschko. Analysis of three-dimensional crack initiation and propagation using the extended finite element method. *International Journal for Numerical Methods*, 63:760–788, 2005.

F. Armero and K. Garikipati. An analysis of strong discontinuities in multiplicative finite strain plasticity and their relation with the numerical simulation of strain localization in solids. *International Journal of Solids and Structures*, 33(20):2863–2885, 1996.

J.L. Asferg, P.N. Poulsen, and L.O. Nielsen. Cohesive crack tip element for X-FEM. In A. Carpinteri, editor, *International Conference on Fracture (ICF 11)*, 2005.

V. Baroghel-Bouny, M. Mainguy, T. Lassabatère, and O. Coussy. Characterization and identification of equilibrium and transfer moisture properties for ordinary and high-performance cementitious materials. *Cement and Concrete Research*, 29:1225–1238, 1999.

N. Barton, S. Bandis, and K. Bakhtar. Strength, deformation and conductivity coupling of rock joints. *International Journal of Rock Mechanics and Mining Sciences & Geomechanics Abstracts*, 22:121–140, 1985.

Z.P. Bažant and L.J. Najjar. Nonlinear water diffusion in nonsaturated concrete. *Materials and Structures*, 5:3–20, 1972.

Z.P. Bažant and W.J. Raftshol. Effect of cracking in drying and shrinkage specimens. *Cement and Concrete Research*, 12:209–226, 1982.

Z.P. Bažant and Y. Xi. Drying creep of concrete: constitutive model and new experiments separating its mechanisms. *Materials and Structures*, 27:3–14, 1994.

Z.P. Bažant, S. Sener, and J.K. Kim. Effect of cracking on drying permeability and diffusivity of concrete. *ACI Materials Journal*, 84:351–357, 1987.

Z.P. Bažant, A.B. Hauggaard, S. Baweja, and F.-J. Ulm. Microprestress-solidification theory for concrete creep. I: Aging and drying effects. *Journal of Engineering Mechanics (ASCE)*, 123:1188–1194, 1997.

J. Bear and Y. Bachmat. *Introduction to Modeling of Transport Phenomena in Porous Media*. Kluwer Academic Publisher, Dordrecht, The Netherlands, 1991.

C. Becker, S. Jox, and G. Meschke. 3D higher-order X-FEM model for the simulation of cohesive cracks in cementitious materials considering hygro-mechanical couplings. *Computer Modeling in Engineering and Sciences*, 57(3):245–276, 2010.

T. Belytschko, N. Moës, S. Usui, and C. Parimi. Arbitrary discontinuities in finite elements. *International Journal for Numerical Methods in Engineering*, 50:993–1013, 2001.

F. Benboudjema, F. Meftah, and J.M. Torrenti. A unified approach for the modeling of drying shrinkage and basic creep of concrete: analysis of intrinsic behaviour and structural effects. In R. de Borst, H.A. Mang, N. Bićanić, and G. Meschke, editors, *Computational Modelling of Concrete Structures*, pages 391–400. Balkema, 2003.

M.A. Biot. General theory of three-dimensional consolidation. *Journal of Applied Physics*, 12:155–165, 1941.

M.A. Biot and D.G. Willis. The elastic coefficients of the theory of consolidation. *Journal of Applied Mechanics*, 24:594–602, 1957.

A.W. Bishop. The principle of effective stress. *Teknisk Ukeblad*, 39:859–863, 1959.

A.W. Bishop. The influence of an undrained change in stress on the pore pressure in porous media of low compressibility. *Géotechnique*, 23:435–442, 1973.

A.W. Bishop and G.E. Blight. Some aspects of effective stress in saturated and partly saturated soils. *Géotechnique*, 13:177–197, 1963.

D. Breysse and B. Gérard. Transport of fluids in cracked media. In H.W. Reinhardt, editor, *Penetration and Permeability of Concrete*, number 16 in RILEM Reports, pages 123–153, London, 1997. E & FN Spon.

K. Brodersen and K. Nilsson. Pores and cracks in cemented waste and concrete. *Cement and Concrete Research*, 22:405–417, 1992.

N. Burlion, F. Bourgeois, and J.F. Shao. Coupling damage - drying shrinkage: experimental study and modelling. In V. Baroghel-Bouny and P.-C. Aïtcin, editors, *Proceedings of the International RILEM Workshop on Shrinkage of Concrete*, number 17 in RILEM Proceedings, 2000.

G.T. Camacho and M. Ortiz. Computational modelling of impact damage in brittle materials. *International Journal for Solids and Structures*, 33:2899–2938, 1996.

J. Carmeliet and K. Van Den Abeele. Poromechanical modelling of shrinkage and damage processes in unsaturated porous media. In V. Baroghel-Bouny and P.-C. Aïtcin, editors, *Proceedings of the International RILEM Workshop on Shrinkage of Concrete*, number 17 in RILEM Proceedings, 2000.

B.D. Coleman and W. Noll. The thermodynamics of elastic materials with heat conduction and viscosity. *Archive for Rational Mechanics and Analysis*, 13:167–178, 1963.

H. Colina and P. Acker. Drying cracks: Kinematics and scale laws. *Materials and Structures*, 33:101–107, 2000.

O. Coussy. Thermomechanics of saturated porous solids in finite deformation. *European Journal of Mechanics, A/Solids*, 8:1–14, 1989a.

O. Coussy. A general theory of thermoporoelastoplasticity for saturated porous materials. *Transport in Porous Media*, 4:281–293, 1989b.

O. Coussy. *Mechanics of Porous Continua*. John Wiley & Sons, Chicester, 1995.

O. Coussy. *Poromechanics*. J. Wiley & SOns, Ltd., 2004.

O. Coussy and F.-J. Ulm. Creep and plasticity due to chemo-mechanical couplings. *Archive of Applied Mechanics*, 66:523–535, 1996.

O. Coussy, R. Eymard, and T. Lassabatère. Constitutive modeling of unsaturated drying deformable materials. *Journal of Engineering Mechanics (ASCE)*, 124:658–667, 1998.

R. de Borst. Some recent issues in computational failure mechanics. *International Journal for Numerical Methods in Engineering*, 52:63–95, 2002.

R. de Borst, N. Bicanic, H.A. Mang, and G. Meschke, editors. *Computational Modelling of Concrete Structures (EURO-C 1998)*, volume 1 & 2, 1998. Balkema, Rotterdam.

P. Dumstorff. *Modellierung und numerische Simulation von Rissfortschritt in spröden und quasi-spröden Materialien auf Basis der Extended Finite Element Method*. PhD thesis, Institute for Structural Mechanics, Ruhr University Bochum, 2005. in german.

P. Dumstorff and G. Meschke. Crack propagation criteria in the framework of X-FEM-based structural analyses. *International Journal for Numerical and Analytical Methods in Geomechanics*, 31:239–259, 2007.

P. Dumstorff, J. Mosler, and G. Meschke. Advanced discretization methods for cracked structures: The Strong Discontinuity approach vs. the Extended Finite Element Method. In *Computational Plasticity 2003*. CIMNE, Barcelona, CD-ROM, 2003.

W. Ehlers and J. Bluhm. *Porous Media: Theory, Experiments and Numerical Applications*. Springer, 2000.

C. Feist. *A Numerical Model for Cracking of Plain Concrete Based on the Strong Discontinuity Approach*. PhD thesis, University Innsbruck, 2004.

C. Feist and G. Hofstetter. An embedded strong discontinuity model for cracking of plain concrete. *Computer Methods in Applied Mechanics and Engineering*, 195(52):7115–7138, 2006.

T.C. Gasser and G.A. Holzapfel. Modeling 3D crack propagation in unreinforced concrete using PUFEM. *Computer Methods in Applied Mechanics and Engineering*, 194:2859–2896, 2005.

D. Gawin, F. Pesavento, and B.A. Schrefler. Modelling of hygro-thermal behaviour of concrete at high temperature with thermo-chemical and mechanical material degradation. *Computer Methods in Applied Mechanics and Engineering*, 192:1731–1771, 2003.

B. Gérard and J. Marchand. Influence of cracking on the diffusion properties of cement-based materials. Part I: Influence of continuous cracks on the steady-state regime. *Cement and Concrete Research*, 30:37–43, 2000.

B. Gérard, D. Breysse, A. Ammouche, O. Houdusse, and O. Didry. Cracking and permeability of concrete under tension. *Materials and Structures*, 29:141–151, 1996.

S. Govindjee, G.J. Kay, and J.C. Simo. Anisotropic modeling and numerical simulation of brittle damage in concrete. *International Journal for Numerical Methods in Engineering*, 38(21):3611–3634, 1995.

L. Granger. *Comportement différé du béton dans les enceintes de centrales nucléaires: Analyse et modélisation*. PhD thesis, ENPC, Paris, 1996.

S. Grasberger. *Gekoppelte hygro-mechanische Materialmodellierung und numerische Simulation langzeitiger Degradation von Betonstrukturen*. Number 186 in Fortschritt-Berichte. VDI Verlag, Düsseldorf, 2002.

S. Grasberger and G. Meschke. Drying shrinkage, creep and cracking of concrete: From coupled material modelling to multifield structural analysis. In R. de Borst, H.A. Mang, N. Bićanić, and G. Meschke, editors, *Computational Modelling of Concrete Structures*, pages 433–442. Balkema, 2003.

S. Grasberger and G. Meschke. Thermo-hygro-mechanical degradation of concrete: From coupled 3D material modelling to durability-oriented multifield structural analyses. *Materials and Structures*, 37:244–256, May 2004. Special Issue on Poromechanis of Concrete.

M. Hassanizadeh and W. G. Gray. General conservation equations for multiphase systems: 1. averaging procedure. *Advances in Water Resources*, 2:131 – 144, 1979.

R. Helmig. *Multiphase Flow and Transport Processes in the Subsurface*. Springer, Berlin, 1997.

G. Hofstetter and H.A. Mang. *Computational Mechanics of Reinforced and Prestressed Concrete Structures*. Vieweg, Braunschweig, 1995.

A.R. Ingraffea and V. Saouma. Numerical modelling of discrete crack propagation in reinforced and plain concrete. In G. Sih and A. DiTommaso, editors, *Fracture Mechanics of Concrete: structural application and numerical calculation*, pages 171–225. Martinus Nijhoff, 1985.

P. Jaeger, P. Steinmann, and E. Kuhl. Modelling three-dimensional crack propagation. a comparison of crack path tracking strategies. *International Journal for Numerical Methods in Engineering*, 76(9):1328–1352, 2008.

M. Jirasek and T. Belytschko. Computational resolution of strong discontinuities. In H.A. Mang, F.G. Rammerstorfer, and J. Eberhardsteiner, editors, *Fifth World Congress on Computational Mechanics*, Vienna, Austria, 2002.

M. Jirašek and T. Zimmermann. Embedded crack model: Part 1: Basic formulation. part 2: Combination with smeared cracks. *International Journal for Numerical Methods in Engineering*, 50:1269–1305, 2001.

J.W. Ju. On energy-based coupled elastoplastic damage theories: Constitutive modeling and computational aspects. *International Journal for Solids and Structures*, 25(7):803–833, 1989.

R. Lackner, Ch. Hellmich, and H.A. Mang. Constitutive modeling of cementitious materials in the framework of chemoplasticity. *International Journal for Numerical Methods in Engineering*, 53:2357–2388, 2002.

R.W. Lewis and B.A. Schrefler. *The Finite Element Method in the Static and Dynamic Deformation and Consolidation of Porous Media*. John Wiley & Sons, Chichester, 1998.

D. Lydzba and J.F Shao. Study of poroelasticity material coefficients as response of microstructure. *Mechanics of Cohesive-Frictional Materials*, 5:149–171, 2000.

H.A. Mang, G. Meschke, R. Lackner, and J. Mosler. *Comprehensive Structural Integrity*, volume 3, chapter Computational Modelling of concrete Structures, pages 1–67. Elsevier, 2003.

S. Mariani and U. Perego. Extended finite element method for quasi-brittle fracture. *International Journal for Numerical Methods in Engineering*, 58:103–126, 2003.

J. Mergheim, E. Kuhl, and P. Steinmann. A finite element method for the computational modelling of cohesive cracks. *International Journal for Numerical Methods in Engineering*, 63:276–289, 2005.

G. Meschke and P. Dumstorff. How does the crack know how to propagate? - a X-FEM-based study on crack propagation criteria. In H. Yuan and F.H. Wittmann, editors, *Nonlocal Modelling of Failure of Materials*, pages 201–218. Aedificatio Publishers, 2007a.

G. Meschke and P. Dumstorff. Energy-based modeling of cohesive and cohesionless cracks via X-FEM. *Computer Methods in Applied Mechanics and Engineering*, 196:2338–2357, 2007b.

G. Meschke and S. Grasberger. Numerical modelling of coupled hygro-mechanical degradation of cementitious materials. *Journal of Engineering Mechanics (ASCE)*, 129(4):383–392, 2003.

G. Meschke, Ch. Kropik, and H.A. Mang. Numerical analyses of tunnel linings by means of a viscoplastic material model for shotcrete. *International Journal for Numerical Methods in Engineering*, 39:3145–3162, 1996.

G. Meschke, R. Lackner, and H.A. Mang. An anisotropic elastoplastic-damage model for plain concrete. *International Journal for Numerical Methods in Engineering*, 42:703–727, 1998a.

G. Meschke, J. Macht, and R. Lackner. A damage-plasticity model for concrete accounting for fracture-induced anisotropy. In R. de Borst, N. Bićanić, H. Mang, and G. Meschke, editors, *Computational Modelling of Concrete Structures*, volume 1, pages 3–12, Rotterdam, 1998b. Balkema.

N. Moës and T. Belytschko. Extended finite element method for cohesive crack growth. *Engineering Fracture Mechanics*, 69:813–833, 2002.

N. Moës, J.E. Dolbow, and T. Belytschko. A finite element method for crack growth without remeshing. *International Journal for Numerical Methods in Engineering*, 46:131–150, 1999.

N. Moës, N. Sukumar, B. Moran, and T. Belytschko. An extended finite element method (X-FEM) for two- and three-dimensional crack modeling. In *European Congress on Computational Methods in Applied Sciences and Engineering*, Barcelona, Spain, 2000.

J. Mosler and G. Meschke. 3D modeling of strong discontinuities in elastoplastic solids: Fixed and rotating localization formulations. *International Journal for Numerical Methods in Engineering*, 57:1553–1576, 2003.

Y. Mualem. A new model for predicting the hydraulic conductivity of unsaturated porous media. *Water Resources Research*, 12:513–522, 1976.

M.B. Nooru-Mohamed. *Mixed-mode Fracture of Concrete: an Experimental Approach*. PhD thesis, Technische Universiteit Delft, 1992.

A. Nur and J.D. Byerlee. An exact effective stress law for elastic deformation of rock with fluids. *Journal of Geophysical Research*, 76:6414–6419, 1971.

B.K. Nyame and J.M. Illston. Relationships between permeability and pore structure of hardened cement paste. *Magazine of Concrete Research*, 33:139–146, 1981.

J. Oliver. A consistent characteristic length for smeared cracking models. *International Journal for Numerical Methods in Engineering*, 28:461–474, 1989.

J. Oliver. Modelling strong discontinuities in solid mechanics via strain softening constitutive equations. Part 1: Fundamentals, Part 2: Numerical simulation. *International Journal for Numerical Methods in Engineering*, 39:3575–3623, 1996.

J. Oliver and A.E. Huespe. On strategies for tracking strong discontinuities in computational failure mechanics. In *Online Proceedings of the Fifth World Congress on Computational Mechanics (WCCM V)*, 2002.

H. Oshita and T. Tanabe. Modeling of water migration phenomenon in concrete as homogeneous material. *Journal of Engineering Mechanics (ASCE)*, 126:551–553, 2000.

Ph.C. Perdikaris and A. Romeo. Size effect on fracture energy of concrete and stability issues in three-point bending fracture toughness testing. *ACI Material Journal*, 92(5):483–496, 1995.

G. Pickett. The effect of change in moisture-content of the creep of concrete under a sustained load. *Journal of the American Concrete Institute*, 13: 333–355, 1942.

T.C Powers. Mechanisms of shrinkage and reversible creep of hardened cement paste. In A.E. Brooks and K. Newman, editors, *The Structure of Concrete and its Behaviour under Load*, International Conference on the Structure of Concrete, pages 319–344, London, 1965. Cement and Concrete Association.

H. Sadouki and F.H. Wittmann. Numerical investigations on damage in cementitious composites under combined drying shrinkage and mechanical load. In R. de Borst, J. Mazars, G. Pijaudier-Cabot, and J.G.M. van Mier, editors, *Fracture Mechanics of Concrete Structures*, volume 1, pages 95–98. Balkema, 2001.

B.A. Schrefler. Thermodynamics of saturated-unsaturated porous materials and quantitative solutions: The isothermal case. In *European Conference on Computational Mechanics ECCM 2001*, pages CD–ROM, Cracow, Poland, 2001.

B.A. Schrefler and X. Zhan. A fully coupled model for water flow and airflow in deformable porous media. *Water Resources Research*, 29:155–167, 1993.

J. Sercombe, Ch. Hellmich, F.-J. Ulm, and H. A. Mang. Modeling of early-age creep of shotcrete. I: model and model parameters. *Journal of Engineering Mechanics (ASCE)*, 126(3):284–291, 2000.

M.J. Setzer. Zum Mikrogefüge des Zementsteins und dessen Einfluß auf das mechanische Verhalten des Betons. *Zement und Beton*, 85/86:29–34, 1975.

J.C. Simo and T.J.R. Hughes. *Computational inelasticity*. Springer, Berlin, 1998.

J.C. Simo, J. Oliver, and F. Armero. An analysis of strong discontinuities induced by strain-softening in rate-independent inelastic solids. *Computational Mechanics*, 12:277–296, 1993.

D.T. Snow. Anisotropic permeability of fractured media. *Water Resources Research*, 5:1273–1289, 1969.

N. Sukumar, N. Moës, B. Moran, and T. Belytschko. Extended finite element method for three-dimensional crack modelling. *International Journal for Numerical Methods in Engineering*, 48:1549–1570, 2000.

N. Sukumar, D.L. Chopp, and B. Moran. Extended finite element method and fast marching method for three-dimensional fatigue crack propagation. *Engineering Fracture Mechanics*, 70:29–48, 2003.

C. Truesdell and R. Toupin. *Handbuch der Physik*, chapter The classical field theories, pages 226–793. Springer, Berlin, 1960.

M.T. van Genuchten. A closed-form equation for predicting the hydraulic conductivity of unsaturated soils. *Soil Science Society of America*, 44: 892–898, 1980.

J.H.M. Visser. *Extensile Hydraulic Fracturing of (Saturated) Porous Materials*. PhD thesis, TU Delft, 1998.

K. von Terzaghi. The shearing resistance of saturated soils and the angle between the planes of shear. In *First International Conference on Soil Mechanics*, volume 1, pages 54–56. Harvard University, 1936.

G.N. Wells and L.J. Sluys. A new method for modelling cohesive cracks using finite elements. *International Journal for Numerical Methods in Engineering*, 50:2667–2682, 2001a.

G.N. Wells and L.J. Sluys. Three-dimensional embedded discontinuity model for brittle fracture. *International Journal for Solids and Structures*, 38:897–913, 2001b.

K. Willam. Experimental and computational aspects of concrete failure. In F. et al. Damjanić, editor, *Computer aided analysis and design of concrete failure*, pages 33–70. Pineridge Press, Swansea, 1984.

B. Winkler. *Traglastuntersuchungen von unbewehrten und bewehrten Betonstrukturen auf der Gunrdlage eines objektiven Werkstoffgesetzes fr Beton*. PhD thesis, Universit"at Innsbruck, 2001.

P.A. Witherspoon, J.S.Y. Wang, K. Iwai, and J.E. Gale. Validity of cubic law for fluid flow in a deformable rock fracture. Technical report, Berkeley Laboratory, LBL-9557, SAC-23, 1979.

F.H. Wittmann. Grundlagen eines Modells zur Beschreibung charakteristischer Eigenschaften des Betons. Technical Report 290, DAfStb, Berlin, 1977.

M. Xie and W.H. Gerstle. Energy-based cohesive crack propagation modeling. *Journal of Engineering Mechanics (ASCE)*, 121(12):1349–1358, 1995.

G. Zi and T. Belytschko. New crack-tip elements for X-FEM and applications to cohesive cracks. *International Journal for Numerical Methods in Engineering*, 57:2221–2240, 2003.

M.D. Zoback and J.D. Byerlee. The effect of microcrack dilatancy on the permeability of westerly granite. *Journal of Geophysical Research*, 80: 752–755, 1975.